The Future of Business and Finance book series features professional works aimed at defining, analyzing, and charting the future trends in these fields. The focus is mainly on strategic directions, technological advances, challenges and solutions which may affect the way we do business tomorrow, including the future of sustainability and governance practices. Mainly written by practitioners, consultants and academic thinkers, the books are intended to spark and inform further discussions and developments.

Thomas Friedli • Jim Cafone
Matteo Bernasconi • Fabian Klinkner
Editors

Striving for Operational Excellence in the Pharmaceutical Industry

Towards Deep Structures

Springer

Editors
Thomas Friedli
Institute for Production and Supply Chain Management
University of St.Gallen
St.Gallen, Switzerland

Matteo Bernasconi
Abbvie
Cham, Switzerland

Jim Cafone
Pfizer
New York, NY, USA

Fabian Klinkner
Institute for Production and Supply Chain Management
University of St.Gallen
St.Gallen, Switzerland

ISSN 2662-2467 ISSN 2662-2475 (electronic)
Future of Business and Finance
ISBN 978-3-032-26286-8 ISBN 978-3-032-26287-5 (eBook)
https://doi.org/10.1007/978-3-032-26287-5

This Springer imprint is published by the registered company Springer Nature Switzerland AG
The registered company address is: Gewerbestrasse 11, 6330 Cham, Switzerland

Contents

Part I

Introduction to "Striving for Operational Excellence in the Pharmaceutical Industry: Towards Deep Structures"

Thomas Friedli, Matteo Bernasconi, and Fabian Klinkner

"Only three things happen naturally in organizations: friction, confusion, and underperformance. Everything else requires leadership." Peter F. Drucker

"Keep working on a plan. Make no little plans. Make the biggest one you can think of, and spend the rest of your life carrying it out." Harry S. Truman

"All of the real heroes are not storybook combat fighters either. Every single man in this Army plays a vital role. Don't ever let up. Don't ever think that your job is unimportant. Every man has a job to do and he must do it. Every man is a vital link in the great chain." George S. Patton

"What is chiefly needed is skill rather than machinery." Wilbur Wright

"We are what we repeatedly do. Excellence, then, is not an act, but a habit." Aristotle

Between 2006 and 2013, we published three books within our research program "Operational Excellence in the Pharmaceutical Industry." We began with a status report based on the first round of the renowned St.Gallen Operational Excellence (OPEX) benchmarking, already featuring selected industry case studies (Friedli, 2006). This was followed by "The Pathway to Operational Excellence" (Friedli, 2010), structured as a journey inspired by Captain James Cook's eighteenth-century expeditions. We outlined a sequence beginning with preparation for the journey and ending with the redefinition of the destination, leading to a continuous cycle of improvement.

In 2013, we released our most comprehensive book to date: "Leading Pharmaceutical Operational Excellence" (Friedli et al., 2013). It remains a source of knowledge and inspiration for managers and OPEX professionals alike, combining leading practical examples and our academic experience gained through collaboration with numerous pharmaceutical companies. The result was a unique collection of state-of-the-art applications that also emphasized the often-underestimated leadership demands and challenges of OPEX.

After two other pharma-related, but not OPEX-focused, books, inspired by research we conducted for the U.S. FDA (Friedli et al., 2018, 2020), we sensed that the time is fitting for the next volume on OPEX. This book should not be viewed as an update or replacement of its predecessors, but as a complementary addition, diving deeper into implementation.

Since 2013, our research on OPEX has continued without interruption. Our team, averaging four members in each of these years, has invested close to 100,000 hours in this research. We collected new data, researched about regulator-industry interactions, supported several companies for over 20 years (offering insights into long-term OPEX sustainability), and built and moderated industrial exchange groups for the pharmaceutical industry[1]. Our OPEX research revolves around cross-industry studies, comparing the pharmaceutical sector to other domains, interface of OPEX and digital in light of the rapid advances in the latter, barriers to continuous improvement, and predictive potential of machine learning. Meanwhile, collaborations with institutions from the pharmaceutical industry on Quality Management Maturity (QMM) remain a vital component of this research.

However, the main reason for writing this book is the positive feedback we received on previous volumes and our belief that, as a university, we should share knowledge broadly. It also reflects our conviction that OPEX will remain essential for all major pharmaceutical companies and that sustainable continuous improvement is not optional—it is a necessity.

The concept for this book aligns with our earlier work, presenting academic and industrial perspectives, side by side to provide direct insights into practice.

The title "Striving for Operational Excellence in the Pharmaceutical Industry: Towards Deep Structures" reflects two decades of research. It emphasizes not an ideal end state, but the ongoing nature of improvement: maintaining momentum, avoiding complacency, leading, motivating, coaching, and never ever giving up. True integration of OPEX into an organization and its culture requires deep structures—not skin-deep actions. Only with these deep structures can OPEX programs and production systems remain resilient, even under pressure, management changes, or other disruptions.

We hope that this book advances the pharmaceutical industry, helping it to become a role model and trailblazer—and finally shedding the reputation that it is a laggard when it comes to production, lean, and OPEX.

To set the stage, our book begins with an introductory part. We define OPEX from a 2026 understanding and investigate triggers and forces over the years. The development is thereafter demonstrated using data from the St.Gallen benchmarking databases. We conclude this part with insights from Pfizer written by one of the co-editors of this book who was the architect of Pfizer's Integrated Manufacturing Excellence System (IMEx).

Part II, "Learnings from Two Decades of OPEX Research," summarizes major findings from our latest OPEX research. After a short introduction into this part, we define OPEX as a program and describe some specifics in three dedicated chapters. After a chapter highlighting the major differences between top and low performers,

[1] In 2025, our global OPEX research group comprised 15 companies, while our QC lab exchange group brought together 20 companies.

derived from a cross-industry benchmarking study, we focus on selected crucial elements of successful OPEX implementations: leadership and the alignment of OPEX and digital as well as of OPEX and quality. This is followed by two chapters dealing with the specifics of highly regulated industries.

Part III provides numerous insights either from practice written by authors who were and are part of managing OPEX in these specific companies or from research and consultancies sharing a common understanding with us. We start again with a short introduction. This is followed by a chapter dealing with quality culture written by a long-term collaborator from TU Dublin. The next chapter builds on this and shares insights from S A Partners, a consultancy collaborating with numerous companies on true excellence. The next chapter provides further insights into Pfizer's IMEx approach, one of the leading OPEX programs in the pharmaceutical industry. Thereafter, Roche describes an approach for an OPEX-digital integration including the organizational integration. This is followed by a personal reflection from a long-term OPEX leader having driven the topic in more than one company. Takeda focuses on the management of dispersed teams, a reality every global company is concerned with. Additionally, two chapters deal with hard implementation, one written by a Boehringer Plant Leader, and the other by the global OPEX lead for Dechra, an animal health company. AbbVie provides insights into data analytics cases before we move again to the quality side of the OPEX journey. We start with a description of what Pfizer calls a quality revolution, changing the way quality supports business. This provides the ground to the next chapter, highlighting today's new requirements for quality leaders. Six traits of quality business leadership are worked out that have the potential to change the industry when built like shown in this chapter. We conclude this chapter with a description of the rollout of OPEX thinking to Pfizer's supply chain(s). This chapter additionally provides deep insights into the potential of artificial intelligence (AI) if used the right way.

Part IV provides a description of the way towards deep structures. The first chapter defines deep structures and how we see them in the context of OPEX. This is followed by a summary of conclusions we draw from the three first parts of the book always with an eye on drivers or reasons for deep structures. We end this book with our outlook for the future of OPEX in the industry.

References

Friedli, T. (Ed.). (2006). *Der Pharmazeutische Betrieb: Vol. 50. Operational excellence in the pharmaceutical industry*. ECV - Ed.-Cantor-Verl.

Friedli, T. (Ed.). (2010). *Der Pharmazeutische Betrieb: Vol. 51. The pathway to operational excellence: Overcoming the internal inertia*. ECV Editio Cantor.

Friedli, T., Basu, P., Bellm, D., & Werani, J. (Eds.). (2013). *Leading Pharmaceutical Operational Excellence*. Springer Berlin Heidelberg. https://doi.org/10.1007/978-3-642-35161-7

Friedli, T., Köhler, S. A., & Calnan, N. (Eds.). (2020). *Quality control lab: A crucial contributor to pharmaceutical value creation and quality system performance*. ECV Editio Cantor Verlag.

Friedli, T., Prabir, K. B., Mänder, C., & Calnan, N. (Eds.). (2018). *21c quality management in the pharmaceutical industry: The journey from compliance to excellence*. Editio Cantor Verlag.

into widespread use in the pharma industry more recently. We therefore use this opportunity to hopefully contribute to a better understanding of terms like "quality," "compliance," "beyond compliance," and "continuous improvement."

1.1 Operational Excellence Defined: The Derivation of Our 2013 Definition (cf. Friedli et al., 2013)

Modern approaches to Operational Excellence (OPEX) have evolved from the understanding of lean production (Friedli & Schuh, 2013) and are generally regarded as part of continuous corporate improvement concepts. However, OPEX programs cannot be viewed as standalone, nor as a new set of methods, as they comprise and rely on several already established manufacturing concepts (Gronauer, 2012).

The systematic discussion of OPEX can be traced back to Hayes and Wheelwright (1984), who published their groundbreaking book "Restoring Our Competitive Edge" and emphasized the relevance of manufacturing in retaining the competitiveness of the American industry. They argued that the competitive advantage of Japanese manufacturers stemmed from superior production capabilities, contrary to an erroneously assumed better product design, marketing ingenuity, or substantial financial strength. Whereas American companies lacked a consideration of manufacturing as a key to success, their Japanese competitors had already realized the benefits of streamlined operations as a competitive factor.[2] As such, the World-Class Manufacturing (WCM) project was initiated in order to identify critical success factors of successful manufacturing companies.

Based on the outcome of the WCM project, two dimensions can be distinguished in describing OPEX. The first dimension refers to the effectiveness of a production system, highlighting the role of manufacturing within an organization. The second dimension embraces the effectiveness of applied approaches and practices and thus considers the utility of selected approaches in terms of their unique combination of different methods.

Discussing the effectiveness of production systems, Hayes and Wheelwright (1984) observed that the most successful companies of their study sample had established more advanced production systems than their competitors. These production systems supported the corporate strategy directly while at the same time providing organizations with the opportunity to develop unique characteristics. The authors considered different success factors—referred to as competitive priorities—as a central aspect for developing these unique characteristics. These competitive priorities can be understood as distinctive dimensions like quality, cost, and time. By stressing dimensions differently, according to an organization's strategy,

grants {#5U01FD005675–02, #1U01FD005675-01}. The views expressed in this report do not necessarily reflect the official policies of the Department of Health and Human Services, nor does any mention of trade names, commercial practices, or organization imply endorsement by the United States Government.

[2] Hayes and Wheelwright (1984, p. 12).

companies distinguish themselves from each other, creating their own competitive advantage. Moreover, Hayes and Wheelwright argue that companies should avoid the pursuit of competitive advantages in every dimension since it is "potentially dangerous, for a company to try to compete by offering superior performance along all of these dimensions simultaneously, since it will probably end up second best on each dimension to some other company that devotes more of its resources to developing that competitive advantage" (Hayes & Wheelwright, 1984, p. 41). Rather, companies should focus on those dimensions in which they intend to develop their unique capabilities. In literature discussed as the idea of "trade-offs," this has been widely acknowledged by various authors.[3]

Hayes and Wheelwright plead that building competitive strength is dependent on a set of approaches and manufacturing practices assigned to six world-class dimensions. If combined in the right way, these practices enable companies to achieve superior performance. These dimensions are categorized as training and qualification of employees at production (workforce skills and capabilities), technical competence of management, organizational understanding of quality (competing through quality), workforce participation, degree of proprietary process and machinery development (rebuilding manufacturing engineering), and finally organizational capability to achieve progress by continuous improvement processes (incremental improvement approaches).

Schonberger (1986) considered 16 manufacturing principles to play a major role in WCM. He enriched the discussion by introducing total quality management (TQM) and just-in-time (JIT), which then evolved to the central aspects of WCM. Hall (1987) brought in another set of criteria to describe excellence; most of these match the criteria introduced by Schonberger. Hall defines "manufacturing excellence" as a system that comprises JIT production, employee participation, standardized tools and machinery, supplier integration, and design-for-manufacturability.[4]

In the early 1990s, the excellence discussion started to change, influenced by one of the most cited approaches in theory and praxis: lean production. First published for a broader audience by Womack et al. (1990), the concept of lean production, describing Toyota's superior production system, took Western manufacturing plants by storm. Numerous empirical studies researching lean production and its constituents (total productive maintenance (TPM); total quality management (TQM); just-in-time (JIT)) revealed a strong correlation between so-called lean practices and operational performance and thus contributed substantially to the concept's global dissemination.[5]

A multitude of studies analyzed the impact of modern manufacturing practices on organizations' operational performance and showed their effectiveness to be out

---

[3] Cf. Skinner (1974), Porter (1985).

[4] The approach "design-for-manufacturability" comprises various methods like quality function deployment (QFD) and house of quality and strives for an engagement of production and other product development-related departments at early stages of the development process.

[5] Cf. Sakakibara et al. (1997), Cua et al. (2001); Shah and Ward (2003), etc.

of the question. In fact, it is the context of their application that became the focus of interest.[6] Hayes and Pisano see the central characteristics of excellent manufacturing companies as follows:[7]

- Excellent manufacturing companies apply modern approaches and manufacturing practices to realize their manufacturing strategy. Their manufacturing strategy is derived from corporate objectives.
- By applying modern approaches and manufacturing practices, excellent manufacturing companies continuously develop new capabilities that provide them with a potential competitive advantage.

Yet, technical aspects summarized in manufacturing practices are only one element of modern OPEX approaches. Social aspects constitute another integral part. They can be considered as a decisive factor in the importance OPEX has gained within manufacturing companies in recent years. Although social aspects in the context of manufacturing had already been acknowledged, it was Spear and Bowen (1999) who highlighted managerial and cultural aspects and their impact on the competitiveness of the manufacturing industry in their article "Decoding the DNA of the Toyota Production System." Loch and Chick (2006) focused on the importance of managerial qualities to explain performance differences in manufacturing plants.

The explanations given above outline that performance of a manufacturing plant or of production in general is multifaceted and may only be described and explained by the interaction of different factors. Unfortunately, many approaches that define OPEX, whether they are practical or theoretical, are lacking a multifaceted and holistic consideration. Moreover, single elements or initiatives are often picked out and treated in isolation, cf. Shah and Ward (2007).

Based on literature and our experience, we see OPEX as the balanced management of cost, quality, and time while at the same time focusing on the customer needs. To achieve this end, OPEX comprises structural and behavioral changes thought to optimally support necessary activities. In order to maintain sustainability, also in changing or volatile environments, OPEX should become part of what we call "deep structures" (cf. Chaps. 26 & 27).

Obviously, OPEX is not only concerned with performance. It also encompasses the way leading to that superior performance and practices that allow an organization to continuously improve itself and fully engrain OPEX into the organization.

Thus, our definition of Operational Excellence is as follows:

> Operational Excellence constitutes the continuous pursuit of improvement of a production plant in all dimensions. Improvement is measured by balanced performance metrics comprising efficiency and effectiveness, thus providing a mutual basis for an improvement evaluation.

---

[6] C.f. Hayes and Pisano (1994), Pilkington (1998).

[7] Hayes and Pisano (1994, p. 78).

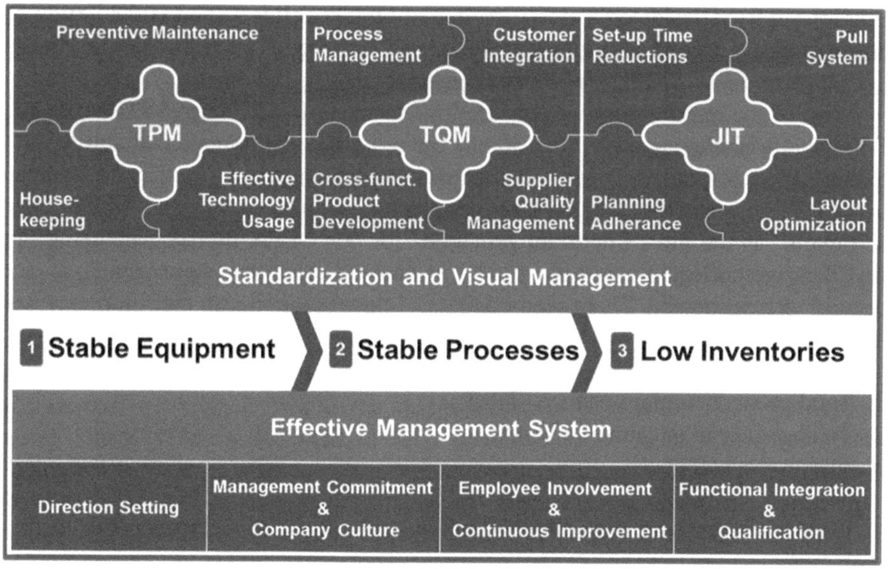

Fig. 1.1 The St.Gallen model for Operational Excellence

1.2 The St.Gallen OPEX Model[8] (cf. Friedli et al., 2013)

Our original model from 2006 is shown in Fig. 1.1. It includes several sub-elements, each of which, in itself, represents an important part that contributes to the overall success. Yet, these distinctive elements reinforce each other.

On the highest level of abstraction, the OPEX reference model can be divided into two larger sub-systems: First, there is a technical sub-system which can be regarded as a tool kit, comprising bundles of practices like total productive maintenance (TPM), total quality management (TQM), and just-in-time (JIT) and structuring them in a consistent manner. Second, there is a "social" sub-system which takes up the quest for an operational characterization of management quality and work organization. This second system focuses on supporting and encouraging people to continuously improve processes.

1.2.1 The Technical Sub-system

The objective of the technical sub-system is to analyze the implementation level of "technical" practices, which can be classified as either (core) principles or techniques/tools. Principles usually span a multitude of techniques and tools. For

[8] cf. for a more detailed description of our model Friedli et al. (2006, p. 47ff), and Friedli et al. (2010, p. 18ff.).

(e.g., unmotivated workforce, high absenteeism). In addition, successful process management is more likely to be achieved by peers working in cross-functional teams than by industrial engineers. TQM specialists suggest that companies should choose vendors primarily on the basis of quality rather than solely on the basis of product price. Moreover, supplier quality management aims to integrate suppliers into the internal quality system to ensure high quality levels. To achieve excellent quality, it is essential to know what customers want and to provide products to meet their needs (customer integration). TQM experts point out that cross-functional product development should help to translate customer requirements into high-quality products.

1.2.4 Just-in-Time (JIT)

Due to the fact that in most industries, the heterogeneity of customer requirements has significantly increased, JIT manufacturing for most companies has become a crucial element to increase flexibility without building up huge inventories. We defined "pull production," "setup time reduction," "layout optimization," and "planning adherence" to be the sub-elements of a JIT production.

While pull production helps to reduce overproduction, inventory and setup time reductions can help to decrease the average lot size. Besides, it can enable a smooth material flow within the manufacturing process. With the need for standardized, stable, reliable processes, we regarded the element "planning adherence" as a further element of JIT. Planning adherence, which means smoothly leveling out the production schedule in both volume and variety, should keep the JIT system stable and allow for minimum inventory. Apart from waste caused by overproduction and excess inventory, an integrated JIT program also endeavors to reduce all kinds of excessive movements caused by material and handling. Hence, layout optimization based on close arrangements of people and equipment in a processing sequence is an additional principle of JIT implementation.

1.2.5 Basic Elements

After structuring the OPEX model according to the three elements TPM, TQM, and JIT, we realized that there are some common practices shared by all three subsystems and that are not unique to each of the programs. These are, for example, cross-functional training, employee empowerment and teamwork, standardization, and visual management. We decided to differentiate between technical practices such as standardization and the more socially oriented aspects such as employee empowerment or cross-functional training (which we will discuss later).

The following two technically oriented practices cannot be solely related to JIT, TQM, or TPM: standardization and visual management. We call them basic elements because they can be regarded as basic prerequisites for successfully implementing TQM, TPM, as well as JIT principles. As Imai (1986) explained in his book on continuous improvement, it is impossible to improve any process before it has been standardized and thus stabilized. Standardization not only refers to processes; it also includes the standardization of technology and equipment. Standardization can be regarded as a common supportive element for TPM, TQM, and JIT. A further basic element is visual management. It provides the workforce with updated information on process and performance data, which assists the deployment of TPM, TQM, and JIT principles (e.g., visual management can provide timely information regarding JIT-related data as, for instance, the actual takt time to enhance flow as well as TQM- or TPM-related information such as process variability or equipment reliability to improve problem-solving). Based on our experience working with numerous pharma companies, we later added problem-solving as an additional category to the basic elements.

1.2.6 The Social Sub-System: Management System

Based on different sources, we developed a management quality model, the objective of which can be summarized as follows: "motivating and aligning people to work for a common goal." To achieve a common goal, employees need autonomy; they need to feel that they have control over their job and belong to a team (people involvement). Targets have to be clear and consistent, as well as challenging (direction setting) and supported by senior management (management commitment). Feedback on progress needs to be given frequently and in a timely manner, and multiple skills should be developed according to individual potential (functional integration and people development).

1.2.7 Direction Setting

We completely agree with Skinner (1974), Hayes et al. (2005), and Loch et al. (2004) that the implementation of certain practices only makes sense if the management has formulated a strategy that is based on clear and consistent objectives. However, we decided to merge the integration variable and the direction setting variable because they are strongly correlated and conceptually interlinked.

1.2.8 Management Commitment

Evidence from several studies supports that management commitment is a crucial element for facilitating change processes, which is a prerequisite for process improvement. Especially TQM literature stresses management commitment as one of the key success factors. A quality improvement process must begin with the management's own commitment to quality. However, management commitment is not just vital for rolling out a TQM program; it is equally important for rolling out a JIT and TPM program. The management has to promote a culture which supports people in doing their work.

1.2.9 Employee Involvement and Continuous Improvement

We also included employee involvement in our management quality model. We strongly believe that one major managerial challenge is to get all employees involved in continuously thinking about how to improve the current situation. This is only possible if process improvement is a common task for everybody and not just for a few smart industrial engineers.

1.2.10 Functional Integration and People Development

A workforce that is eager to contribute to the goals set by the management but lacks proper know-how in doing so will fail in achieving them. If complex decisions are to be delegated, as it is the case in the concept of autonomous problem-solving, this can only succeed if employees are given the chance to acquire new knowledge. Furthermore, the flexibility of the technical system we have introduced (especially the concept of mixed model production) requires a multiskilled workforce that can perform different tasks. Consequently, functional integration and employee development are basic pillars for the achievement of goals set by the plant management. We also introduced the aspect of feedback and reward as an enabler for motivation. Organizational scientists argue that rewards go beyond money. The important point is that positive or negative reinforcement ensues as quickly as possible after the action.

1.2.11 Excellence in Pharma Manufacturing

The above section aimed at explaining our OPEX model and our understanding of excellence in general. In the following, we will transfer this understanding to excellence in pharmaceutical manufacturing. The performance of a manufacturing plant can be described in several ways. Obviously, high performance in a single Key Performance Indicator (KPI) can be derived from a mere focus on that respective KPI. However, this is not the case for excellence. Our understanding of an excellent manufacturing site is influenced by the St.Gallen school of systems theory and cybernetics.

Thinking in systems determines the holistic consideration of the system (organization) itself. The systemic approach acknowledges that an organization is embedded in economy and society (Ulrich, 1984). The complexity of situations and problems is recognized, and it is sought to consider them in their full context. Isolated consideration of single aspects and inexpedient problem definitions should be avoided (Ulrich & Krieg, 1974). As there is already only a limited number of opportunities to monitor and interpret surrounding events and happenings, it is difficult, sometimes even impossible, to reflect interrelations objectively (Bleicher, 1995). Thus, it might be misleading to artificially narrow down the horizon. The conscious management of those social systems and therefore an adaptation of structures and people's behavior is an approach to mitigate these shortcomings. Moreover, evoking self-organized adaptations from the system itself requires approaches that are based on an evolutionary organizational development in contrast to the steering and directing of just a few managers. As such, also Bleicher (1995) concludes to strive for a holistic consideration in order to satisfy the connectivity of problems and interrelations within the social system.

That said, we rely on our model's entire technical sub-system in order to distinguish excellent plants from weaker-performing ones. Transparent manufacturing operations of an organization are achieved by assigning a distinctive set of KPIs to TPM, TQM, and JIT. This, however, implies that excellent manufacturing sites have high performance on a holistic level instead of high performance on a single KPI only. If only a single KPI is analyzed, even average or weak-performing sites might excel. But due to the balanced approach of effectiveness and efficiency, overall performance of the manufacturing system is critical to qualify as operationally excellent.

1.3 Further Key Terminology Defined

Under a contract with the US FDA from 2023 to 2025, a comparison of terminology from the literature with insights from over 80 interviews with OPEX and quality experts in the pharmaceutical and MedTech sectors was conducted to establish consolidated definitions of key terms with a relevance for the FDA/CDER Quality Management Maturity Program (cf. FDA, 2025). This approach integrates academic and industry perspectives, providing a shared foundation for understanding key quality-related concepts.

Quality has evolved from a product-focused concept to an enterprise-wide perspective, emphasizing organizational accountability and patient safety. It encompasses both technical requirements and broader obligations to deliver consistent, reliable outcomes.

> Quality is about consistently meeting defined requirements and specifications to ensure compliance, product integrity, and patient safety through effective and efficient processes.

While technical measures alone are insufficient to achieve superior outcomes, quality culture has emerged as a crucial enabler for embedding quality values and behavior across the organization. A strong quality culture fosters ownership and empowers employees to make quality-driven decisions.

> Quality culture is a shared commitment—driven by leadership and embraced across the organization—to consistently uphold high standards through accountability, proactivity, and ethical behavior in everyday work.

In highly regulated industries, compliance reinforces quality by providing the framework necessary to operate safely and legally. It ensures that organizations meet external requirements while supporting internal quality objectives.

> Compliance is about adhering to regulations and requirements set by authorities to meet standards, ensure safety, and support the license to operate.

Triggers and Forces 2

Thomas Friedli

Abstract

Researching Operational Excellence (OPEX) in the pharmaceutical industry, we observed that there must be reasons to engage in serious efforts to become an "operationally excellent company." In this chapter, we provide an overview of what has triggered OPEX over the years and which forces have made companies move in this direction but also what has worked against a deep implementation. We will first define trigger and forces before describing the selected ones we have witnessed over time.

What became obvious to us researching Operational Excellence (OPEX) in the pharmaceutical industry for 20 years is that there must be reasons to engage in serious efforts to become an "operationally excellent company". If this is not given or it's not transparent, the whole program will remain skin-deep and fade away over time. What we intend to do in this chapter is to provide an overview of what triggered OPEX over the years and what forces made companies move in this direction but also what worked against a deep implementation. We will first define trigger and forces before describing selected ones we have witnessed over time.

T. Friedli (✉)
Institute for Production and Supply Chain Management, University of St.Gallen, St.Gallen, Switzerland
e-mail: thomas.friedli@unisg.ch

T. Friedli et al. (eds.), *Striving for Operational Excellence in the Pharmaceutical Industry*, Future of Business and Finance,
https://doi.org/10.1007/978-3-032-26287-5_2

2.1 Definitions

The Cambridge Dictionary defines a trigger as "an event or situation, etc. that causes something to start".

A force is defined there as "physical power or strength", "to force" as "to make an object move by pushing or pulling it hard".

2.2 Historical Development of OPEX in Pharma

For the early history of OPEX in the pharmaceutical industry we will first provide a summary from our previous book chapter describing it until 2012 (cf. Friedli et al., 2013, Sect. 1.3). After this we will discuss some more recent developments and draw some conclusions.

2.2.1 The History up to 2012 (cf. for a more Extended Version, Friedli et al. (2013), Sect. 1.3)

"As pharmaceutical manufacturing evolves from an art to a science- and engineering-based activity, application of this enhanced science and engineering knowledge in regulatory decision-making, establishment of specifications, and evaluation of manufacturing processes should improve the efficiency and effectiveness of both manufacturing and regulatory decision-making."[1]

"..., industry's hesitancy to broadly embrace innovation in pharmaceutical manufacturing is undesirable from a public health perspective. Efficient pharmaceutical manufacturing is a critical part of an effective U.S. health care system. The health of our citizens (and animals in their care) depends on the availability of safe, effective, and affordable medicines."[2]

Compared to other industries, the pharmaceutical industry was rather late to adopt OPEX programs. There were only a few isolated actions with a rather limited scope by the late 1990s, mostly generated in single plants and not part of a real global program. In the first decade of the 2000s there was suddenly an appetite for embracing OPEX, mostly driven by the USA FDA which expressed concerns about the absence of science in manufacturing and the hesitancy in industry to embrace innovations (see the two quotes above). In the early beginning the US FDA can be seen as the trigger as well as the force that made the industry move. Additionally, the industry had to react to increased economic pressure stemming from the rise of the generics industry and challenges with R&D productivity. Pioneering companies were, for example, Pfizer, launching a huge Six Sigma-like program and sending the

[1] FDA, Final Report "Pharmaceutical CGMPs for the 21st Century – A Risk-Based Approach", September 2004.

[2] Guidance for Industry PAT – A Framework for Innovative Pharmaceutical Development, Manufacturing, and Quality Assurance, FDA, September 2004, Page 3.

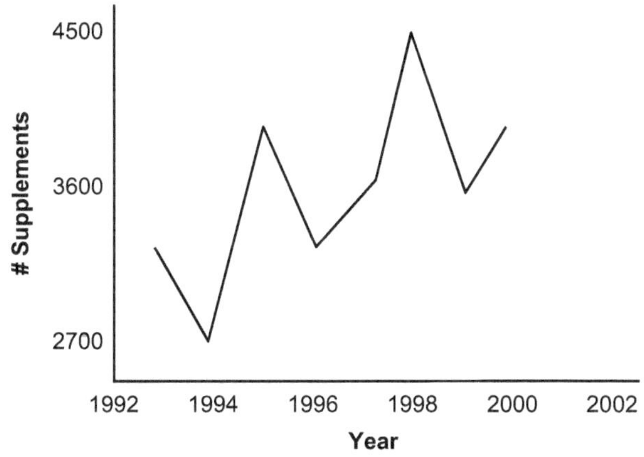

Fig. 2.1 Post-approval manufacturing supplements (Cf. Presentation from Janet Woodcock, Meeting of the FDA science board, November 16th, 2011)

company on a trajectory to remain at the forefront in pharmaceutical OPEX ever since (e.g. Chaps. 4, 16, 23, and 25).

The discussion about a more scientific approach to pharmaceutical manufacturing was seriously launched in a meeting of the scientific advisory board of the FDA end of 2001. One of the difficulties the agency faced was the increasing number of post approval manufacturing supplements (see Fig. 2.1).

This situation led to major problems for the FDA to fulfill its inspection obligations and also made clear that the industry faced major difficulties with the scientific mastering and understanding of its production processes. It was constated that Good Manufacturing Practice (GMP) worked empirically rather than science-based and the industry as well as the regulators were risk-averse.[3]

Doug Dean and Francis Brutton from PricewaterhouseCoopers (PwC) presented at the same meeting a rather bleak analysis about the status of pharmaceutical manufacturing. They came to the following conclusions:[4]

- The status quo is untenable
- Pharmaceutical manufacturing—lots of room for improvement
- Traditional metrics hide poor performance
- Compliance infrastructures are not economic
- Technologies are critical enablers—but not in isolation
- Huge potential for industry and regulators to create a win-win

They further identified some of the reasons for this situation, for example: Processes are transferred that are neither fully understood or capable at commercial scales, lengthy & elaborate new product introduction exercises that generate data but fail to provide critical information, 50% of production costs locked in before

[3] Cf. Presentation from Janet Woodcock, Meeting of the FDA science board, November 16th, 2011.

[4] Cf. Presentation from Dean and Bruttin, Meeting of the FDA science board, November 16th, 2011.

Phase III begins, process inefficiencies "institutionalized" and no scientific basis for trading-off time in return for deeper process understanding.

Therefore, data became another critical trigger, making managers aware of the waste and numerous other issues in pharmaceutical production.

Both the industry and the FDA were well aware of the deficiencies in pharmaceutical manufacturing. To move forward they jointly encouraged the use of innovative technologies to enhance process understanding and to establish science and risk-based approaches to quality and regulatory processes. Process analytical technology (PAT) was selected by the FDA as a pilot to evaluate how they could further promote this thought. The PAT team and manufacturing science work group stated in the executive summary of a report: "Pharmaceutical manufacturing operations are inefficient and costly. The cost of low efficiency is generally not understood or appreciated (e.g., manufacturing costs far exceed those for research and development operations). Low efficiency is predominantly due to "self-imposed" constraints in the system (e.g., static manufacturing processes, focus on testing as opposed to quality by design, approach to specifications based on discrete or the so called "zero tolerance" criteria, a less-than-optimal understanding of variability, etc.). These constraints keep the system in a corrective action mode. Continuous improvement is an essential element in a modern quality system, and it aims at improving efficiency by optimizing a process and eliminating wasted efforts in production. In the current system continuous improvement is difficult, if not impossible."[5] In response to these findings, the FDA changed its position of focusing on product purity and potency as its measure of quality, in favor of a regime spending more time on trying to address issues dealing with actual physical manufacturing processes. For example, what effects, if any, do small changes in the reactor vessel, blending, drying, pressing, coating or other manufacturing steps have on the final dosage form?[6] The main objective was to obtain a more thorough understanding of pharmaceutical manufacturing processes and, with this understanding, more predictable and efficient manufacturing. PAT should be thought of as a system for designing, analyzing and controlling manufacturing through timely measurements, taken during the process or processes, of critical quality and performance attributes, and of raw and in-process materials and processes, with the goal of ensuring final product quality. Associated benefits such as faster development of new products; shorter manufacturing cycle times; higher yields; reduced waste materials; and fewer product recalls should also be achieved by the greater understanding of the process gained by using this methodology.[7] Therefore, a focus on process understanding can be added to the list of early triggers for OPEX.

The PAT team and the initiative preceded the broader pharmaceutical Current Good Manufacturing Practices (cGMP) initiative by about a year. In August 2002, the FDA announced this significant new initiative for the 21st century to enhance and modernize the regulation of pharmaceutical manufacturing and product

[5] Cf. Report of the PAT Team and Manufacturing Science Working Group 2004.

[6] Cf. Clark, G. (2004): FDA's PAT initiative, in: Pharmaceutical Technology Europe.

[7] Clark (2004).

quality—to bring a 21st century focus to this critical FDA responsibility. The main objectives of this initiative were: *(1) encourage the early adoption of new technological advances by the pharmaceutical industry, (2) base regulatory review and inspection policies on state-of-the-art pharmaceutical science, (3) facilitate industry application of modern quality management systems, (4) use risk-based approaches that focus both industry and agency attention on critical areas*; and *(5) incorporate enhanced quality system approaches into the agency's business processes.*[8]

The initiative also stated a so-called desired state for pharmaceutical manufacturing:

- Product quality and performance achieved and assured by design of effective and efficient manufacturing processes
- Product specifications based on mechanistic understanding of how formulation and process factors impact performance
- Continuous improvement approaches, with innovative use of new technology as desired
- Continuous "real time' assurance of quality

The PAT activities became part of cGMP. The interesting aspect of cGMP and the current thinking of the agency was and is that economic considerations impacted the approach. Quality and productivity surfaced on the agenda of the agency, giving industry new opportunities under this changed paradigm: "Quality and productivity improvement share a common element—reduction in variability through process understanding (e.g., application of knowledge throughout the product lifecycle). Reducing variability provides a *win-win* opportunity from both public health and industry perspectives. And, since manufacturing technologies and practices are generally similar between both innovator and generic companies, facilitating efficiency improvements provide opportunities for both sectors of the pharmaceutical industry."[9]

Later FDA activities were all based on the same underlying idea: to modernize the scientific base for pharmaceutical manufacturing and pharmaceutical quality management. This could be observed as well in the documents about the "Critical Path Initiative" as in the work along of ICH Q8-Q10 and the [Quality by Design (QbD)] initiative as well as in the process validation guideline released in 2011.

Words like "continuous improvement"[10] became common when thinking about manufacturing but there was also an awareness that changing a paradigm that had lasted for so long in this industry will take time and efforts: Continuous improvement is an essential element in a modern quality system. Its aim is to improve efficiency by optimizing a process and eliminating wasted efforts in production.

---

[8] Cf. Report of the PAT Team and Manufacturing Science Working Group 2004.

[9] FDA, Final Report "Pharmaceutical CGMPs for the 21st Century – A Risk-Based Approach", September 2004.

[10] The current understanding is discussed in Chap. 12.

Dimension	Definition	Examples of Activities
Assessing Safety	Show that product is adequately safe for each stage of development	• Preclinical: show that product is safe enough for early human testing Eliminate products with safety problems early • Clinical: show that product is safe enough for commercial distribution
Demonstrating Medical Utility	Show that the product benefits people	• Preclinical: Select appropriate design (devices) or candidate (drugs) with high probability of effectiveness • Clinical: Show effectiveness in people
Industrialization	Go from lab concept or prototype to a manufacturable product	• Design a high-quality product - Physical design - Characterization - Specifications • Develop mass production capacity - Manufacturing scale-up - Quality control

Fig. 2.2 The three dimensions of the Critical Path (FDA (2004): Challenge and Opportunity on the Critical Path to New Medical Products)

Improvement efforts are carried out in a structured manner with appropriate predefined protocol and oversight. These efforts are primarily directed towards reducing variability in a process and product quality characteristics and are not for changing the fundamental design of a manufacturing process. Generally, the term *continuous improvement* is broadly used for all improvement efforts including those that result from *corrective actions*. In the regulatory setting a distinction between corrective action and continuous improvement is essential. Need for corrective actions occur when product quality characteristics are in question (e.g., out of specification). Such a situation can require urgent risk assessment and sound quality decisions to prevent any adverse impact on patients. In the current state corrective actions are the dominant mode for improvement and continuous improvement is difficult."[11]

The Critical Path Initiative from 2003 had its own industrialization perspective (cf. Figs. 2.2 and 2.3) and QbD was the logical next step: making sure that introduced processes were better understood from the launch.

In summary, the industry built as well on the broader use of manufacturing science as on a related push to lower manufacturing costs. As regulators were part of this development, it was clear from the start that this had to be done without sacrificing quality. The majority of pharma companies therefore started with a more

[11] FDA, Final Report "Pharmaceutical CGMPs for the 21st Century – A Risk-Based Approach", September 2004.

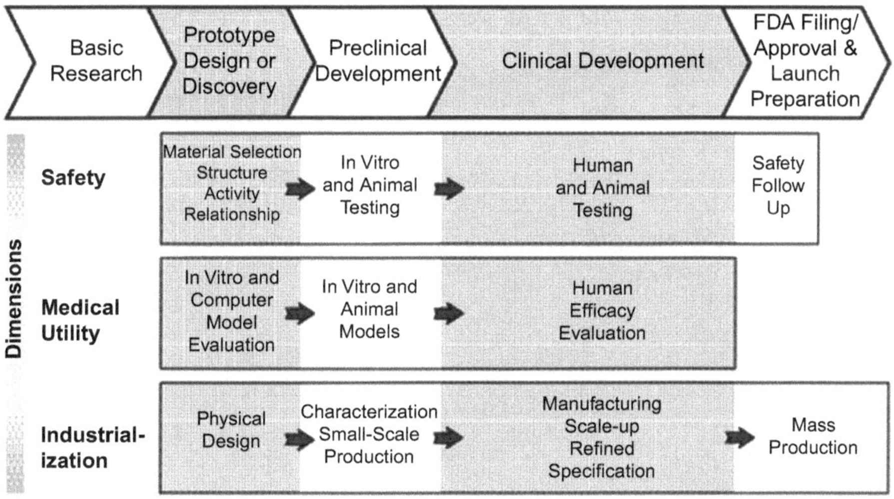

Fig. 2.3 The industrialization perspective along the product lifecycle (FDA (2004): Challenge and Opportunity on the Critical Path to New Medical Products)

sustainable approach for OPEX than many other industries, following the learnings from the sand cone model (cf. Chap. 6), which demands that cost improvements can only be sustained when built on-top of a solid foundation in quality.

In summary we can identify the following triggers/forces in the first decade:

- Regulators and especially the US FDA requiring more science when it comes to manufacturing
- Data demonstrating waste and issues across manufacturing
- A new process focus going away from "only" looking at the product specifications
- The quest for cost improvements

The introduction happened in stages and is described in more detail in Gronauer et al. (2010). The first phase was the "pre-OPEX" phase which lasted until the late 1990s, followed by a "best-practice transfer" phase which gave way to today's "transformation" phase. Already in 2010, we predicted another phase an "integrated operations systems" phase as the coming dominating pattern. Whereas some pharmaceutical companies are on the threshold of entering the fourth phase, the big majority is still far away from this stage. Therefore, this book makes integration an integral part of numerous chapters and will have a dedicated chapter about deep structures, that have to be built to make this move sustainable. The pathway to OPEX in the pharmaceutical industry and its four phases are illustrated in Fig. 2.4.

Even the most advanced OPEX programs today, i.e., the ones targeting the "integrated operations systems" phase, did not evolve over night. They went through all the other phases as well. The pre-OPEX phase is characterized by isolated

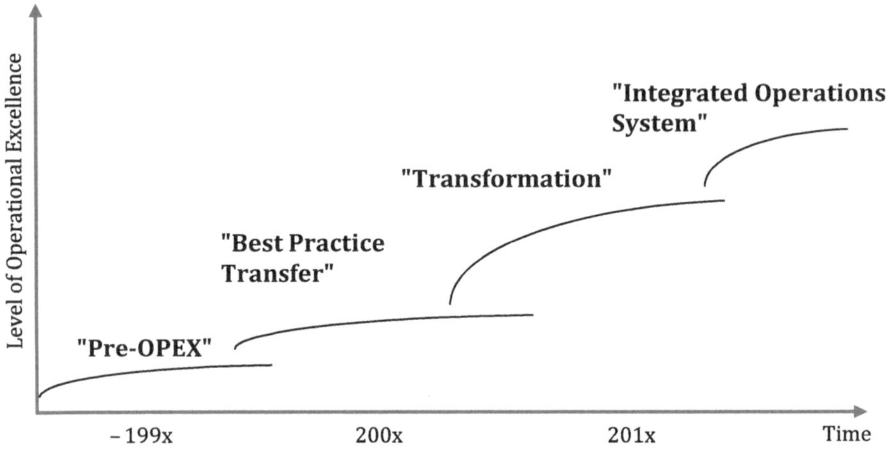

Fig. 2.4 The pathway to OPEX (Gronauer et al., 2010)

improvements in manufacturing that are not the result of a structured and carefully designed approach. Changes are only introduced reluctantly, the underlying culture is a "no-changes"-culture. With the described intentions by the FDA and the rising cost pressure a new approach was needed. Pharmaceutical production managers visited plants from BMW, Audi, Toyota etc. The main intention was to copy successful practices and apply them on the pharmaceutical production floor. After some initial successes there was a sudden realization that simply copying training programs is not enough to get a buy-in from the employees. The next phase was therefore consistently focused on people and designed as huge change management approaches. This transformation stage is where most of the more advanced companies have been back in 2013. We foresaw in the future more integrated approaches, on the one hand bringing together preventive with reactive OPEX (e.g. QbD and OPEX combined) and on the other hand the alignment of all improvement initiatives at the top-management level. This is a necessity to make sure that people will understand why OPEX does help at the end of the day.

2.2.2 Why There Is Still a Long Way to True Integration

The OPEX programs we described back in 2013 certainly had an impact on the respective companies and their employees. However, there were new triggers and forces coming into play in recent years and the pressure to further reduce costs never loosened up. The cost pressure in some cases brought with it the reduction of the people in central OPEX teams as well as in travel budgets. Therefore, it was no longer possible to support OPEX at every single site. The COVID-19 pandemic additionally demonstrated that quite some supply chains were just not robust what brought back higher inventories across the board (see Chap. 3). Finally, the push to invest in digitalization, often poorly aligned with OPEX, took away additional resources from OPEX and reduced the credibility of the programs. Whereas in the first decade of OPEX in the industry almost all triggers and forces pushed the companies towards deeper implementation and integration the second decade brought detrimental forces and triggers with it. For about two years now we are back again in a noticeable drive towards more integration and deeper implementation. This is partly driven again by the US FDA pushing programs like Quality Management Maturity (QMM), recognizing advanced quality management practices and quality culture. Nevertheless, the current drive is rather based on a technical and not a human-centric approach. We will describe later what we believe would be the "way to go" to finally implement deep structures (cf. Chap. 27).

Triggers/forces in the second decade:

1. Working against deeper and more integrated implementation:
   - Cost pressure leading to the reduction of OPEX resources and travel budgets
   - Experiences during the pandemics
   - Digitalization poorly aligned with OPEX and taking away resources
   - Resulting loss in credibility
2. Working towards deeper and more integrated implementation:
   - The realization that programs have to be aligned to have a sustainable impact
   - A recognition that OPEX first is the way towards impactful digitalization
   - New US FDA programs like Quality Management Maturity (QMM) recognizing the importance of advanced quality management practices and quality culture

2.3 Conclusions

A combination of different reasons led to the rise of OPEX in the pharmaceutical industry in the first decade. The second decade did not fully build on this and in some cases even led to a regression on the way to sustainable OPEX implementations. However, this helped to understand the real requirements for a sustainable implementation of OPEX in the pharma industry. We come back to this in Part IV of this book.

References

Clark, G. (2004). FDA's PAT initiative, in: pharmaceutical technology Europe. https://www.pharmtech.com/pharmtech/Analytical/article/detail/129539. Retrieved February 2026.

FDA. (2004). Challenge and opportunity on the critical path to new medical products. https://www.fda. gov. (Archive), Retrieved February 2026.

Friedli, T., Basu, P., Bellm, D., & Werani, J. (Eds.). (2013). *Leading pharmaceutical operational excellence*. Springer Berlin Heidelberg. https://doi.org/10.1007/978-3-642-35161-7

Gronauer, T., Friedli, T., & Götzfried, M. (2010). The roadmap to operational excellence – pattern and elements of OPEX programs. In: Friedli et al. (Eds.), The pathway to operational excellence – overcoming the internal inertia. Editio Cantor Verlag, Aulendorf.

Reassessing the State of Pharmaceutical Operational Excellence: A (R)evolution in Figures

3

Fabian Klinkner and Thomas Friedli

Abstract

After a decade of global pandemics, geopolitical conflicts, economic turbulences, and the advent of new technologies, we feel obligated to revisit and update our assessment on the state of Operational Excellence (OPEX) in the pharmaceutical industry. We perform our analyses on the same premise as our previous assessment (cf. Friedli et al., 2013), capturing the state of pharmaceutical OPEX through both maturity and performance elements across the building blocks of our St.Gallen OPEX model. Further, we identify overarching trends and developments the industry faces before drawing conclusions from our reassessment.

3.1 Introduction

In our 2013 book publication "Leading Pharmaceutical Operational Excellence" (Friedli et al., 2013), we assessed the state of Operational Excellence (OPEX) approximately 10 years after its emergence in the pharmaceutical industry. Now, another decade later, we feel obligated to revisit and update our assessment. While the initial push for OPEX in pharmaceutical manufacturing was motivated by increasing cost pressure, productivity crises, and regulatory aspirations, more disruptive forces have reshuffled the industrial landscape in recent years (cf. Chap. 2). Global pandemics, geopolitical conflicts, economic turbulences, and the advent of new technologies present unprecedented challenges and opportunities. In light of

F. Klinkner (✉) · T. Friedli
Institute for Production and Supply Chain Management, University of St.Gallen, St.Gallen, Switzerland
e-mail: fabian.klinkner@unisg.ch; thomas.friedli@unisg.ch

T. Friedli et al. (eds.), *Striving for Operational Excellence in the Pharmaceutical Industry*, Future of Business and Finance,
https://doi.org/10.1007/978-3-032-26287-5_3

these developments, being operationally excellent becomes inevitable to maintain a competitive stance as a pharmaceutical manufacturer. This raises the following question: To which extent has the pharmaceutical industry evolved in OPEX over the last decade?

To answer this question, we perform our analyses on the same premise as our previous assessment (cf. Friedli et al., 2013). Since 2004, our OPEX research team at the University of St.Gallen has been conducting an international benchmarking project on the implementation of OPEX in the pharmaceutical industry. Having grown our database by over 75% since our previous assessment, as of November 2025, the St.Gallen OPEX database includes data of 438 pharmaceutical manufacturing sites (drug substance and drug product) from over 160 different companies, ranging from small- and medium-sized organizations to global enterprises. This chapter presents results from our analysis of more than 20 years of benchmarking data (2003–2024). This comparison is presented in reference to the two sub-systems of the St.Gallen OPEX model (see description in Chap. 2a): (1) the technical sub-system, consisting of the building blocks' total productive maintenance (TPM), total quality management (TQM), and just-in-time (JIT) and (2) the social sub-system, represented by the effective management system (EMS). Further, these sub-systems are complemented by a fifth building block, basic elements (BEs) (cf. Chap. 2a).

The state of OPEX is often mischaracterized and oversimplified, either as traces of practice implementation or as measurable operational results. However, only if companies excel in both these domains at the same time, they reach a state of world-class OPEX (Voss et al., 1995). This belief is deeply anchored in our benchmarking approach. Therefore, we determine the state of OPEX in the pharmaceutical industry by two indices: (1) maturity, which is assessed by the implementation of specific enablers (defined as "methods and tools leading to better performance"), and (2) performance, which is calculated through key performance indicators (KPIs) (defined as "metrics for tracking progress and comparing with other companies").[1] Our conceptualization of OPEX and consequently our approach to assess its state are based on well-established scientific models, specifically the Voss Matrix (Voss et al., 1995) and the sand cone model (Ferdows & de Meyer, 1990). While the former outlines the interplay between maturity and performance, the latter describes how quality, time, and costs can be synchronized. Proposing an alternative to trade-off theory, which states that excelling in these three dimensions simultaneously is not feasible, the sand cone model understands quality as the foundation to building manufacturing capabilities. To improve in dependability, speed, and cost, this foundation must be expanded first—like building a sand cone, which cannot be built from the top but relies on a solid base to grow in height.

The remaining chapter is structured as follows: Sections 3.2 and 3.3 assess the state of OPEX in maturity and performance, whereas Sect. 3.4 identifies

[1] Enablers and KPIs featured in the St.Gallen OPEX benchmarking are listed in the appendix of this chapter.

overarching trends and developments in relation to the aforementioned disruptive forces. Finally, we draw our conclusions in Sect. 3.5.

3.2 The State of Pharmaceutical Operational Excellence Maturity

The assessment of maturity is based on companies' self-reported degree of implementation of 136 enablers on a Likert scale from 1 (*not implemented*) to 5 (*fully implemented*). Corresponding maturity sub-scores are calculated for the respective elements of the St.Gallen OPEX model and aggregated into an overall maturity score, ranging from 0% (*not implemented*) to 100% (*fully implemented*). As of November 2025, the median level of pharmaceutical OPEX maturity stands at 66.6%, indicating an increase by 3.5%pts. since 2013. In the same sense, main elements of the St.Gallen OPEX model display a positive delta from the baseline year. The strongest increase is recorded in JIT (+10%pts.) and TPM (+7.6%pts.), whereas TQM (+1.2%pts.) and EMS (+2.4%pts.) showcase the lowest increment. For TQM, this is in line with expectations, as the regulatory environment made pharmaceutical companies focus on at least some quality-related enablers already before the advent of OPEX. Meanwhile, BE (+5.8%pts.) splits the field, displaying moderate growth.

Detailed observations on the evolution of maturity can be made by applying a more granular lens to these elements. While the baseline delta is positive for most sub-components, enabler implementation has also regressed in certain aspects. Figure 3.1 displays the maturity growth relative to the baseline year. Further,

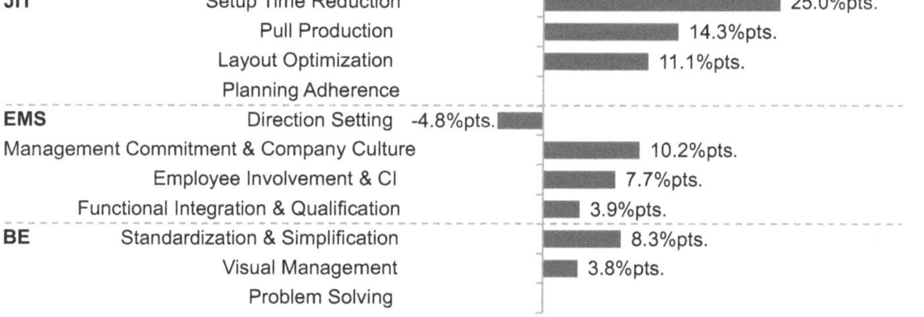

Fig. 3.1 Overall maturity evolution (baseline year: 2013)

distinct changes in trajectory compared to the first decade of pharmaceutical OPEX emerge. All these observations are elaborated in the following sections.

3.2.1 Total Productive Maintenance

The strong growth in TPM maturity since 2013 can primarily be attributed to a surge in technology assessment and usage (+29.2%pts.) and housekeeping (+16.7%pts.). Meanwhile, growth in preventive maintenance (+4.7%pts.) has remained rather steady since our previous assessment. The maturity evolution across all three elements is displayed in Fig. 3.2.

Preventive maintenance programs help companies to maintain continuity and stability in their manufacturing processes. At the point of our current analysis, TPM maturity stands at 78.1%, having grown from 73.4% since 2013. This boost is linked to a stronger focus on the documentation of maintenance plans and checklists (D02: +25.0%pts.) and the continuous improvement of maintenance programs based on failure analyses (D05: +25.0%pts.). Growth in maturity is, however, dampened by decreasing emphases on maintenance as a strategy for quality and planning compliance (D03: −12.5%pts.). Further, companies provide less assistance to operators in performing preventive maintenance tasks independently (D06: −25.0%pts.).

Previously, we noted a decrease in technology assessment and usage between 2004 and 2013. Although pharmaceutical companies used new technologies and state-of-the-art machines, they neither expanded their focus on proprietary process technologies, nor did the number of patents on established processes increase. This trend has reversed since, as companies now rely on proprietary equipment and

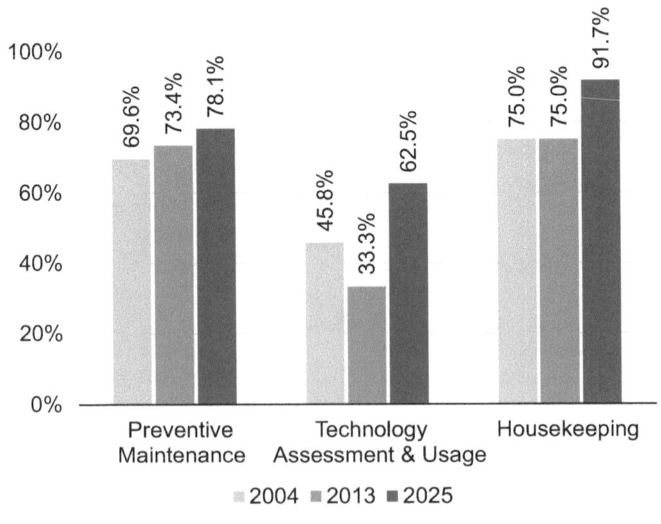

Fig. 3.2 TPM maturity evolution

process technology much more to gain competitive advantages (D13: +100.0%pts.; D14: +50.0%pts.). Thus, maturity for this category has climbed to 62.5%.

Due to regulatory requirements, housekeeping, concerned with maintaining an orderly production environment, has always been relatively mature in pharmaceutical manufacturing. Nevertheless, since our previous assessment of the status quo, its implementation has risen further to 91.7%. This increase is tied to further emphasis on cleanliness (D15: +25.0%pts.) and more thorough documentation via checklists (D17: +25.0%pts.).

3.2.2 Total Quality Management

TQM maturity is marked by a more volatile development since 2013. While an incline is recorded in cross-functional product development (+10.0%pts.) and supplier quality management (+3.6%pts.), a decline emerges in process management (−1.7%pts.) and customer involvement (−4.2%pts.). Therefore, the universal growth across all four components, which we highlighted in 2013, has partially stagnated (see Fig. 3.3).

The lower enabler implementation in process management is tied to a lower utilization of statistical process control (SPC) (D23: −37.5%pts.) and process analytical technology (PAT) (D26: −25.0%pts.). Instead of relying on these conventional methods, companies increasingly turn to other solutions. By leveraging artificial intelligence to track and analyze multiple process parameters and their interdependencies simultaneously, companies overcome limitations of SPC, which only considers individual parameters. Thus, process management maturity drops to 63.9%.

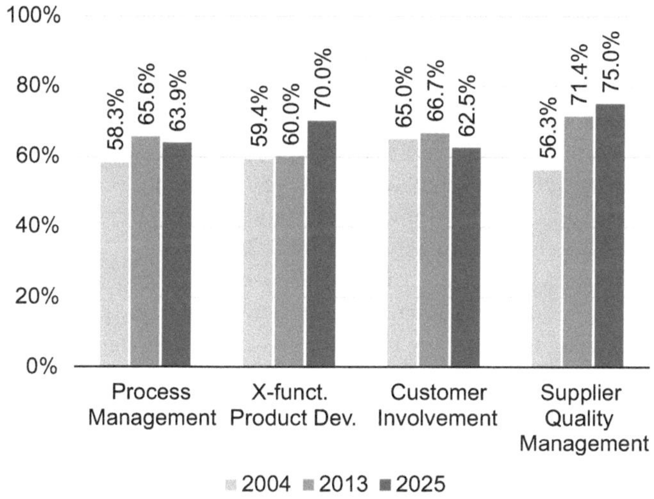

Fig. 3.3 TQM maturity evolution

Increasing maturity in cross-functional product development is caused by a more sophisticated integration of product and process development (D28: +25.0%pts.), including greater involvement of manufacturing engineers in the development of new drug formulations and respective production processes (D27: +25.0%pts.). Thus, timeliness of product launches has improved (D30: +25.0%pts.). However, the collaboration between R&D and manufacturing departments has worsened, dropping to 2004 levels (D30: −25.0%pts.).

Maturity in customer involvement has decreased due to companies paying less attention to surveying customer requirements (D34: −25.0%pts.) and collecting feedback on quality and delivery performance (D33: −25.0%pts.). This stands in contrast with an intensified on-time-delivery philosophy (D36: +25.0%pts.). Such contradictions provide evidence for the struggle companies face to maintain consistency in their OPEX efforts.

A change in philosophy, and a partial departure from previous developments, can also be observed for supplier quality management. Compared to 2013, the focus has shifted away again from selecting suppliers on the premise of quality (D38: −25.0%pts.). Instead, greater efforts are put into inspecting incoming shipments of parts and materials (D41: +75.0%pts.; D43: +25.0%pts.) to detect below-par deliveries from suppliers. Thus, enabler implementation in supplier quality management reaches an all-time high at 75.0%.

3.2.3 Just-in-Time

JIT records the greatest boost in maturity since our previous assessment, as three out of its four sub-components exhibit a positive delta from the baseline year. Setup time reduction accounts for the largest growth (+25.0%pts.), followed by pull production (+14.3%pts.) and layout optimization (+11.1%pts.). While general planning adherence maturity has remained unchanged (±0%pts.), variations at the enabler implementation level are apparent. The evolution of maturity across all JIT elements is displayed in Fig. 3.4.

The improvement in setup time reduction can be attributed to two developments, First, companies put greater emphasis on the optimization of batch sizes (D48: +50.0%pts.) and production schedules (D49: +50.0%pts.). Second, machine operators receive more support in minimizing setup and cleaning times (D46: +25.0%pts.) and practice these activities extensively (D47: +25.0%pts.).

Pull production, which stands at a maturity of 64.3%, is also affected by this optimization in scheduling. Allowing to catch up on delays, production schedules are increasingly designed to compensate for production stoppages (D51: +25.0%pts). Further, companies have increased the share of long-term supplier associations (D55: +25.0%pts.). Although not regressing, the implementation of pull-driven production control has stagnated since 2013 (D52: ±0%pts.).

The increase in layout optimization maturity is linked to stronger implementation of multiple enablers. In this sense, products are more frequently grouped in accordance with their processing and routing requirements to enhance flow (D59:

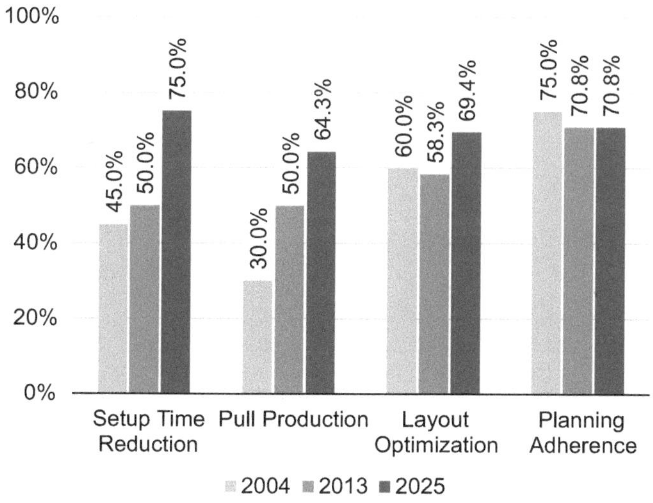

Fig. 3.4 JIT maturity evolution

+25.0%pts.; D60: +50.0%pts.). This is complemented by shop floor layouts being purposefully designed to support inventory reduction and throughput acceleration (D61: +25.0%pts.). Finally, the improved synchronization of processes contributes to the maturity growth (D63: +25.0%pts.). On the contrary, the application of value stream mapping (VSM) as a tool to visualize and optimize processes has declined (D65: −25.0%pts.). This can be linked to the introduction of new hard- and software solutions, which enable live visualizations of process flows, whereas traditional VSM does not offer such real-time capabilities.

While planning adherence maturity has remained at 70.8% over the past decade, its underlying elements have evolved over time. On the one hand, daily adherence to production plans has improved (D67: +25.0%pts.) as companies increasingly share data with customers and suppliers to enhance planning adherence (D69: +25.0%pts.). On the other hand, flexibility of shift models as a method to adjust production capacities has decreased (D71: −25.0%pts.), while companies prioritize capacity utilization over level production schedules (D72: −25.0%pts.).

3.2.4 Effective Management System

EMS accounts for the second-lowest growth in maturity over the past decade, even displaying maturity regression in one of its sub-elements. In this sense, direction setting has dropped below the 2013 mark (−4.8%pts.), countering the growth across the remaining elements. The largest increase is seen in management commitment and company culture (+10.2%pts.), followed by employee involvement and continuous improvement (CI) (+7.7%pts.) and functional integration and qualification (+3.9%pts.). Figure 3.5 depicts the development of EMS maturity over time.

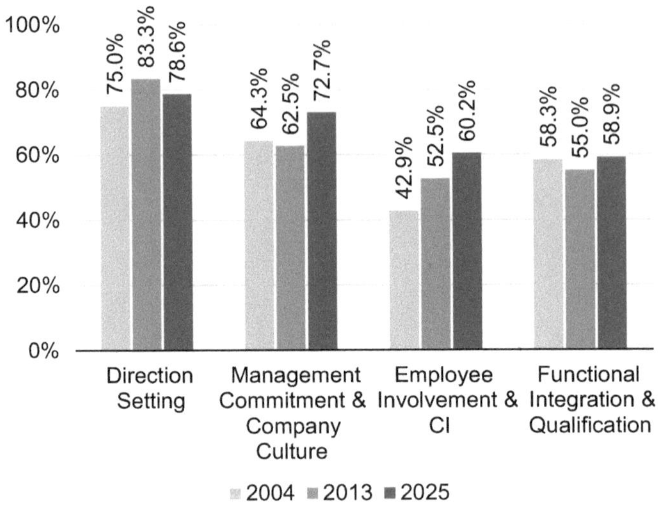

Fig. 3.5 EMS maturity evolution

The drop in direction setting is caused by manufacturing managers' decreasing familiarity with crucial manufacturing priorities, such as low costs, delivery, or quality (D78: −25.0%pts.). As a result, direction setting maturity drops from 83.3% to 78.6%. The growth, previously reported in our 2013 assessment, is thus mitigated.

A change in trajectory is also apparent for management commitment and company culture, however, in an upward direction. The previous drop in maturity from 2004 to 2013 has been turned around, due to a stronger implementation of three enablers. First, we observe enhanced interactions across departments with the intention of better aligning actions with the strategic aspirations (D81: +50.0%pts.). Second, stronger alignment can also be observed regarding quality standards (D86: +50.0%pts.). Third, companies report improvements in the clarity of leadership approaches (D88: +50.0%pts.). The upward trend from 2013 is, however, hampered, as less attention is paid to continuously reducing waste in processes (D87: −25.0%pts.).

Rising to 60.2%, employee involvement and CI maturity have improved further since 2013, continuing its positive development. In part, this is caused by the wider deployment of tools and methods to display CI progress (D89: +25.0%pts.). Further, work teams' responsibilities are better aligned with wider management goals (D92: +25.0%pts.), and employees receive greater authority to address problems they observe (D93: +12.5%pts.). Meanwhile, we observe a stagnation regarding problem-solving, as supervisors have not intensified their efforts to support their subordinates' efforts, while cross-functional teams are also not leveraged further (D95: ±0%pts.; D96: ±0%pts.).

Finally, we observe improved implementation of two functional integration and qualification enablers. In this sense, companies increasingly install formal programs to increase their production workers' flexibility to operate different processes

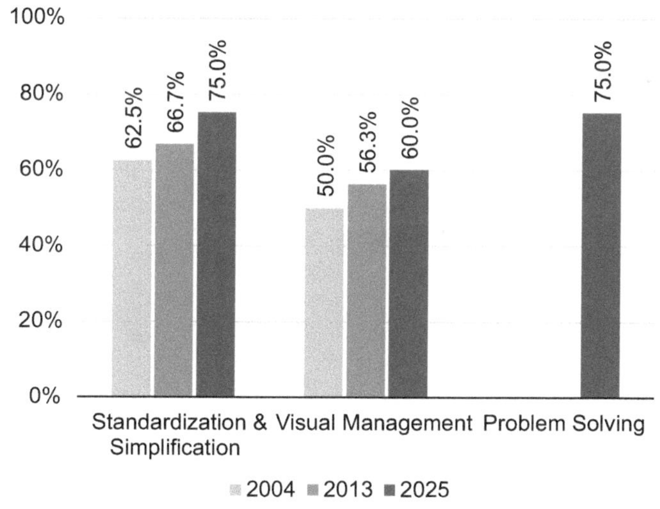

Fig. 3.6 BE maturity evolution

(D 101: +12.5%pts.). Further, companies appear to rely more heavily on open feedback to drive maturity in this category (D102: +25.0%pts.).

3.2.5 Basic Elements

Complementing the St.Gallen OPEX model's technical and social sub-systems, BEs record growth in both standardization and simplification (+8.3%pts.) and visual management (+3.8%pts.). Newly introduced as a dedicated component, problem-solving maturity stands at 75.0%.[2] The evolution of BE maturity is captured in Fig. 3.6.

Enhanced standardization and simplification maturity is linked to better documentation of operating procedures (D106: +25.0%pts.) and standardized functional descriptions to shorten vocational trainings (D108: +25.0%pts.).

Meanwhile, visual management maturity growth results from an intensified utilization of charts to display objectives (D111: 25.0%pts.) and current performances (D113: +25.0%pts.) on the shop floor. On the contrary, takt times and schedule compliance are illustrated less frequently (D114: −25.0%pts.).

Completing the set of enablers, problem-solving maturity is evaluated by companies' efforts to make performance gaps (D134) and necessary actions (D135) visible as well as the structuration of problem escalations (D136). All these elements stand at a maturity of 75.0% at the time of our current assessment.

[2]While other elements of the St.Gallen OPEX model also touch on problem-solving, its general maturity is assessed under BE, whereas in other instances, the focus is on the implementation of other enablers to foster problem-solving.

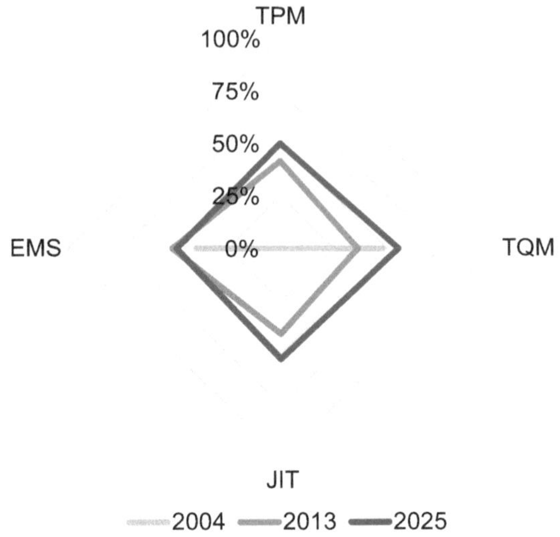

Fig. 3.7 Overall performance evolution

3.3 The State of Pharmaceutical Operational Excellence Performance

The calculation of performance is based on companies' measurement of 40 KPIs in relation to their manufacturing operations. Data points are normalized against the St.Gallen OPEX database, and corresponding performance sub-scores are calculated for the respective elements of the OPEX model. These are then aggregated into an overall performance score. While sub-scores are calculated for TPM, TQM, JIT, and EMS, the latter element does not contribute to the overall performance score.[3] As of November 2025, the median level of pharmaceutical OPEX performance stands at 52.9%, indicating an increase by 12.3%pts. since our last assessment. The delta from the baseline year is positive for all elements that feed into the overall score. The greatest improvement occurs in TQM (+19.1%pts.), followed by JIT (+12.0%pts.) and TPM (+8.4%pts.). Meanwhile, EMS emerges as the only element where performance has regressed over the past decade (−2.0%pts.). Overall, we observe a steady expansion of pharmaceutical OPEX performance since the initiation of our benchmarking project (see Fig. 3.7).[4]

However, as in the case of maturity, the picture is more diverse in the underlying layers. As this chapter presents an update to our previous analysis, the following depictions focus on the same selection of KPIs as our 2013 assessment, representing some of the most relevant figures for pharmaceutical performance measurement.

[3] No performance score is calculated for BE.

[4] TPM and JIT performances were initially not measured when we began our benchmarking project. Thus, no corresponding scores are reported for 2004.

3.3.1 Total Productive Maintenance

TPM aims at maximizing equipment effectiveness in production at moderate costs, revolving around creating stability and reliability in relation to equipment and maintenance. Thereby, TPM provides the basis for improvements in efficiency. At the time of our updated assessment, the pharmaceutical industry displays a median TPM performance of 50.0%.

Equipment breakdowns and losses[5] interrupt the production flow, resulting in waste and avoidable costs. To detect such disturbances, Nakajima (1988) introduces overall equipment effectiveness (OEE). As an indicator of TPM, OEE analyzes the interrelated effects of plant availability, performance, and quality, ensuring a valid and distinct measurement of equipment usage and stability. For the OPEX benchmarking, we calculate OEE as the product of availability, performance,[6] and quality (see also Nakajima, 1988). While OEE captures the general extent to which equipment runs steadily without interruptions, unplanned maintenance emphasizes maintenance-related interruptions. Thus, it provides further insights into TPM performance. In the OPEX benchmarking, unplanned maintenance is measured as the proportion of unforeseen maintenance work. Since 2013, the median OEE value in pharmaceutical manufacturing has increased by 2.5%pts., now standing at 57.4%. The delta is, however, less significant than the one we observed between 2013 and 2004 (+18.9%pts.). Accompanying the improvement in OEE, unplanned maintenance has decreased further (−2.1%pts), as pharmaceutical manufacturing companies exhibit a median value of 15.4%.[7] Again, the delta between 2025 and 2013 falls short of the delta between 2013 and 2004 (−7.5%pts.). We conclude that, despite a positive trend in individual KPIs (see Fig. 3.8), more comprehensive maintenance programs are still required to further improve this performance dimension.

3.3.2 Total Quality Management

TQM is based on the assumption that costs for corrective quality activities and complaint management exceed costs for preventive quality activities (Hackman & Wageman, 1995). Thus, TQM provides the foundation to retain an operating license without excessive costs. Recovering from a drop between the initiation of our

[5] Robinson and Ginder (1995) define the following losses: (1) downtime due to machine breakdown; (2) time required for setup and adjustments; (3) time or cycles lost to inefficient start-up; (4) time or cycles lost to tooling; (5) time or cycles lost to minor stoppages; (6) operating at less than ideal speed; and (7) producing defective or off-spec product that is rejected, requires rework or repair, or is sold at a lower price.

[6] The reader is advised not to confuse performance as a component of OEE with performance as an overall measurement.

[7] The reader is advised to take note that an improvement may occur as a positive or negative delta from the baseline year, depending on the KPI. For unplanned maintenance, for example, a decrease is favorable.

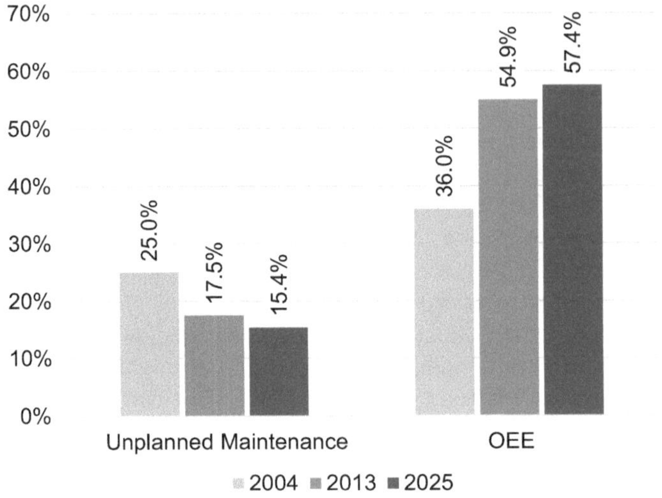

Fig. 3.8 TPM KPI evolution

benchmarking project and our 2013 assessment, median TQM performance now stands at 55.3%.

Central to OPEX and quality management philosophy is a stringent focus on customer value (Flynn et al., 1995). Thus, surveying customer satisfaction is essential to maintaining this focus. Simultaneously, it provides insight whether TQM efforts are effective. Vice versa, tracking whether suppliers meet their quality criteria prevents compromising own manufacturing operations through the usage of poor-quality supplies and materials. Therefore, we measure complaint rates on both the customer and supplier sides in our assessment of pharmaceutical OPEX for an end-to-end perspective. To prevent nonconforming products to reach markets and endanger patient lives, regulatory requirements mandate a review of each production batch before release and shipment. As another effectiveness measure for TQM, rejected batches therefore not only capture a company's ability to manufacture to quality standards but also its capability to detect, correct, and prevent quality issues during production. We therefore measure rejected batches in relation to all production batches. While overall TQM performance has increased over the past decade, developments in the aforementioned KPIs are more volatile. We observe only a minor improvement by −0.1%pts. in customer complaint rates, compared to the delta between 2013 and 2004 (−0.4%pts.). As of November 2025, the median value for customer complaint rates is 0.5%. Meanwhile, supplier complaint rates have slightly increased to 1.2%. This does, however, not necessarily imply that supplier quality has become worse. As stated in Sect. 3.2.2, the share of inspections for inbound deliveries has risen drastically since 2013. Thus, companies have simply become more effective at detecting insufficient quality on the supply side. While the share of rejected batches had remained stable between 2004 and 2013, it has since risen above 1.3%. The evolution of these selected TQM KPIs is depicted in Fig. 3.9.

Fig. 3.9 TQM KPI evolution

3.3.3 Just-in-Time

A guiding principle of JIT is the limitation of overproduction, inventories, and inconsistencies through pull. To achieve this, companies rely on setup time reduction and factory layout optimization. JIT performance, which we measure along related elements, stands at 52.7% at the point of our revisited assessment.

Inventory levels can hardly be compared between different companies, due to the specifics of their respective products and processes. However, as inventories are lowered, the number of turns over a certain period increases, regardless of the product or process type. We therefore consider inventory turns for raw material and finished goods to be an insightful metric for JIT performance. We define raw material turns as the annual cost of raw materials purchased divided by the average raw material inventory. Equally, we define finished good turns as the annual cost of goods sold divided by the average finished goods inventory. While, in part, customer value relates to product quality (see Sect. 3.3.2), it is also determined by delivery timeliness and completeness. Therefore, the St.Gallen OPEX benchmarking features on-time-in-full delivery (OTIF) as another measure of JIT performance. We define the metric as the percentage of orders shipped on time from the manufacturing site (±1 day of the agreed shipment date), in the right quantity (±3% of the agreed quantity) and in the right quality to the customer. Since 2013, raw material turns have dropped from 5.4 to 3.5. Meanwhile, finished good turns have risen from 7.0 in 2013 to 8.5 in 2025. Finally, OTIF has dropped by 3.8%pts. to a median value of 92.9%, inversing the upward trend we observed at the time of our previous assessment. The evolution of JIT KPIs is displayed in Figs. 3.10 and 3.11.

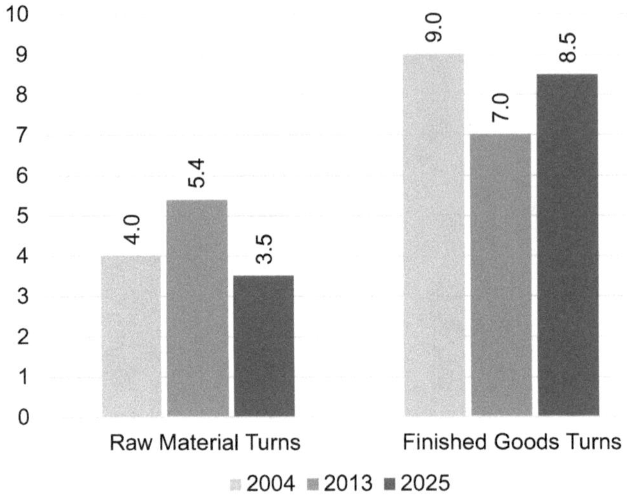

Fig. 3.10 JIT KPI evolution (1/2)

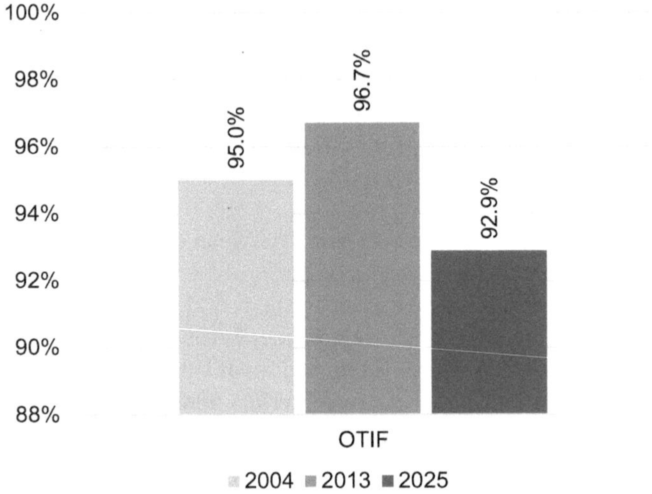

Fig. 3.11 JIT KPI evolution (2/2)

3.3.4 Effective Management System

Finally, EMS performance complements the qualitative assessment of cultural elements through quantitative indicators. Contrary to the previous three performance dimensions, EMS has regressed since 2013 to a median value of 48.9%.

While culture cannot be measured directly, selected KPIs can serve as a proxy in this domain. OPEX relies on a work environment where every employee is valued, heard, and understood, as well as involved in the continuous improvement efforts. If companies fail to create such a setting, OPEX cannot flourish. A poor company culture often results in employees remaining absent or leaving the company. Thus, sick leave (defined as total time of employees absent as a percentage of total working time) and employee fluctuation (defined as employees exiting the company as a percentage of all employees) hint at whether an appropriate cultural foundation exists. In the context of pharmaceutical manufacturing, these two metrics bear additional relevance in relation to knowledge management and expertise. As pharmaceutical manufacturing relies on highly skilled and knowledgeable operators, extensive levels of sick leave or employee fluctuation hinder companies from exploiting and retaining their employees' know-how. This can have negative effects on quality and delivery performance. In the same sense, investing in employees' capabilities is an important factor to secure competitive advantages. Thus, training (captured by the number of yearly training days per employee) represents another measure of EMS performance. While employee fluctuation has improved by 2.5%pts. to 5.0% in 2025, sick leave has stagnated at a median value of 3.3%. Dropping by 1.5 days per employee per year, training stands at a median value of 6.1 days. The evolution of these figures is depicted in Figs. 3.12 and 3.13.

3.4 Trends and Developments

As highlighted in the introduction of this chapter, disruptive forces have reshuffled the industrial landscape in recent years. The impact these are having on the pharmaceutical industry can be seen clearly in our benchmarking data.

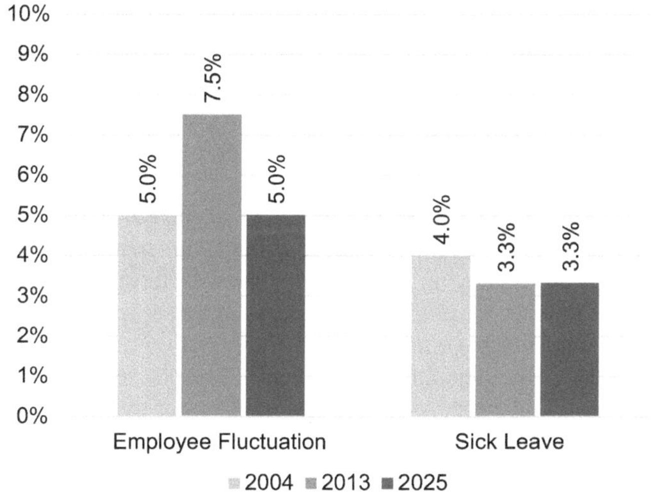

Fig. 3.12 EMS KPI evolution (1/2)

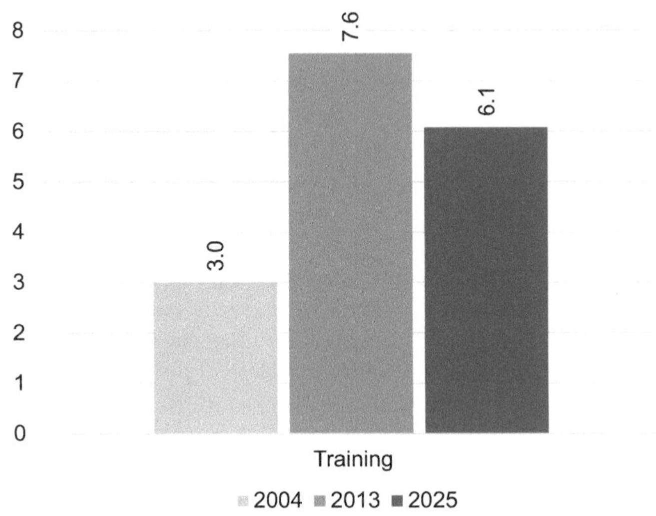

Fig. 3.13 EMS KPI evolution (2/2)

Recent supply chain disruptions, caused, for example, by the COVID-19 pandemic, protective tariff policies, and geopolitical tensions, have led to a state of caution across many industrial sectors. In such scenarios, increased safety stock levels help to maintain lead times and to prevent product supply shortages. The described drop in raw material turns provides substantiating evidence that pharmaceutical manufacturers have taken similar actions to counter these supply chain disruptions. Simultaneously, they have faced a drastic surge in product demand, especially in the second half of the past decade (IQVIA Institute for Human Data Science, 2024). Effects of this phenomenon are reflected in the incline of finished good turns. Thus, manufacturers sell through their finished products at a quicker rate than before. Finally, the drop in OTIF shows that pharmaceutical companies increasingly struggle to satisfy demand in a timely manner. In combination, these developments pose economic and financial risks for the industry. If not managed and contained, these trends could spiral out of control and create dangerous bullwhip effects along pharmaceutical supply chains.

In our assessment of pharmaceutical OPEX, we also consider structural factors, which shed light on contextualities (e.g., company/site size, production volume, distribution structure), allowing to compare manufacturing operations beyond maturity and performance. Ever since the industrialization, regardless of the sector, manufacturing companies have invested in the automation of machines and equipment, seeking advantages in quality and efficiency. The pharmaceutical industry has not been immune to this trend. A century ago, tablets, for example, were produced manually and packed by hand. Nowadays, automated and integrated production lines are capable to press, blister, and pack thousands of pills per minute, before industrial robots then load the finished packages onto pallets ready for shipment. However, such solutions have yet to become the norm. As of November 2025, only

20.5% of pharmaceutical manufacturing processes are fully automated. Instead, companies still rely on operators to perform certain or all steps. Thus, 46.9% of manufacturing is characterized as semi-automated, whereas 32.6% is still manual. While we observe no statistically significant difference in relation to overall performance, we see variations for different product types, especially for tablet manufacturing. Here, semi-automation exceeds the industry average by 20.6%pts., standing at 67.5%, whereas most other technologies display values similar to the industry average.

Finally, we also find evidence for the cost pressure pharmaceutical companies have faced over the past decade, as inflation has plagued the global economy. In an effort to stay afloat, companies thus relied on drastic cost-cutting measures. Especially in and after 2021, when the economic effects of the COVID-19 pandemic began to unfold, we see clear signs for reduced spending. For example, companies limited their plant, property, and equipment costs (see Fig. 3.14) and direct labor costs (see Fig. 3.15). While such measures present an ad hoc solution, they are not sustainable, do not reflect the goal of OPEX, and bear additional threats. Reducing expenditures for maintaining or upgrading equipment may destabilize machines and thus processes at the expense of output quality and quantity. Such short-sighted approaches to cost improvement thus contradict our school of thought. Following the sand cone model's logic (see also Ferdows & de Meyer, 1990), we advocate focusing on quality first to build a solid foundation for sustainable improvements in speed and costs. Meanwhile, having fewer operators available to perform the same production tasks can have similar effects, if pressure for productivity rises. Further, the loss of expertise when cutting employees must not go overlooked, as restoring know-how and building capabilities are lengthy and complex processes.

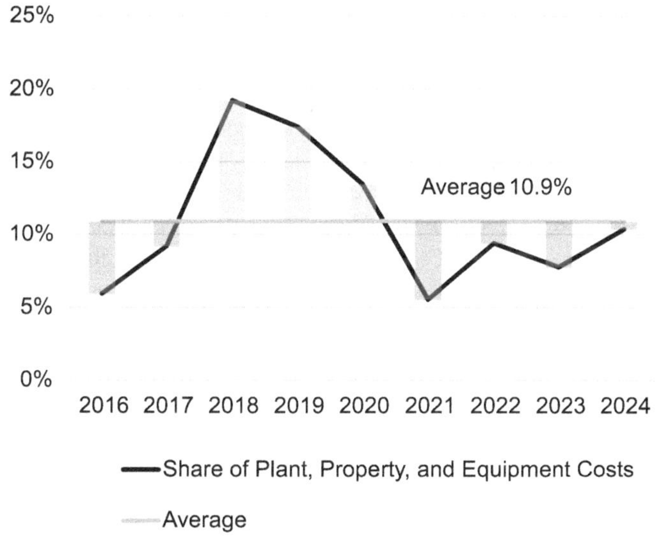

Fig. 3.14 Share of plant, property, and equipment costs in total costs of goods sold

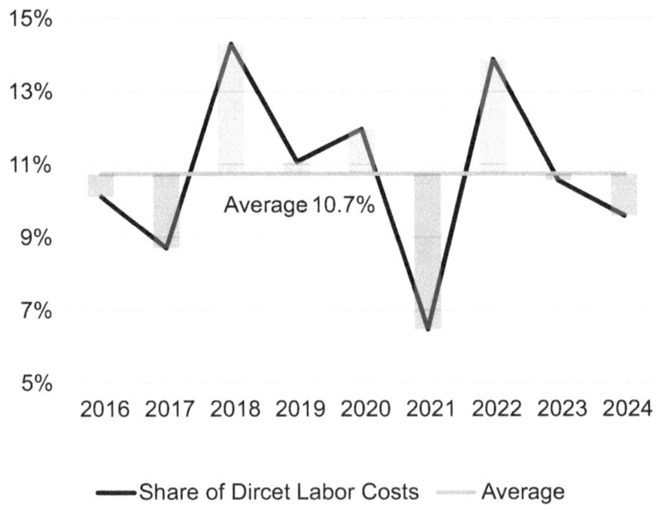

Fig. 3.15 Share of direct labor costs in total costs of goods sold

3.5 Conclusion

Given the disruptive forces the pharmaceutical industry has been subject to over the past decade, we assumed that the state of OPEX must have progressed since 2013. Given the growth in both maturity and performance, we can confirm this hypothesis. However, compared to sectors where OPEX has become ubiquitous (e.g., automotive), the pharmaceutical industry still faces an uphill battle. While maturity levels range between 60% and 70% overall as well as for the building blocks of the St.Gallen OPEX model, respective performance scores rarely exceed the 55% mark. Thus, pharmaceutical manufacturers need to intensify their efforts in both domains, expanding the processual foundation of their operations, while transferring and translating practices into measurable performance outcomes.

There will also be a need for a real focus on the social system of the OPEX model and therewith on people and culture. After initial progress during the first decade, the industry has reverted to a rather technical implementation and brute force. This can have a positive impact on the culture but only if the priorities remain on OPEX. A clearer shift to aiming at behavior and culture would make a lot of sense and is the way for a sustainable implementation. This could be supported again by regulatory programs like quality management maturity (cf. Chap. 12). Additionally, there is already a visible impact from digitalization on enablers. Some traditional approaches example (e.g., statistical process control) lose in importance. However, digitalization only works if it is aligned with OPEX and built on simplification and standardization (cf. Chap. 11).

Further, uncertainties and disruptions have confronted pharmaceutical companies with unprecedented challenges and will continue to do so in the future. Although companies have found ways to manage these, they need to apply caution. Current ad hoc and short-term solutions that do not embody the core values and principles of OPEX and bear a risk of erosion, threatening to jeopardize the progress that has been made in the pharmaceutical industry. Thus, companies will have to put even greater effort into finding a sustainable approach to their OPEX journey. With the research insights and practical examples provided in this book, we strive to provide guidance on this journey.

Appendix

Enabler

Total productive maintenance (TPM)	
Preventive maintenance	
D01	*To what degree is there a formal program for maintaining machines and equipment?*
D02	*To what degree are maintenance jobs documented and maintenance plans and checklists posted close to machines?*
D03	*To what degree is good maintenance emphasized as a strategy for increasing quality and planning for compliance?*
D04	*To what degree are potential bottleneck machines identified and supplied with additional spare parts?*
D05	*To what degree is the maintenance program continuously optimized based on a dedicated failure analysis?*
D06	*To what degree does the maintenance department focus on assisting machine operators perform their own preventive maintenance?*
D07	*To what degree are machine operators actively involved in the decision-making process when buying new machines?*
D08	*To what degree are machines maintained internally* vs. *externally?*
Technology assessment and usage	
D09	*To what degree is the site situated at the leading edge of new technology?*
D10	*To what degree do you screen the market for new production technology and assess new technology concerning its technical and financial benefit?*
D11	*To what degree is the site effectively using new technology?*
D12	*To what degree does the site rely on vendors for its equipment?*
D13	*What proportion of equipment is proprietary?*
D14	*To what degree is proprietary process technology and equipment used to gain a competitive advantage?*
Housekeeping	
D15	*To what degree do employees strive to keep the site neat and clean?*
D16	*To what degree are plant procedures in place, regarding putting all tools and fixtures in their place?*
D17	*To what degree are housekeeping checklists used to continuously monitor the condition and cleanness of our machines and equipment?*

Total quality management (TQM)	
Process management	
D18	*To what degree are direct and indirect processes documented?*
D19	*To what degree is process quality continually measured?*
D20	*To what degree is process quality continually measured using process metrics* (e.g., *on-time-in-full delivery rate)?*
D21	*To what degree are process measures directly linked to site objectives?*
D22	*To what degree are dedicated process owners responsible for planning, managing, and improving their processes?*
D23	*What proportion of the equipment on the shop floor is currently under statistical process control (SPC)?*
D24	*To what degree is statistical process control utilized to reduce variances in processes?*
D25	*To what degree are standardized tools in place for root cause analysis, to get a deeper understanding of the influencing factors* (e.g., *DMAIC)?*
D26	*What proportion of equipment makes use of process analytical technology (PAT), for real-time process monitoring and controlling?*
Cross-functional product development	
D27	*To what extent are manufacturing engineers* (e.g., *industrial engineers) involved in the development of a new drug formulation and the development of the necessary production processes?*
D28	*To what degree are product and process development linked?*
D29	*To what degree is there close collaboration between the R&D and the manufacturing department to shorten launch times?*
D30	*In the last 2 years, what proportion of product launches have been on time (launch by planned launch date)?*
D31	*To what degree are standardized procedures in place for product and process transfers between different units or sites?*
Customer involvement	
D32	*To what degree are you in close contact with your customers* (e.g., *email, phone, e-room)?*
D33	*To what degree do your customers give you feedback on quality and delivery performance?*
D34	*To what degree do you survey your customer's requirements?*
D35	*To what degree do you conduct customer satisfaction surveys?*
D36	*To what degree is on-time delivery part of your philosophy?*
D37	*To what degree do you use joint improvement programs with your customers to increase your performance?*
Supplier quality management	
D38	*To what degree is supplier selection oriented around quality?*
D39	*To what degree do you conduct supplier qualifications and audits in order to rank your suppliers?*
D40	*What proportion of your suppliers have you validated?*
D41	*For what proportion of your suppliers do you perform inspections of the incoming parts/materials?*
D42	*To what degree are inspections of incoming materials usually performed in proportion to the past quality performance or type of supplier?*
D43	*What proportion of incoming shipments do you inspect?*

D44	*To what degree do you use joint improvement programs with your suppliers to increase your performance?*
D128	*To what extent do you employ a holistic supplier management process incorporating lean/OPEX tools (*e.g.*, map/visualize supplier relationships and performance metrics)?*
Just-in-time (JIT)	
Setup time reduction	
D45	*To what degree do you continuously work to lower setup and cleaning times in your plant?*
D46	*To what degree does the production department focus on assisting machine operators to lowering setup and cleaning times?*
D47	*To what degree do crews practice setups to reduce the time required?*
D48	*To what degree do you put a high priority on optimizing batch sizes in your site?*
D49	*What proportion of setups are scheduled so that the regular uptime of our machines is not affected?*
D50	*To what degree are optimized setup and cleaning procedures documented as best practices and rolled out throughout the whole site?*
Pull production	
D51	*To what degree is the production schedule designed to allow for catching up due to production stoppings because of problems (*e.g.*, quality problems)?*
D52	*To what degree is production control driven by a "pull" system (kanban squares, containers, or signals)?*
D53	*To what degree is your production aligned with S&OP planning?*
D54	*To what degree are suppliers integrated and to what degree do vendors fill kanban containers, rather than filling purchasing orders?*
D55	*What proportion of supplier associations are long term (>3 years)?*
D56	*To what degree do you enforce your suppliers to deliver on time (>70% of deliveries on time)?*
D57	*What is your manufacturing approach (make-to-order* vs. *make-to-stock)?*
Layout optimization	
D58	*To what degree are your processes located close together so that material handling and part storage are minimized?*
D59	*What proportion of products are classified into groups with similar processing requirements to reduce setup times?*
D60	*What proportion of products are classified into groups with similar routing requirements to reduce transportation time?*
D61	*To what degree does the layout of the shop floor facilitate low inventories and fast throughput?*
D62	*To what degree can your shop floor layout be characterized as separated into "mini plants," if products have been classified based on their specific requirements?*
D63	*If applicable, to what degree are manufacturing processes synchronized over all steps (*e.g.*, within 1 takt)?*
D64	*To what degree can your current manufacturing process be described as a full continuous flow, involving almost no interruptions?*
D65	*To what degree are you working to reach the status of a full continuous flow?*
D66	*To what degree do you use "value stream mapping" as a methodology to visualize and optimize processes?*

Planning adherence	
D67	*To what degree do you meet your daily production plans?*
D68	*To what degree do you know the root causes of variance in your production schedule and continuously try to eliminate them?*
D69	*To what degree do you share data with customers and suppliers based on a rolling production plan to increase planning adherence?*
D70	*To what degree do you work on achieving smoothly leveled production capacity throughout the whole production process?*
D71	*To what degree does your site have flexible working shift models in order to easily adjust production capacity according to current demand changes?*
D72	*To what degree do you prefer a smoothly leveled production schedule to a high level of capacity utilization of single machines?*
Effective management system (EMS)	
Direction setting	
D73	*To what degree is your site's strategy and vision clear and linked to your corporate mission statement?*
D74	*To what degree is your vision, mission, and strategy communicated to employees?*
D75	*To what degree are the focus, goals, and objectives of your site's manufacturing unit closely linked to and consistent with corporate objectives?*
D76	*To what degree are the overall objectives of the site closely linked to the team or personal objectives of your shop floor teams and employees?*
D77	*To what degree do manufacturing managers (head of manufacturing, site leader* etc.*) have a good understanding of how the corporate/ divisional strategy is formed?*
D78	*To what degree do manufacturing managers know the most important criteria for manufacturing jobs* (i.e., *low costs, delivery, quality)?*
D129	*To what extent are maturity standards set and assessed by an external party* (e.g., *central team, peers)* vs. *by the site?*
Management commitment and company culture	
D79	*To what degree does leadership promote and develop the necessary behaviors and standards required for operational excellence at your site?*
D80	*To what degree is site management (committee) personally involved in improvement projects?*
D81	*To what degree do you align your approach to departmental interaction with your wider management goals* (e.g., *encouraging competition/cooperation)?*
D82	*To what degree is communication made* via *official channels?*
D83	*To what degree does your site have an open communication culture and encourage the flow of information between the departments and the different management levels?*
D84	*To what degree are you informed early enough about innovations* (e.g., *new products, new SKUs, new technologies)?*
D85	*To what degree are problems* (e.g., *complaints) traced back to their origin to identify root causes?*
D86	*To what degree do you align your approach to quality standards with your wider management goals* (e.g., *shared responsibility or primarily the task of QA/QC)?*
D87	*To what degree do your employees continuously strive to reduce waste in processes* (e.g., *waste of time, waste of production space)?*
D88	*To what degree do you have a clear leadership approach (command and control* vs. *open culture)?*
D130	*To what extent do shop floor leaders seek to implement a "change culture" built on lean/OPEX principles?*

Employee involvement and continuous improvement	
D89	*To what degree have you implemented tools and methods to deploy a continuous improvement process?*
D90	*To what degree are your employees involved in writing policies and procedures (concerning site vision down to standard operating procedures)?*
D91	*To what degree do shop floor employees actively drive suggestion programs?*
D92	*To what degree do you align your work team responsibility with wider management goals* (e.g., *work teams taking autonomous action or requiring management approval)?*
D93	*To what degree do your employees have the authority to correct problems when they occur?*
D94	*To what degree do you align supervisor problem-solving responsibility with wider management goals* (e.g., *problems always solved by supervisors or solved by teams)?*
D95	*To what degree do supervisors focus on assisting employees to perform their own problem-solving?*
D96	*To what degree does your plant form cross-functional project teams to solve problems?*
D97	*To what degree does the company take care of the employees?*
D98	*We have organized production employees into teams in production areas. For each team, there is one dedicated team member that is responsible for supervisory tasks.*
D99	*We have organized production employees into teams in production areas. For team leadership, we have an additional supervisory level in our organization.*
Functional integration and qualification	
D100	*What proportion of your production employees are cross-trained?*
D101	*To what degree has your plant implemented a formal program to increase the flexibility of your production workers?*
D102	*How frequently does your company hold open feedback meetings?*
D103	*To what degree is the information from these official feedback meetings used in further training?*
D104	*To what degree does your site invest in the training and qualification of your employees?*
D131	*To what degree are lean/OPEX trainings rolled out in a structured program tied to business needs?*
D132	*To what degree do shop floor leaders participate in and facilitate lean/OPEX trainings?*
Basic elements (BEs)	
Standardization and simplification	
D105	*To what degree is standardization emphasized as a strategy for continuously improving processes, machines, and products (closely relates to JIT segment)?*
D106	*To what degree are documented operating procedures used to standardize processes* (e.g., *setups) (closely relates to JIT segment)?*
D107	*To what degree are optimized operating procedures* (e.g., *shortened setups) documented as best practice processes and rolled out throughout the whole site (closely relates to JIT segment)?*
D108	*To what degree are standardized functional descriptions used to reduce the period of vocational training for new employees (closely relates to EMS segment)?*

D109	*To what degree are standardized machines and equipment* (e.g., *machine design, spare parts) used to achieve a high uptime of our machines (closely relates to TPM segment)?*
D110	*To what degree are standardized machines and fixtures used to lower material costs for spare parts (closely relates to TPM segment)?*
Visual management	
D111	*To what degree do you utilize performance charts to show objectives across production processes on the shop floor?*
D112	*To what degree are technical documents* (e.g., *maintenance documents) and workplace information* (e.g., *standardized inspection procedures, team structures) visible on the shop floor?*
D113	*To what degree do you utilize charts showing the current performance status* (e.g., *current scrap rates, current uptimes) on the shop floor?*
D114	*To what degree do you utilize charts showing current takt times and schedule compliance* (e.g., *Andon boards) on the shop floor?*
D133	*To what extent do you make use of real-time data collection and visualization?*
Problem-solving	
D134	*To what degree do you ensure that performance gaps are clear* (e.g., *make conditions visible to identify nonstandard performance)?*
D135	*To what degree do you ensure that, where necessary, actions are underway to close performance gaps* (e.g., *monitor timelines and ensure responsibility for actions underway)?*
D136	*To what extent do you have a structured approach to escalating problems* (e.g., *identifying critical escalation points, clarity on how to escalate well)?*
Reasons for launching OPEX initiatives	
D115	*To meet FDA regulations*
D116	*To implement process analytical technology (PAT)*
D117	*To change from functional organization to process organization*
D118	*To increase cost awareness*
D119	*To increase employee involvement*
D120	*To increase employee empowerment*
D121	*To reduce lead times and inventory*
D122	*To change the quality focus from final product to process quality*
D123	*To initiate a cultural change for continuous improvement*
D124	*To introduce standardized methodologies for problem-solving*
D125	*To launch a broader cost-cutting program*
D126	*To improve final product quality*
D127	*To fulfill site targets between corporate and plant management*

KPIs

Total productive maintenance (TPM)	
Setup and cleaning	*The total time spent for setup and cleaning as a percentage of total scheduled time*
Dedicated equipment	*The percentage of your equipment that is dedicated to one product*
Unplanned maintenance	*Proportion of unplanned maintenance work as a percentage of the overall time spent for maintenance works*

Shift model	*Number of shifts per day*
Shift length	*Average length of one shift in hours*
Line staffing	*FTEs per shift*
Loading	*Scheduled time/calendar time*
Availability (OEE)	*(Scheduled time − Downtime)/Scheduled time*
Performance (OEE)	*(Amount produced × Ideal cycle time)/Available time*
Quality (OEE)	*(Input − Defects)/Input*
OEE	*Availability (OEE) × Performance (OEE) × Quality (OEE)*
Total quality management (TQM)	
Yield	*Real achieved output in pharmaceutical production (input minus material losses, weighting, sediments* etc.*)*
Release time	*Average time from sampling to release of finished products including all waiting times*
Deviations	*Number of deviations per month that arise from raw materials purchased, production components (equipment), and product/ process specifications*
Deviation closure time	*Average deviation closure time in days*
Recurring deviations	*A deviation, during the defined review period, in which at least in one other occasion, the same event has occurred with the same root cause within the same process and/or work area. If redundant/duplicative processes or equipment exist, consider deviation events common to the grouping/work center as recurring*
CPKs	*Average CPK score across the top five products*
Customer complaint rate	*Number of justified complaints as a percentage of all customer orders delivered*
Supplier complaint rate	*Number of complaints as a percentage of all deliveries received (from your supplier)*
Right-first-time (RFT)	*Total number of batches produced without document errors or exception reports as a percentage of the total number of batches produced*
Rejected batches	*Number of rejected batches as a percentage of all batches produced*
Mean-time-between-failure (MTBF)	*Average time between breakdowns*
Scrap rate	*Average difference between 100% and real achieved output in packaging operations*
Number of tests	*Number of quality tests per month*
Critical-to-quality (CTQ)	*Is "critical to quality" analysis utilized?*
Just-in-time (JIT)	
Order lead time	*Time between a customer placing an order and receiving delivery*
Production lead time	*Time in days from start of production (SOP) to release of finished products.*
... thereof waiting time	*Sum of waiting times, for example during setup, or before and after QA/QC*
... thereof production time	*Value-adding time of the production process not including setup time or waiting time in between two process steps*
...thereof QA/QC	*Time for release testing of finished goods. If different release tests have to be performed, it also includes waiting in between two tests*

Replacement time to customer	*Response time for short-term delivery to the customer for goods not on stock (delivery time supplier and your production time)*
Days on hand (DOH)	*Average inventory less write downs × 365 divided by the cost of goods sold (COGS)*
On-time-in-full delivery (OTIF)	*Perfect order fulfillment (percentage of orders shipped in time from your site (± 1 day of the agreed shipment day) and in the right quantity (± 3% of the agreed quantity) and right quality to your customer*
On-time-in-full supply	*Perfect order fulfillment (percentage of orders shipped in time to your site (± 1 day of the agreed shipment day) and in the right quantity (± 3% of the agreed quantity) and right quality from your supplier*
Forecast accuracy	*Actual orders received compared to the annual sales forecast*
Production schedule adherence	*Number of orders finished in the correct week and volume (± 10%)/Total number of orders planned for completion that week*
Priority orders	*Number of priority orders as a percentage of all orders produced*
Production freeze period	*Freezing period in which you do not allow any changes of your production schedule*
Number of stockouts	*Number of out-of-stock events per month*
Raw material turns	*Annual cost of raw materials purchased divided by the average raw material inventory*
Work-in-progress (WIP) turns	*Annual cost of raw materials purchased plus annual cost of conversion divided by the average work in process inventory*
Finished goods turns	*Annual cost of goods sold divided by the average finished goods inventory*
Changeover time	*Time in minutes spent between different SKUs for setting up and cleaning the equipment (distinguished by product, format, and country)*
Number of changeovers	*Average number of changeovers performed per month including changing lots and changing formats (distinguished by product, format, and country)*
Effective management system (EMS)	
Management layers	*Number of management levels between production workers and the highest-ranking manager at the site (*e.g., *worker–supervisor–manager of the department–site leader = 4 levels)*
Management span of control	*The average number of employees directly reporting to supervisors*
Group work	*Percentage of production workers that are organized in self-directed teams in terms of,* e.g., *holiday planning and team meetings*
Functional integration	*Number of production workers that are qualified to work on three or more technologies/functional areas as a percentage of all workers*
Suggestions (quantity)	*Average number of suggestions per employee in the last year*
Suggestions (financial impact)	*Estimated total savings due to suggestions that were implemented*
Employee fluctuation	*Employees leaving your site due to terminations, expired work contracts, retirements,* etc. *as a percentage of all employees*

Sick leave	*Total time of employees absent* (e.g., *sick leave) as a percentage of the total working time*
Overtime	*Hours worked in paid overtime (excludes the overtime which is compensated with free time) in the last year as a percentage of the overall working time*
Training	*Number of training days per employee (all kinds of training off and on the job) in the last year*
Level of qualification	*Number of workers with prior work-related qualification/ education as a percentage of the total number of workers at your site*
Level of safety	*Reportable incidents due to accidents and safety on average per month that are internally (on site) reported*

References

Ferdows, K., & de Meyer, A. (1990). Lasting improvements in manufacturing performance: In search of a new theory. *Journal of Operations Management, 9*(2), 168–184. https://doi.org/10.1016/0272-6963(90)90094-T

Flynn, B. B., Schroeder, R. G., & Sakakibara, S. (1995). The impact of quality management practices on performance and competitive advantage. *Decision Sciences, 26*(5), 659–691. https://doi.org/10.1111/j.1540-5915.1995.tb01445.x

Friedli, T., Basu, P., Bellm, D., & Werani, J. (Eds.). (2013). *Leading pharmaceutical operational excellence*. Springer Berlin Heidelberg. https://doi.org/10.1007/978-3-642-35161-7

Hackman, J. R., & Wageman, R. (1995). Total quality management: Empirical, conceptual, and practical issues. *Administrative Science Quarterly, 40*(2), 309. https://doi.org/10.2307/2393640

IQVIA Institute for Human Data Science. (2024). *Global use of medicines 2024: Outlook to 2028.* IQVIA Institute for Human Data Science. https://www.iqvia.com/insights/the-iqvia-institute/reports-and-publications/reports/the-global-use-of-medicines-2024-outlook-to-2028

Nakajima, S. (1988). *Introduction to TPM: Total productive maintenance*. Productivity Press.

Robinson, C. J., & Ginder, A. P. (1995). *Implementing TPM: The North American experience.* Productivity Press. https://doi.org/10.4324/9781003075547

Voss, C., Blackmon, K., Hanson, P., & Oak, B. (1995). The competitiveness of European manufacturing — A four country study. *Business Strategy Review, 6*(1), 1–25. https://doi.org/10.1111/j.1467-8616.1995.tb00169.x

Reinventing OPEX in Pfizer: Why and How We Did It?

4

History: Pfizer's History and Strategic M&A Impact

Jim Cafone

Abstract

This chapter explains why and how Pfizer changed the way it works to deliver medicines more efficiently. Pfizer grew from a small company into a global leader by internally developing important drugs and acquiring other company's scientific innovation. At first, Pfizer let the companies it purchased keep their own ways of working. This helped avoid disruptions but made it hard to work as one team. Over time, Pfizer saw that different systems slowed things down, raised costs, and made it harder to share best practices. So, in 2019, Pfizer created one standard way of operating all sites, inspired by the Toyota Production System. This new approach, called Integrated Manufacturing Excellence (IMEx), uses common processes, clear goals, and teamwork at every level—from the shop floor to top leaders. IMEx helped Pfizer react quickly during the COVID-19 crisis by uniting efforts to make and distribute vaccines. Today, this system guides how Pfizer's global supply chain operates, making the company more flexible, consistent, and ready to serve patients worldwide.

Pfizer was founded in 1849 in Brooklyn, NY, by Charles Pfizer and Charles Erhart. Initial achievements included the development of Santonin (an antiparasitic compound) and subsequent citric acid production, positioning Pfizer as a leader in fermentation technology. During World War II, Pfizer was successful in its efforts to mass-produce penicillin and became the world's largest producer of the "miracle drug," evolving from a chemical company to a pharmaceutical innovator. Postwar,

J. Cafone (✉)
Pfizer, New York, NY, USA
e-mail: jim.cafone@pfizer.com

T. Friedli et al. (eds.), *Striving for Operational Excellence in the Pharmaceutical Industry*, Future of Business and Finance,
https://doi.org/10.1007/978-3-032-26287-5_4

the company expanded its scientific capabilities through the introduction of antibiotics like Terramycin and subsequently diversified its portfolio to include cardiovascular, anti-inflammatory, and mental health therapies. The 1980s and 1990s saw Pfizer's research engine deliver several landmark medicines: Lipitor (cholesterol lowering), Zoloft (antidepressant that revolutionized mental health treatment), and Viagra (created an entirely new therapeutic category). These breakthroughs propelled Pfizer into the top tier of global pharmaceutical companies.

As R&D and associated disease states and epidemiology became more complex, Pfizer adapted its strategy, accordingly, combining organic innovation with targeted mergers and acquisitions. This approach enabled the company to leverage emerging science and expand its therapeutic expertise in order to sustain its leadership within the global pharmaceutical industry. A brief overview of the companies acquired by Pfizer includes the following:

- **Warner Lambert (2000):** Secured rights to the blockbuster cholesterol medicine Lipitor
- **Pharmacia (2003):** Expanded oncology and anti-inflammatory portfolios
- **Wyeth (2009):** Added vaccines (Prevnar) and enhanced biologic capabilities
- **Hospira (2015):** Strengthened sterile injectables and biosimilar platforms
- **Medivation (2016):** Advanced oncology leadership with prostate cancer therapies
- **Array BioPharma (2019):** Expanded targeted cancer therapies
- **Seagen (2023):** Added antibody-drug conjugate (ADC) technology
- **Metsera (2025):** Added next generation of medicines for obesity and cardiometabolic diseases

The integration of both internal and external innovation strategies has enabled Pfizer to diversify its pipeline and accelerate the delivery of medicines to patients by investing in advanced manufacturing platforms—including vaccines, biologics, antibody-drug conjugates, and obesity treatments. This approach has established an unparalleled global manufacturing and supply chain infrastructure for the company. By transforming internal manufacturing expertise into modular technology capabilities accessible to acquired entities, Pfizer has built a **world-class pharmaceutical supply chain platform** that supports the **development, scaling, and innovation of science delivery worldwide**. Today, Pfizer has multiple manufacturing technology platforms:

- **Small-molecule manufacturing**: API production (chemical synthesis at flexible scales, from grams to tons) and oral solid dose (tablets and capsules through granulation, coating, and compression; modular continuous manufacturing)
- **Large-molecule biologics**: Manufacturing facilities for biologics, vaccines, and rare disease categories
- **Sterile injectables and aseptic processing**: Advanced sterile fill-finish using barrier, robotic, and modular aseptic facilities
- **Specialized capabilities**: Antibody drug conjugation, lyophilization, and high-potency drug handling

Through a manufacturing network defined by **speed, flexibility, and innovation**, Pfizer sustains a powerful competitive edge in the following:

- **Rapid scale-up**: Continuous and modular systems enable rapid response to market demand and emergencies.
- **End-to-end manufacturing technology capability**: Covers the full spectrum of modern therapeutic modalities, including small molecules and biologics.
- **End-to-end supply chain integration**: Advanced analytics and automation deployed across its supply chain to reduce costs, improve quality, and accelerate time to market. This spans across demand and supply planning, procurement, manufacturing, quality release, customer order management, and distribution.
- **Global reach**: Strategically designed network of internal sites and external contract manufacturing organizations around the world, ensuring supply chain resilience and regional responsiveness.
- **Future-readiness**: Targeted investments in mRNA and ADC technologies position Pfizer for leadership in next-generation therapies.

However, each acquisition requires the integration of corporate cultures—a process that can ultimately determine the success of each transaction. Corporate cultures differ across organizations in their underlying values, communication styles, decision-making processes, and organizational structure, making effective cultural alignment a critical component of post-merger integration.

4.1 History: M&A Impact on Manufacturing and Supply Chain

In the manufacturing and supply space, Pfizer's initial approach to acquired companies was to retain their localized and decentralized operational excellence (OPEX) cultures, while certain functions such as quality, regulatory, HR, and IT were quickly integrated. This deliberately nonintegrated OPEX approach was intended to preserve distinctive cultures, specialized processes, and bespoke workflows. At the time, Pfizer believed that integrating an acquired company into a new production system could risk jeopardizing its core strengths. Additionally, because speed to market is critical and transitioning to a new production system can be time-consuming, Pfizer was reluctant to enforce this move. To expedite market entry and generate immediate cash flow with minimal disruption, the company aimed to maintain ongoing operations and reduce integration risk.

However, over time, it became clear that sustainable value is realized through standardization and process discipline. In its absence, Pfizer's global operations, which at the time included over 60 internal manufacturing sites and 300 contract manufacturers, employed varied management systems, metrics, definitions of world-class performance, and operational languages. This decentralized approach resulted in network inefficiencies related to cost, service, cycle time, and overall efficient performance. Over time, operating multiple distinct production systems hindered Pfizer's ability to leverage the network and establish a truly world-class operational system. What the company needed was a standardized set of processes, built on a strong culture, shared infrastructure, and coordinated systems to achieve scalability and local responsiveness, for customers.

In 2018, it was decided that a shift in strategy was needed, and Pfizer developed a new pharmaceutical production system that was modeled after the Toyota Production System. This single **integrated** production system across the end-to-end supply chain allowed for one **common operating model** through **one language of performance— Integrated Manufacturing Excellence (IMEx)**. From the shop floor to the CEO, IMEx established one "operating model" through standardized work, shared metrics, and a common set of continuous improvement tools. This also allowed for easier oversight, faster best practice transfer, and consistent leadership development.

To this day, there are still ongoing debates around the merits of a fully integrated production system versus decentralized, entropic models that allow processes to diverge without centralized control. While decentralization can foster entrepreneurship (like in many biotechnology companies), it is less suited for organizations needing **global platforms to support customer operations across regions and markets**. So, the debate within most operating organizations is the order vs. disorder, or standardization vs. autonomy, within organizations' discussion. In the context of large-scale, global manufacturing operations, the case for integration is compelling.

4.2 Integration: What Is It and Why Is It Critical?

OPEX: In the modern business world, Operational Excellence (OPEX) is the process where a business unit relentlessly executes operational processes better than competitors, with less waste and fewer mistakes, so customers get more value. It brings together and harmonizes every level of the company—from frontline staff to executive leadership—into a unified force working in concert.

Companies that excel at Operational Excellence have figured out that it is not just a set of individual tools (Lean, Six Sigma, Kaizen, Business Process Management, Balanced Scorecard, ISO, etc.) but an **integrated management system** across processes and systems that makes the difference.

The management system drives **integration of work** through (1) standardization of work methods, (2) standardized metrics, (3) common digital systems, and (4) relentless opportunity and loss prevention via continuous improvement. But integration does not equal unification. Integration is the act of making people work together more easily but leveraging their individual cultures, beliefs, and diversity. Pfizer deployed this integrated system from the front line to the executives—much different than deploying individual tools. This approach integrated everyone to become greater than the sum of the parts, precisely built for serving our patients.

Before moving forward, let us explore the benefits of integration and harmonization from different angles—(1) historical, (2) business, and (3) the world-famous Toyota Production System. **A historical integration reference**: In the aftermath of

World War II, European leaders aligned a shared enterprise objective—long-term stability and prosperity—and redesigned how critical systems worked together. By integrating foundational industries and later harmonizing markets, rules, and governance, they replaced fragmentation with interdependence. Over time, this operating-model mindset scaled across borders, enabling seamless flows, standardized rules, and shared accountability. With a clear purpose and aligned execution, integration became the natural outcome rather than a forced initiative. The result was sustained performance, resilience, and one of the longest periods of stability and collaboration in modern Europe.

A modern business reference: In modern times, companies and industries that excel harmonize and blend cultures, systems, and values to create shared prosperity rather than conflict or domination. When people feel that their work contributes to something larger—whether that is a social mission, an innovation idea, or even just pride in quality and output—the enterprise naturally transcends individual silos. With a shared purpose, you get **integration**.

The Toyota Production System: The Toyota Production System (TPS) is one of the clearest modern examples of harmony—"people working together harmoniously" to transform manufacturing into a collaborative, human-centered process—not just a mechanical one. The Toyota Production System is a model of unity and harmonization in manufacturing, a management philosophy and manufacturing method developed by Toyota to produce vehicles efficiently by eliminating waste, reducing costs, and improving quality.

4.3 IMEx: Integrated Manufacturing Excellence (IMEx) at Pfizer

4.3.1 Building a Culture of Integrated Excellence

Over the past 15 years, Pfizer has strategically transformed itself through a series of high-impact acquisitions, which have not only expanded its therapeutic portfolio but also reinforced its position as a global biopharmaceutical leader. Through these acquisitions, Pfizer has been harmonizing and integrating cultures, ideas, and systems to create a work environment that is greater than the sum of their parts. This integration has enabled the company to drive cutting-edge science, expand therapeutic capabilities, and streamline operations. Pfizer achieved this integration through the deployment of its production system—**Integrated Manufacturing Excellence (IMEx)**. IMEx serves as Pfizer's global "way of working" designed to drive end-to-end supply chain integration, continuous improvement, and operational excellence. It is not optional; it is the standard way of working across Pfizer Global Supply (PGS).

The model is built on the foundational principles:

- Standardized work
- Visual management
- Problem-solving/Lean and Six Sigma methodologies

- Empowerment of teams built around processes
- Outcome-focused tracking
- Coaching-centric leadership behaviors

Our IMEx journey represents a transformative shift in how the company approaches Operational Excellence across its global supply chain. IMEx is not just a framework; it is the standard operating model for Pfizer Global Supply (PGS), designed to drive end-to-end integration, continuous improvement, and data-driven decision-making.

The journey began with foundational efforts in 2019, focusing on the deployment of a mandated "management system." The IMEx leadership program defines five core behaviors—strategic vision, simplicity, people focus, Lean leadership, and long-term commitment—that guide PGS leaders in fostering a culture of excellence. These behaviors are reinforced through the IMEx capability and learning environment, which offers targeted training and resources to support deployment and sustainment.

4.3.2 Executing at Unprecedented Scale During COVID-19

In response to one of the biggest public health crises of modern times, Pfizer unified work methods across the enterprise to drive quality, speed, and innovation from "science to supply." Pfizer (with its partner BioNTech) aligned on a single, urgent mission: create a safe and effective vaccine to curb the COVID-19 pandemic. The results were very clear—harmonization and integration of expertise to deliver groundbreaking mRNA technology, research agility, large-scale manufacturing, regulatory experience, and a global distribution network. During this period, decisions were made rapidly and collectively, guided by shared priorities of quality, trust, and mutual respect—individual interests were replaced with the larger goal of public health for the world.

In this time, Pfizer Global Supply combined people, process, and purpose. The organization initiated its **IMEx production system and prioritized system-level outcomes over individual or departmental gain**. Like a civilization unifying for the better, Pfizer showed that coordinated execution through a harmonized management system can outperform isolated efforts.

The IMEx harmonization took place both vertically and horizontally across Pfizer Global Supply. Vertical harmonization connected colleagues across tiers from the shop floor to the executive leadership team through shared goals, clear metrics, and a common set of operating standards. Horizontal harmonization connected functions within sites, across sites, and globally across the network. IMEx became Pfizer's new way of working.

Part II

Introduction and Overview: Part II

Thomas Friedli, Matteo Bernasconi, and Fabian Klinkner

"Success is not final, failure is not fatal: it is the courage to continue that counts." Winston Churchill

The following chapters present key insights from our research on pharmaceutical Operational Excellence (OPEX). The University of St.Gallen has been an academic pioneer in this field, launching the first dedicated research project in 2005. This early start gave us a unique vantage point: we were able to accompany an entire industry on its journey towards catching up with the global state of the art in OPEX. Over more than two decades, each project and every benchmarking initiative has helped us remain at the forefront of industry knowledge and practice.

This section of the book does not aim to provide a comprehensive overview of all St.Gallen research in this domain (for that, see also our other publications such as Friedli et al. (2006, 2010, 2013). Instead, it highlights recent research outcomes that form the foundation for the main conclusions presented in Part IV.

We begin with three chapters dedicated to OPEX programs. Chapter 5 introduces the phenomenon itself and outlines the key success factors for effectively managing such programs. The following chapter examines how the results of OPEX programs can be measured, and the subsequent one explores effective rollout strategies. Chapter 8 then contrasts leading and lagging companies, offering a wealth of insights into the factors that differentiate high performers.

Given that both our research and our practical experience consistently underscore the critical role of leadership, we continue with a historical case from outside the pharmaceutical context to illustrate the profound impact that strong leadership can have. Chapter 9 revisits the 1948 Berlin Airlift and traces the actions of its chief architect, General William H. Tunner.

One of the persistent challenges in sustaining results lies in aligning OPEX efforts with the multitude of other programs and initiatives typically active in global organizations. Mismanaged interfaces—particularly with the quality function or with digital and AI initiatives—can lead to recurring conflicts. To address this, we include one chapter on each of these alignment challenges (Chaps. 10 and 11).

We conclude this part of the book with a closer examination of the specific requirements of highly regulated industries and their implications for managing OPEX. Chapter 12 discusses continuous improvement in regulated environments, followed by Chapter 13, which focuses on interactions between industry and regulators.

Taken together, these chapters offer a rich set of evidence-based insights and practical guidance, enabling readers to identify meaningful improvement opportunities for their own OPEX programs and activities.

References

Friedli, T., Kickuth, M., Stieneker, F., Thaler, P., Werani, J. (Eds.). (2006). *Der Pharmazeutische Betrieb: Vol. 50. Operational excellence in the pharmaceutical industry*. ECV Editio Cantor.

Friedli, T., Basu, P., Gronauer, T., Werani, J. (Eds.). (2010). *Der Pharmazeutische Betrieb: Vol. 51. The pathway to operational excellence: Overcoming the internal inertia*. ECV Editio Cantor.

Friedli, T., Basu, P., Bellm, D., & Werani, J. (Eds.). (2013). *Leading Pharmaceutical Operational Excellence*. Springer Berlin Heidelberg. https://doi.org/10.1007/978-3-642-35161-7

Operational Excellence Programs: The Phenomenon

5

Fabian Klinkner and Thomas Friedli

Abstract

In this chapter, we explore the phenomenon of Operational Excellence (OPEX) programs. For this purpose, we elaborate where these originated from and how they have evolved, especially in the pharmaceutical industry. We define the phenomenon conceptually in line with our understanding of OPEX programs and investigate their typical life cycle. Further, we review common characteristics and key success factors for the management of OPEX programs.

5.1 Origins

After its defeat in World War II, Japan had lost its industrial and economic strength, battling resource scarcity, government debt, and hyperinflation. Companies were forced to explore new avenues in the pursuit of competitive advantages or simply to survive. These circumstances led to the creation of the Toyota Production System (TPS). Drawing inspiration, for example, from historic shipbuilding techniques and American supermarkets, Toyota developed a novel production management approach that revolved around two principles, namely just-in-time and Jidoka. While the former element was concerned with producing the right parts, in the right

Parts of this chapter are based on the works of Grothkopp (2023) and Klinkner (2026), who conducted their research as members of the St.Gallen OPEX research team.

F. Klinkner (✉) · T. Friedli
Institute for Production and Supply Chain Management, University of St.Gallen, St.Gallen, Switzerland
e-mail: fabian.klinkner@unisg.ch; thomas.friedli@unisg.ch

T. Friedli et al. (eds.), *Striving for Operational Excellence in the Pharmaceutical Industry*, Future of Business and Finance,
https://doi.org/10.1007/978-3-032-26287-5_5

quantity, at the right time, the latter aimed at making problems in production visible. Stable and standardized processes provided the foundation for these components, while the idea of continuous improvement (CI) stood at the center of the TPS (Liker, 2004). Through this approach, Toyota was able to overcome the industrial and economic challenges it faced and established itself as one of the most renowned global car manufacturers.

Toyota's success initiated the rise of Operational Excellence (OPEX). Propelled to global recognition by the seminal works of Ohno (1988), Krafcik (1988), and Womack et al. (1990), many companies in the 1990s and early 2000s shifted their attention to this new production paradigm. Manufacturers quickly began to systematize their OPEX efforts, copying and later reinterpreting the TPS and creating their own approaches. At the forefront of this movement stood the automotive industry, which has since become a role model for many sectors in their respective OPEX transformations. The popularity of OPEX in the automotive industry can, to a considerable extent, be attributed to cost pressure caused by fierce competition and thin margins.[1] At the same time, high production volumes offered a favorable setting for process optimization through standardization and reduced variation. Compared to the automotive sector, the pharmaceutical industry found itself in a more comfortable position at the turn of the millennium. Stellar margins and patent protection created a sense of security among manufacturers. Consequently, they felt less compelled to pursue OPEX at first. Ultimately, regulator expectations, market dynamics, and increased competition due to the rise of generics caused an awakening and led companies to embark on their own OPEX journeys. As a result, despite initial delay, formalized OPEX programs have now become a ubiquitous phenomenon in the pharmaceutical industry (cf. Chap. 2). In the following sections, we establish the necessary theoretical foundation to conceptualize these. Therefore, we first define what OPEX programs are. Subsequently, we describe their life cycle and common characteristics as well as key success factors for their management.

5.2 Definition

Over time, a plethora of different definitions around OPEX have emerged. Many scholars have debated what it is in general and what it entails. While some argue OPEX to be a collection of tools for process optimization, others find it to be a management philosophy. In fact, no one perspective can be dismissed, as OPEX draws from both domains. A consensus exists, however, that OPEX relies on longevity (Bhasin & Burcher, 2006). Therefore, OPEX must be managed as a lasting program rather than a temporary project (cf. Pellegrinelli, 1997). The TPS, which Toyota began developing in the mid-twentieth century, embodies and exemplifies this perfectly, as it has remained in place ever since.

[1] Here, we would like to reiterate that OPEX does not serve mere cost- or employee-cutting purposes, but offers efficiency and profitability improvements through various means.

Scholars and practitioners have used different terminologies to refer to OPEX programs. Some simply use the term *lean*.[2] Others tie their wording directly to Toyota, speaking of *lean production systems*. As manufacturers began imitating and reinterpreting the TPS, the term *company-specific production system* later emerged (Netland, 2012). In practice, companies often personalize their OPEX program, giving it a company-specific name. Prominent examples featured in this book are Pfizer's Integrated Manufacturing Excellence System (IMEx) or Dechra's One Dechra Production System (ODPS).

In general, an OPEX program is defined as a corporate process improvement program with the goal of driving competitiveness (Netland, 2012). Often emerging in production, OPEX programs are, however, not limited to manufacturing in their applicability. In fact, their purview extends to the entire organization with the intention of transforming processes across all departments and functions.

Conceptually, OPEX programs may be described in three dimensions, namely content, process, and structure (Netland, 2012). The content dimension relates to the tools and philosophical principles that companies employ to improve their processes. OPEX programs often resemble the TPS, drawing from similar content. However, tools and principles must be relevant to the specific company. Only if custom-selected and tailored to the respective requirements will the content unfold its desired effects. Further, to foster identification and awareness, an OPEX program's content is typically visualized as a distinct model, such as a house or circle. This practice was inspired by the TPS, which was also visualized as a house (see Fig. 5.1). Meanwhile, the process dimension addresses the implementation of the OPEX program across an organization's network of manufacturing sites. This presents a highly complex and context-dependent undertaking, as these networks can be quite comprehensive and may even encompass hundreds of sites. As no one-size-fits-all approach to installing an OPEX program exists, companies must identify the right path for their specific circumstances. Our research shows that companies must, for example, consider resource availabilities, corporate and site-specific needs, as well as preexisting OPEX capabilities when managing the process dimension. Further, we have found that success can only be obtained through comprehensive engagement of employees and manufacturing sites. Lastly, the structure dimension refers to organizational and technical aspects. While the former revolves around the allocation of roles and responsibilities across the corporate hierarchy, the latter relates to information, knowledge management, and learning systems for the OPEX program. Commonly, the program is owned by a corporate entity, which oversees it globally. Tasked with the local management of the OPEX program, site entities complement this organizational structure. Typically, site OPEX teams report to the general manager of their site, rather than the corporate team. Through this circumstance, these general managers become gatekeepers for the success of an OPEX program and, as such, key

[2] Although lean and OPEX are often used interchangeably, we make a distinction between the two. While lean is directly associated with the TPS and a core part of OPEX, OPEX itself is an umbrella term for production management approaches that build on CI to create competitive advantages.

Fig. 5.1 The TPS house (Liker, 2004)

figures to involve. Finally, divisional or regional OPEX teams may also be installed for further support, depending on the company's requirements.

5.3 Life Cycle

When first embarking on their OPEX journey, companies typically see isolated improvement initiatives emerge in different parts of the organizations. Motivated by initial successes from these still uncoordinated activities, companies then usually harmonize their efforts by developing a corporate-wide approach—often supported through external consultancies or academic institutions. Once the development is completed, this formalized OPEX program is subsequently rolled out across the entire organization and maintained thereafter. Drawing on learnings made during the introduction and application of the program, companies then typically optimize their approach further, introducing updates to the current program or even entirely new versions. Our recent cross-industrial study on the management of OPEX programs shows that companies who update their program frequently and proactively

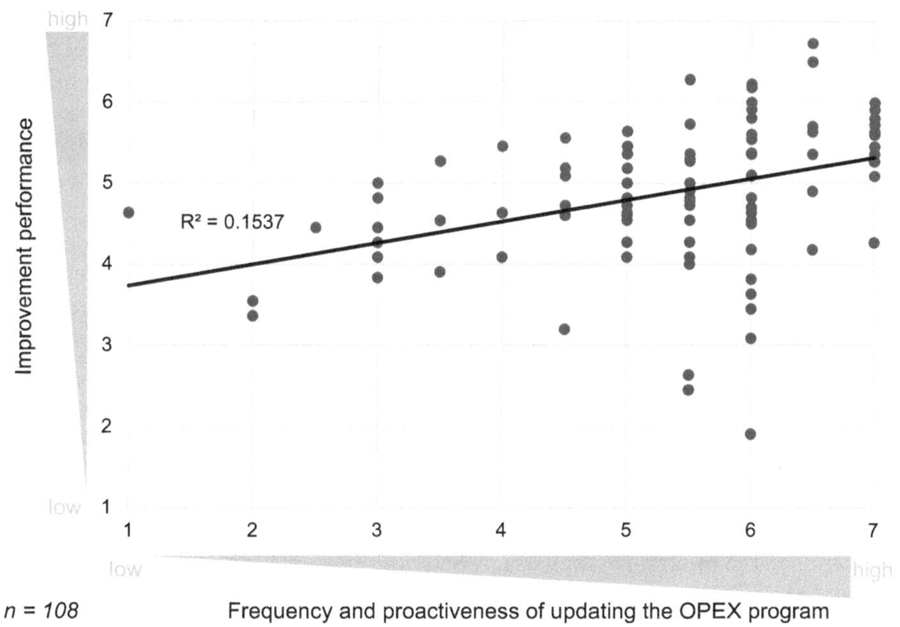

Fig. 5.2 Impact of updating the OPEX program

tend to achieve far greater improvements than those who do not (see Fig. 5.2). In this sense, many of the successful OPEX programs we feature in this book represent second, third, or even fourth iterations of the respective program, all building on learnings from their predecessors.

Simultaneously, OPEX programs are also subject to corporate dynamics. Although they should persevere and outlast disruptions in the sense of longevity, their stability is often jeopardized by different forces pulling on the organization. At a given point, an OPEX program may rest in an equilibrium state, where it has, for example, just been put in place. Certain triggers may now punctuate this equilibrium and cause significant shifts. These triggers can come in the form of M&A activities, strategic reorientations, or management changes. Especially the latter element is a frequent cause for destabilization. Top-level managers are often brought in to turn around an organization's trajectory. To do so, they may replace the previous program with a new approach that matches their personal OPEX understanding more clearly or they demand to see immediate improvements from a still young program when, in reality, the effects require significant time to fully unfold. We are not saying that new management cannot bring the sought turnaround. However, as we have witnessed many times in our industry and research work, OPEX programs that would have had the potential to yield great improvements, given enough time, fall victim to impatience and short-term thinking. Then, companies either reset or completely halt their OPEX efforts. This is until a new manager or a new strategy breathes fresh life into it and the procedure starts all over. Therefore, we believe that the truly sustainable OPEX programs are the ones that had the time to develop deep structures that were less vulnerable to punctuations (see Part IV). A novel trigger that we have seen to emerge more recently revolves around digitalization and artificial intelligence (AI). As these technologies have already become part of daily life for many, we also sense great excitement and eagerness among manufacturers to integrate these technologies into their OPEX programs. Creating opportunities for next-level process improvements, the capabilities these technologies possess carry unprecedented potential. Consequently, companies have already begun to update their OPEX programs' content accordingly. However, to succeed when entering this emerging life cycle iteration, they must rethink their approach holistically across all dimensions. Given the relevance of this topic, we explore the reconciliation of OPEX and digitalization at length in Chap. 11.

In general, the life cycle of an OPEX program can be described as three subsequent and iterative phases.[3] The design of the program, where its content, process, and structure dimensions are defined, represents the first phase. This is followed by the deployment. Here, the OPEX program is rolled out to all manufacturing sites, departments, and functions. This includes the diffusion of knowledge and practices in relation to the program as well as any activities that enable this deployment. This phase therefore directly relates to the process dimension. In addition, the process dimension also addresses the sustainment of the program, which embodies the third life cycle phase. Here, the organization is focused on substantiating the implementation by maintaining and expanding its OPEX maturity, striving for the establishment of deep structures. Further, the program is enhanced through frequent updates. A typical OPEX program life cycle is displayed in Fig. 5.3. While the schematic depiction suggests that life cycle phases are distinct, in reality, companies often find themselves in transition between these phases. In that, they frequently circulate between designing and deploying updates to certain elements of their program while simultaneously sustaining others.

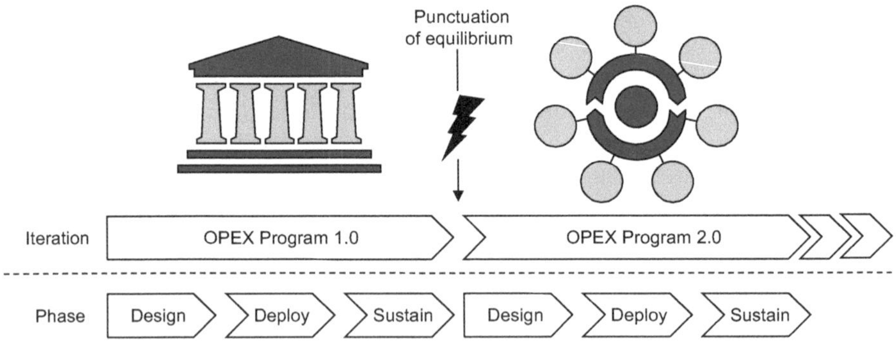

Fig. 5.3 OPEX program life cycle (adapted from Grothkopp, 2023)

[3] Szulanski (1996), Netland (2012), Schmidt and Zahn (2015), and Danese et al. (2017) describe slightly refined yet compatible life cycle phases. Their works provide the basis for our generalization.

5.4 Characteristics

We identify five distinct characteristics for OPEX programs that determine how these must be approached and managed. We have already touched on some of these features in this chapter as they relate to the definition and life cycle. We derive others from our research on OPEX programs. All are summarized hereafter.

First, OPEX programs represent an interrelated, socio-technical system of synergetic practices. In that, their content builds on tools and principles that, when combined and employed correctly, unfold positive effects that lead to operational improvements and competitive advantages. The key challenge, especially when designing an OPEX program, is to translate practices to the specifics of the respective company. Blindly copying content from the likes of Toyota usually fails. Instead, it must be reinterpreted to fit the setting and purpose.

Second, OPEX programs are corporate and strategic undertakings. Thus, their scope ranges across the entire manufacturing network, not just selected locations. Thereby, OPEX programs become multiplant endeavors. This bears additional challenges. Given the vastness of many manufacturing organizations, managing an OPEX program with consistency and precision becomes a highly complex task. Sociocultural differences, site-specific circumstances, and limited resources at the global and local levels add to this complexity and must be accounted for.

Third, OPEX programs evolve over time and must be updated frequently and proactively. Following the idea of CI, companies must not become complacent with initial successes. As corporations are highly dynamic, what works today may not work tomorrow. Evolving circumstances, new challenges, and strategic reorientations must be accounted for by the OPEX program. Scrutinizing the content, process, and structure dimensions frequently and thoroughly therefore becomes inevitable to stay on top.

Fourth, OPEX programs are contingent. Their management is, for example, influenced by the maturity that sites have already reached in the OPEX transformation. This aspect may determine how much support the corporate team must provide. The corporate OPEX team often also considers the strategic role a site plays and what it contributes to the overall business. Finally, contextualities, such as the OPEX history of a site or its national environment, influence the OPEX program.

Fifth, OPEX programs are path dependent and equifinal. This means that no one-best-way or one-size-fits-all approach to their management exists. Rather, as OPEX programs are highly company specific, what works, and what does not, can depend on contextualities (cf. fourth characteristic). At the same time, companies that face similar circumstances may still find success by employing different approaches to managing their program. We find evidence for this, for example, in our research on the deployment of OPEX programs (cf. Chap. 7).

5.5 Key Success Factors

OPEX programs promise great improvements and competitive advantages, but many manufacturers struggle to reap these benefits. Reportedly, three in four companies progress too slowly in their OPEX implementation, while only a quarter is able to achieve considerable results (Pay, 2008). Consequently, what truly yields success has been studied at length by many scholars, resulting in a plethora of success factors (cf. Netland, 2015). Instead of listing every single factor that has ever been identified, we want to address those we find to be most relevant and essential.

As with any corporate undertaking, OPEX programs require significant resources to succeed. Although the TPS and eventually OPEX thinking emerged from a sense of thriftiness, comprehensive change cannot be achieved without the right financial investments. These can, of course, be justified if their return is positive and savings outweigh costs. As mentioned before, managers often expect to see immediate payoffs. However, in the sense of longevity, OPEX programs will come to fruition only after some time. For example, when implementing preventive maintenance procedures, improvements will not show in instantly doubled production capacities or outputs. Instead, the benefits will be evident when the lifespan of machinery can be extended and costly equipment purchases can be put off. Thus, investments have to be not only made but also given enough time to deliver the desired returns. In addition to monetary means, sufficient personnel resources are key to managing an OPEX program across all its dimensions and over its entire life cycle. In theory, dedicated OPEX teams become obsolete once a corporation has truly internalized OPEX thinking. In reality, though, corporate dynamics often require dedicated teams to be kept in place for the program's management. Companies typically aim for a ratio of 1:200 to 1:100 for site OPEX full-time equivalents (FTEs) to total site FTEs. For corporate OPEX FTEs to total company FTEs, the average ratio is approximately 1:1700. However, depending on the company size, this figure can vary. We find that, to a certain point, corporate OPEX FTEs grow disproportionately compared to total company FTEs. Beyond this point, companies then again create denser ratios to maintain company-wide alignment (see Fig. 5.4).

Despite the complexity of managing an OPEX program and the necessity to invest in it, resources are rarely abundant. Thus, to succeed, companies must set clear priorities and focus their means on what truly matters. Identifying the most

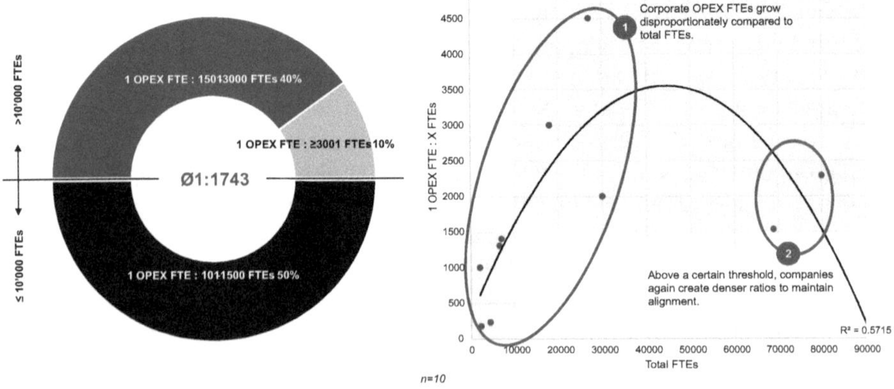

Fig. 5.4 Corporate OPEX FTEs to total FTEs in pharmaceutical manufacturing

value-adding improvement opportunities can, however, become a tricky task. Especially in the pharmaceutical industry, where operations must be validated and approved by regulatory bodies, transforming entire production processes to exploit the largest upside potentials is not always feasible. As such changes require lengthy reapproval procedures, which would freeze revenue streams, pharmaceutical companies are hindered and thus reluctant to pursue these opportunities. As a result, the notion that CI does not work in the pharmaceutical industry may sometimes be sensed.[4] Essentially, CI relies on small incremental improvements that, in combination, create considerable value added. Thus, companies who have truly mastered OPEX excel at identifying those opportunities that are incremental yet highly valuable and focus their efforts on these. Given the limited resources, identifying what truly adds value also relates to integrating digitalized or AI-based solutions into the OPEX program. While newer and shinier innovative solutions are often tempting, not all improve processes. Thoroughly scrutinizing digital use and business cases and having a sophisticated approach to reconciling OPEX and digitalization therefore become critical for success (cf. Chap. 11). Finally, we have witnessed in our research that, especially when rolling out their OPEX program to the manufacturing sites, successful companies deliberately and consciously prioritize certain sites over others. This is inevitable as corporate OPEX deployment managers are usually responsible for multiple sites at once. To avoid skin-deep results where it matters the most, they therefore concentrate their efforts on these sites.

To become operationally excellent, corporate and site OPEX teams cannot rely exclusively on their internal competencies. In this sense, successful companies seek inspiration for successful practices from external cases. At the same time, successful companies also disseminate proven approaches that emerged locally, proactively across the global manufacturing network. As previously noted, blindly copying practices does not, however, lead to success. Thus, these practices must be evaluated in detail and adapted carefully to fit the environment where they should be applied. In addition to sharing as well as sourcing practices internally and externally, successful companies benchmark their OPEX achievements against others. This allows them to identify and close performance as well as maturity gaps. Further, they receive reference values for world-class results, which helps them to set attainable targets for their own OPEX roadmap. While benchmarking is highly useful, finding the right peer group is crucial to avoid meaningless comparisons. As companies and sites are unique, finding the perfect match to compare is virtually impossible. However, based on contextualities, such as company or site size, production volume, and product type, reliable peers can be identified to benchmark the OPEX program against.[5]

---

[4]We disagree with this notion, as we believe that CI is highly applicable to the pharmaceutical industry. Therefore, we address such hurdles in Chap. 12 and present solutions to overcome these.

[5]Given the importance of successful practice sharing as well as benchmarking, the St.Gallen OPEX research team supports pharmaceutical companies in these activities through benchmarking projects and industrial exchange groups. For further information, please contact us.

Finally, OPEX programs demand high organizational commitment and engagement. Only if the company as a whole believes in the approach and pulls in the right direction can and will the program succeed. Given their role and exposure in the organization, top managers are particularly important in this regard. If they are not committed to the cause, where should the rest of the organization draw its motivation from to contribute to the program? As OPEX programs are resource intensive, these decision-makers must also be willing to allocate the necessary means to the program for it to succeed. Further, OPEX programs can only flourish if all employees, especially frontline workers, are encouraged and enabled to contribute to the program. As they possess crucial process expertise, leveraging their know-how is essential to improve operations. Further, they are the ones directly affected by process changes. If not involved in the transformation, they likely will not feel valued and therefore obstruct the improvement efforts. Ultimately, the OPEX program must become part of a company's identity. In this sense, truly successful companies do not view their OPEX program as an abstract strategic aspiration or checkbox exercise but as the way to conduct their work.

References

Bhasin, S., & Burcher, P. (2006). Lean viewed as a philosophy. *Journal of Manufacturing Technology Management, 17*(1), 56–72. https://doi.org/10.1108/17410380610639506

Danese, P., Romano, P., & Boscari, S. (2017). The transfer process of lean practices in multi-plant companies. *International Journal of Operations & Production Management, 37*(4), 468–488. https://doi.org/10.1108/IJOPM-12-2014-0571

Grothkopp, M. (2023). *Managing production systems: A corporate-wide improvement program perspective* [Dissertation]. University of St. Gallen.

Klinkner, F. (2026). *Rolling out company-specific production systems: A network perspective* [Dissertation]. University of St. Gallen.

Krafcik, J. F. (1988). Triumph of the lean production system. *Sloan Management Review, 30*(1), 41–52.

Liker, J. K. (2004). *The Toyota way: 14 management principles from the world's greatest manufacturer* (1st ed.). McGraw-Hill Education, McGraw Hill.

Netland, T. H. (2012). Managing strategic improvement programs: The XPS program management framework. *Journal of Project, Program & Portfolio Management, 3*(1), 31–44. https://doi.org/10.5130/pppm.v3i1.2556

Netland, T. H. (2015). Critical success factors for implementing lean production: The effect of contingencies. *International Journal of Production Research, 54*(8), 2433–2448. https://doi.org/10.1080/00207543.2015.1096976

Ohno, T. (1988). *Toyota production system: Beyond large-scale production* (1st ed.). Productivity Press. https://doi.org/10.4324/9780429273018

Pay, R. (2008). *Everybody's jumping on the Lean Bandwagon, but many are being taken for a ride*. IndustryWeek. https://www.industryweek.com/leadership/companies-executives/article/21946187/everybodys-jumping-on-the-lean-bandwagon-but-many-are-being-taken-for-a-ride

Pellegrinelli, S. (1997). Programme management: Organising project-based change. *International Journal of Project Management, 15*(3), 141–149. https://doi.org/10.1016/S0263-7863(96)00063-4

Schmidt, S., & Zahn, T. (2015). Einführung Ganzheitlicher Produktionssysteme. In U. Dombrowski & T. Mielke (Eds.), *VDI-Buch. Ganzheitliche Produktionssysteme:*

Aktueller Stand und zukünftige Entwicklungen (pp. 171–187). Springer Vieweg. https://doi.org/10.1007/978-3-662-46164-8_3

Szulanski, G. (1996). Exploring internal stickiness: Impediments to the transfer of best practice within the firm. *Strategic Management Journal, 17*(S2), 27–43. https://doi.org/10.1002/smj.4250171105

Womack, J. P., Jones, D. T., & Roos, D. (1990). *The machine that changed the world.* Rawson.

Operational Excellence Programs: Measuring and Improving Results

6

Fabian Klinkner and Thomas Friedli

Abstract

Operational Excellence (OPEX) programs are implemented to deliver process improvements and competitive advantages. Given the immense resource investment these programs require, corporate management is therefore rather emphatic – or sometimes even impatient – to see their effects unfold. Thus, to steer an OPEX program in the right direction, monitoring whether it delivers the sought results is essential. In this chapter, we explore four complementary perspectives for measuring these results. Further, we share insights on the use of Key Performance Indicators (KPIs) and the development of elaborate KPI systems in the pharmaceutical industry. Especially when an OPEX program is still in its infancy, it may not yet deliver stellar results. Therefore, we close this chapter by looking at how improvement potentials can be tackled through a well-structured approach.

As outlined before, Operational Excellence (OPEX) programs are implemented with the intent to achieve process improvements and competitive advantages. Given the immense resource investment these programs require, corporate management is therefore rather emphatic—or sometimes even impatient (cf. Chap. 5)—to see their effects unfold. Thus, to steer an OPEX program in the right direction, monitoring whether it delivers the sought results is essential. In our industry and research

F. Klinkner (✉) · T. Friedli
Institute for Production and Supply Chain Management, University of St.Gallen, St.Gallen, Switzerland
e-mail: fabian.klinkner@unisg.ch; thomas.friedli@unisg.ch

T. Friedli et al. (eds.), *Striving for Operational Excellence in the Pharmaceutical Industry*, Future of Business and Finance,
https://doi.org/10.1007/978-3-032-26287-5_6

Table 6.1 Maturity assessment typology

Frequency	Quarterly	Yearly	Biyearly		>Biyearly
Duration	<1 day	1–2 days	3–5 days		>5 days
Scope	Entire site	Certain product(s)	Certain process(es)/ value stream(s)		Certain function(s)
Level of detail	All sites		Selected sites		
Scale	0–4	1–5	Other scale		Text based
Number of questions	0–10	11–50	51–100		>100
Capture	Assessment handbook		Questionnaire		
Participants	Global OPEX team	Local OPEX team	Site leadership/ line functions	Internal/ external third parties	Customer(s)
Mode	Self-assessment	By corporate team		Combination	
Format	On-site	Remote		Hybrid	

activities, we observe different approaches to measuring the effects of OPEX programs, yielding four complementary perspectives.

The first perspective revolves around the OPEX program's maturity, which is defined as "*the breadth and depth of its implementation*" (Netland & Ferdows, 2016, p. 1107). In other words, maturity reflects how far the content has already been disseminated across the organization and how rigorously it is being applied (Klinkner, 2026). To capture this, companies rely on dedicated assessments. Typically, these are led by the corporate OPEX team and conducted at frequent time intervals across the organization. Variations between companies can occur in the extent of the assessment as well as its execution. While some organizations, for example, conduct their assessment on a yearly basis in an on-site 2-day format, others may revert to a less frequent, 1-day remote approach. Based on our practical insights, we have developed a typology to classify different maturity assessment approaches (see Table 6.1). Given the many options companies can choose from when configuring their assessment, the question arises whether a one-best-way exists. While we have found no supporting evidence for this in our research, we have nevertheless identified two success factors. First, as a maturity assessment aims to uncover areas of insufficient OPEX implementation, exhaustiveness and consistency in the approach are crucial. Companies must therefore be thorough and assess the OPEX implementation across the entire organization and all elements of the program, as well as with sufficient granularity. Further, they must not let too much time pass between assessments to prevent any deterioration of previous successes. By conducting detailed and frequent assessments, companies can counter *acting* and *avoiding* behavior, where entities only pretend to implement the OPEX program or bypass it completely (Netland & Aspelund, 2014). Second, a maturity assessment must yield tangible outputs. Maturity reports as well as improvement plans and road maps allow to translate identified gaps into concrete actions, thus helping to build upward momentum. However, equally important to defining and taking suitable measures is following up on their implementation and effectiveness. Thereby, companies can monitor whether the necessary progress is made to close any maturity gaps and that the upward momentum is maintained.

The second perspective relates to operational performance. As processes are optimized and inefficiencies eliminated, results translate to key performance indicators (KPIs).[1] Positive effects become, for example, evident in improved on-time-in-full (OTIF) deliveries, fewer deviations, or higher right-first-time (RFT) rates. To monitor such performance elements, companies develop elaborate KPI systems. These systems must be designed to reflect the corporate strategy. This link between performance measurement and business needs can be created through strategy maps, balanced scorecards, or Hoshin Kanri procedures. Further, regular reviews of the system help to uphold its strategic alignment. As we will elaborate later, which KPIs to measure and how to interpret them are highly contextual. Therefore, companies must put considerable effort into selecting the right KPIs to create a holistic view. However, as holistic does not mean exhaustive, a cautious differentiation between key indicators and other metrics is necessary when configuring these systems. In addition to measuring *what* matters, companies must do so *where* it matters. In this sense, successful KPI systems include intricate cascading and aggregation logics, which connect KPIs across the entire organization. Drawing on previous research and industry activities, we shed further light onto KPI systems in later parts of this chapter. Finally, effective performance measurement requires company-wide harmonization. In this sense, establishing corporate standards for the definition and measurement of performance eliminates island solutions, while granting certain autonomy to sites and departments produces successful practices. Typically, a delay can be observed between emerging maturity and translation into performance.

The third perspective revolves around financial performance. This viewpoint is closely linked to the previous lens as financial figures may also be considered as KPIs. As such, these are commonly integrated in performance measurement systems. Thus, the success factors we elaborated above also apply here. However, the operational performance and financial performance perspectives represent alternate, but not necessarily contradictory, schools of thought to assess results. Figures around cost savings and cost avoidance embody key indicators to assess improvements in a firm's financial situation and are therefore essential for managing an OPEX program. However, as established by Ferdows and Meyer's (1990) sand cone model (see Fig. 6.1), improvements in financial performance build on previous improvements in operational performance. In the long run, companies do not face a trade-off between quality, cost, and time, but can only excel in all these elements simultaneously if tackled in the right sequence. Thus, the effects of implementing an OPEX program translate to financial performance later than to operational

[1] Improper program design and insufficient engagement or commitment, among many other factors, can prevent an OPEX program from unfolding its desired effects. Here, we assume that the prerequisites for a successful OPEX implementation are in place, resulting in improved performance.

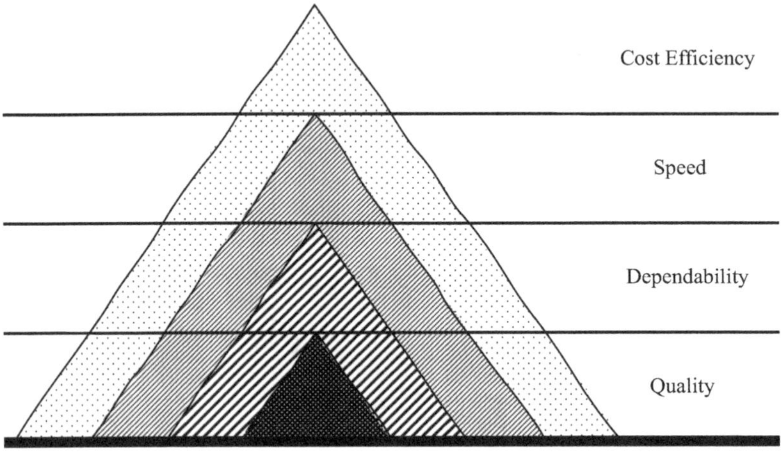

Fig. 6.1 Sand cone model (based on Ferdows & Meyer, 1990)

performance. Consequently, measuring results only through financial figures would be rather short-sighted and is often characteristic of impatient OPEX program management.

Finally, the fourth perspective captures a company's continuous improvement (CI) mindset. As CI is a pivotal element to any OPEX program, its philosophical components serve the purpose of establishing a culture where every individual is directly engaged in the improvement efforts and actively contributes to these. Contrary to maturity and performance, the CI mindset is less tangible and harder, if not impossible, to determine directly. To capture results from this meta-perspective, companies therefore revert to proxy measures. One approach may be to monitor the number of improvement ideas submitted by employees. The underlying hypothesis is that in companies with more suggestions, the CI mindset is higher. However, this does not necessarily provide sufficient insight into how far this mindset has truly spread, as these suggestions could originate from a limited number of sources in the organization. Thus, companies should not only track how many suggestions are submitted but whether all departments, functions, and sites are contributing ideas. Further, simply tracking suggestions by quantity reveals little about their value. Successful practice companies therefore also evaluate the quality of improvement ideas, which can be captured through operational and financial performance figures. However, linking suggestions and KPI improvements can be cumbersome, requiring close collaboration between OPEX and controlling teams. The level of OPEX-related training across an organization can serve as another proxy measure for the CI mindset. Here, the underlying hypothesis is that the more employees have been trained, the higher the CI mindset must be. Besides simply calculating the rate of employees that have received training in relation to the OPEX program, companies also take into account the number of lean or six sigma belt certifications. However, a common struggle companies face when trying to increase these training and certification ratios is that the built know-how is not always applied in day-to-day activities, failing to yield actual improvements. As a countermeasure, companies may therefore require employees to draft and conduct a concrete improvement project as part of their training or certification. Finally, cultural surveys offer another common tool to assess the current CI mindset but may fail to shed light on how and where to improve it, if off course.

While all four perspectives measure certain effects and results from implementing an OPEX program, no perspective is superior to or better equipped than the others. In fact, companies must not focus on one perspective alone to track whether they are achieving their goals. Instead, they must combine these four perspectives to have a holistic picture of their improvement efforts.

Measuring a system's performance entirely by a single indicator or through the aggregation of only a few metrics lacks balance and fails to provide a holistic perspective, evoking narrow-minded and misguided behavior among employees (Friedli et al., 2013). At the same time, companies must be careful not to create KPI overload by including too many metrics in their performance measurement approach, losing sight of what truly matters. The KPI selection funnel provides a structured approach to resolve this dilemma. When designing their system, companies turn to different sources to make their KPI selection. Typically, they consider which metrics are already being tracked in the organization. Further, metrics that are common in their respective industries or even in other sectors provide a sufficient starting point for the selection process. Finally, state-of-the-art research insights provide additional guidance on which KPIs to potentially include in the system. This vast set of performance indicators serves as input to the selection funnel. At the start of the selection process, the purpose of the KPI system must be defined. This creates the necessary link between performance measurement and strategic management. Subsequently, four filters narrow down the list of potential metrics to a holistic set of strategically aligned KPIs. The first filter, *expressiveness*, assesses the internal and external relevance of a certain metric and whether it describes key characteristics of process performance. Second, *sensitivity* takes into account whether an indicator is responsive and granular enough to capture the necessary process performance parameters. Third, *apprehendability* considers how well and easily a metric can be understood as well as how accepted it is in the organization. Finally, *ascertainability* relates to how practical the capture of a certain metric is in relation to time expenditure or automation. The KPI funnel and its filter criteria are depicted in Fig. 6.2.

To illustrate the application of the KPI selection funnel, we review overall equipment effectiveness (OEE) as a performance indicator. OEE is widely used in pharmaceutical manufacturing contexts and is typically tracked at the shop floor level, making it a natural candidate for inclusion in an OPEX performance measurement system.

Expressiveness: OEE performs well on the expressiveness criterion, as it captures key dimensions of equipment-related process performance by integrating availability, performance, and quality into a single indicator. Internally, it provides a concise representation of losses related to downtime, speed, and defects, while externally it is a well-established benchmark across manufacturing industries. As

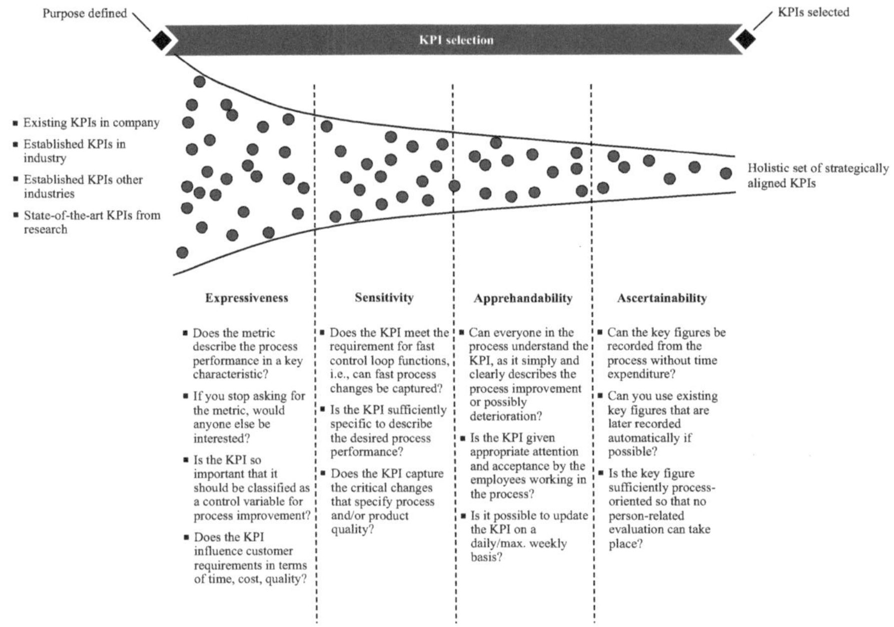

Fig. 6.2 KPI funnel

such, OEE effectively describes central characteristics of production process performance.

Sensitivity: From a sensitivity perspective, OEE is sufficiently responsive and granular to reflect operational improvements or deteriorations over short time horizons. Changes in setup times, minor stoppages, scrap rates, or cycle-time adherence are quickly reflected in OEE values, allowing managers to link improvement initiatives directly to measurable performance outcomes. This makes OEE suitable for CI environments where frequent feedback is required.

Apprehendability: OEE is generally well understood by operational staff and management alike, particularly in organizations with prior exposure to lean or TPM practices. Its decomposition into availability, performance, and quality supports transparency and facilitates communication across hierarchical levels. While the aggregated nature of OEE may initially obscure root causes, its widespread acceptance and interpretability contribute positively to employee buy-in and usage.

Ascertainability: Finally, OEE scores high in terms of ascertainability. The required data—machine states, production counts, and quality losses—are often already available through manufacturing execution systems or shop floor data collection tools. Where digital infrastructure is in place, OEE can be calculated automatically with limited manual effort, making it practical to capture on a regular basis without excessive time expenditure.

Passing all four filters of the KPI selection funnel, OEE represents a balanced, strategically aligned performance indicator. While it should not be used in isolation, its inclusion as part of a broader KPI system provides meaningful insight into equipment-related performance and supports targeted OPEX initiatives.

Although KPI systems are company specific, corporations often end up selecting similar indicators. Such interorganizational resemblances may be motivated by regulatory activities and industry-specific contexts as well as by certain metrics' relevance for general production management. While a universal KPI system would contradict the aforementioned company specificity, the apparent similarities provide insight into which indicators are particularly important for certain industries. In this sense, the set of KPIs featured in our St.Gallen OPEX benchmarking embodies a representative sample of such successful practice metrics for the pharmaceutical industry. A detailed list of these metrics, including their definitions, can be found in our reassessment of the current state of pharmaceutical OPEX.

As stated before, intricate hierarchies aggregate or cascade KPIs across the entire organization and allow to measure performance at different levels of abstraction. In relation to these levels, reporting structures typically change. This creates a virtual pyramid, where at the top the overall manufacturing contribution is monitored in retrospect, providing a more strategic lens. Thus, such "lagging" indicators are typically reviewed at upper management levels and less frequently. At the bottom, operational KPIs provide a prospective view on performance. Review cycles for such "leading" indicators are much shorter, usually daily. In between operational KPIs and manufacturing contribution, aggregated diagnostics combine these perspectives. Which metrics may be considered most relevant at which level as well as reporting structures are again company specific. Thus, the hierarchy depicted in Fig. 6.3 only provides an illustrative example.

As hinted before, KPI systems and performance measurement are highly contingent. In this sense, KPIs must always be viewed in light of contextualities, as these bear significant implications. In our reassessment of the current state of OPEX in Chap. 3, we consciously limited our analyses to the pharmaceutical industry. We are

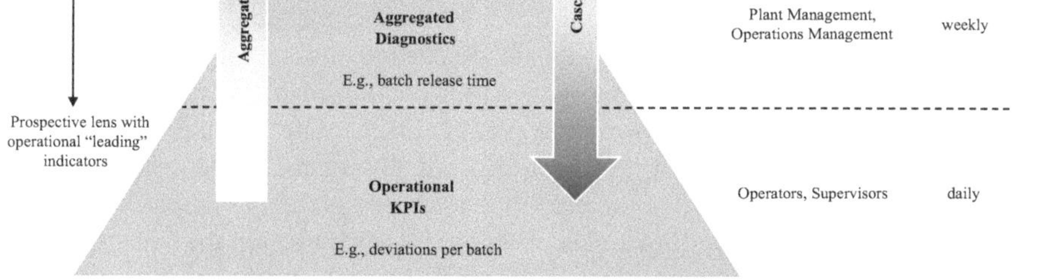

Fig. 6.3 Exemplary KPI hierarchy

not motivated to do so because we believe that the pharmaceutical industry is superior to other sectors or better suited for such exercises. Instead, we are strong advocates of industry-specific performance benchmarks because different industrial sectors set different baselines for performance. In the automotive industry, for example, an OEE rate between 70% and 85% is considered acceptable, whereas in pharmaceutical manufacturing, values between 10% and 60% are much more common. Sticking to this example, we see that lead times in pharma may well rise to 120–180 days, while automotive manufacturers strive for values between 1 and 7 days. As both industries are distinctly different and "work" in their own way, what characterizes high performance varies significantly.

In a similar vein, comparisons between companies or even processes within the same industry must also be made with great caution, as baseline performance levels may again differ. For example, comparing the number of deviations per batch between a sterile liquid manufacturing unit and a tablet manufacturing unit is senseless, as the formulation and packaging processes of the former drug product are naturally much more prone to GMP-relevant deviations. Likewise, the OEE rate of a tablet packaging line, which blisters millions of pills annually, will usually exceed the OEE rate of a bioreactor for drug substance manufacturing.

Finally, even for the same process type, performance levels must be assessed under consideration of the respective circumstances. In this sense, context factors, such as production volumes, number of stock-keeping units (SKUs), new product launches, equipment age, or automation degrees, among many other aspects, can have a drastic impact on the operational metrics. If, for example, a production unit excels in the proportion of setup and cleaning work, it can really only be compared to another unit of the same process type if the number of changeovers is similar in both cases. Otherwise, the unit with fewer changeovers may display better numbers for this particular metric despite not necessarily being operationally more excellent.

In conclusion, when measuring the results of an OPEX program along KPIs, implications and interpretations must not be derived blindly. We urge OPEX managers to always pay careful attention to any contextualities that may affect their performance indicators to avoid ill-advised decision-making.

When implementing an OPEX program, at certain steps of the transformation journey, results will fall short of expectations, requiring additional efforts to be made to overcome any maturity or performance gaps. A well-structured approach to tackle such improvement potential is therefore essential.

Assuming that the current state has been assessed thoroughly and root causes have been identified, these insights must be translated into actions next. Companies therefore develop systematic improvement plans that outline appropriate resolutions, thus driving their transformational journey. Such plans typically consist of four elements: (1) a detailed description of the current state, including a problem statement regarding the challenge to be addressed; (2) a detailed description of the solution, i.e., improvement idea, to tackle the respective issue; (3) a precise definition of the target state, including a collection of metrics by which the success of the solution can be measured; and (4) a comprehensive description of the approach, including milestones, responsibilities, and methodology.

While all four elements are equally important, often such plans do not deliver actual improvements due to insufficiently defined target states or unclear approach descriptions, such as missing to set measurable targets or outlining activities too vaguely. A successful practice to overcome the former issue may be, for example, to set so-called SMART goals, meaning that these should be **s**pecific, **m**easurable, **a**chievable, **r**elevant, and **t**ime bound. To overcome the latter issue, for example, work-breakdown and Gantt charts specify timelines, task sequences, and key stages, as well as the relationships and roles of all involved stakeholders. Thereby, the necessary level of detail for a successful approach description can be achieved.

Finally, improving results requires thorough governance. In this sense, OPEX program managers must track the implementation of improvements as much as they track their effectiveness. Further, they are tasked with maintaining oversight across the entire manufacturing network, which is a prerequisite for the program to work in synchronicity and harmony. More so, OPEX managers must be well informed about all corporate improvement activities to be able to create visibility for successful practices that could potentially be transferred across the network.

References

Ferdows, K., & de Meyer, A. (1990). Lasting improvements in manufacturing performance: In search of a new theory. *Journal of Operations Management, 9*(2), 168–184. https://doi.org/10.1016/0272-6963(90)90094-T

Friedli, T., Basu, P., Bellm, D., & Werani, J. (Eds.). (2013). *Leading pharmaceutical operational excellence*. Springer Berlin Heidelberg. https://doi.org/10.1007/978-3-642-35161-7

Klinkner, F. (2026). *Rolling out company-specific production systems: A network perspective* [Dissertation]. University of St.Gallen.

Netland, T. H., & Aspelund, A. (2014). Multi-plant improvement programmes: A literature review and research agenda. *International Journal of Operations & Production Management, 34*(3), 390–418. https://doi.org/10.1108/IJOPM-02-2012-0087

Netland, T. H., & Ferdows, K. (2016). The S-curve effect of lean implementation. *Production and Operations Management, 25*(6), 1106–1120. https://doi.org/10.1111/poms.12539

Operational Excellence Programs: Network-Level Rollout

7

Fabian Klinkner and Thomas Friedli

Abstract

When Operational Excellence (OPEX) programs fail to deliver substantial improvements, the root causes often lie in the program's implementation. Depending on how the rollout is managed, the achieved outcome can be vastly different. Deploying an OPEX program is therefore a highly context-dependent matter. Already proving to be a daunting endeavor at the site level, managing this rollout successfully is evermore complex at the network level. Drawing on our research in the pharmaceutical industry, we present two distinct rollout strategies to successfully roll out an OPEX program from headquarters (HQ) to the manufacturing network. Further, we shed light on how the HQ supports its sites at different stages during this process.

7.1 Introduction

When Operational Excellence (OPEX) programs fail to deliver substantial improvements, the root causes often lie in the program's implementation (Secchi & Camuffo, 2019). Depending on how the rollout, which we define as the process of deploying an OPEX program from the corporate headquarters (HQ) to the manufacturing

The following chapter is based on the work of Klinkner (2026), who conducted his research as a member of the St.Gallen OPEX research team.

F. Klinkner (✉) · T. Friedli
Institute for Production and Supply Chain Management, University of St.Gallen, St.Gallen, Switzerland
e-mail: fabian.klinkner@unisg.ch; thomas.friedli@unisg.ch

T. Friedli et al. (eds.), *Striving for Operational Excellence in the Pharmaceutical Industry*, Future of Business and Finance,
https://doi.org/10.1007/978-3-032-26287-5_7

network, is managed, the achieved outcome can be vastly different (Secchi & Camuffo, 2019). Deploying an improvement program is therefore a highly context-dependent matter (Shah & Ward, 2003). Already proving to be a daunting endeavor at the site level, managing this rollout successfully is evermore complex at the network level (Demeter & Losonci, 2019). Here, companies are tasked with rolling out their program to many different sites, in some cases more than a hundred, adding to the intricacy of the undertaking (Yadav et al., 2017). During the rollout, the HQ must support its sites, for example, through coaching and training, by assessing the progress of implementation or by collaborating closely with the sites on deployment-related activities (cf. Boscari et al., 2016; Danese et al., 2017; Linderson, 2024; Linderson et al., 2025; Netland, 2012; Netland & Ferdows, 2016). Generally, HQ support can be provided through both financial and personnel resources and varies during the rollout (Netland, 2015; Netland & Ferdows, 2016). Consequently, the HQ's support role evolves over the course of the deployment.

Drawing on our research in the pharmaceutical industry, we present two distinct rollout strategies to successfully roll out an OPEX program to the manufacturing network. Further, we shed light on how the HQ supports its sites at different stages during this process.

7.2 Rollout Strategies

To successfully deploy their OPEX program, pharmaceutical manufacturing companies can choose from two strategies, namely a deliberate sequence (DS) and an emergent sequence (ES). In the former approach, the rollout is purposefully staggered, and the OPEX program is deployed in waves of descending priority across the manufacturing network. In the latter approach, the rollout is initiated simultaneously but becomes gradually staggered as sites deviate in their progression rates. Both strategies are detailed hereafter.

7.2.1 Deliberate Sequence

Following the DS, the HQ differentiates its sites by their contribution, which it assesses by their strategic importance (i.e., size, manufactured products, served markets), revenue streams, or business needs. Based on this contribution, the HQ defines a priority ranking for its sites and partitions the manufacturing network into corresponding, structurally equivalent entities. The rollout is purposefully staggered across these entities, deploying the OPEX program in waves of descending priority and creating a deliberate sequence. Therefore, we label this strategy as such. The number of waves can vary, but primary, secondary, and tertiary waves are not uncommon. The sequence is defined by the HQ, which considers two additional determinants: first, feasibility, which is captured by site resource availability, and second, context, which is assessed through previous OPEX exposure and current site performance. The differentiation by contribution is accompanied by discrimination in relation to HQ support. While prioritized sites can usually draw from extensive HQ resources during the rollout, deprioritized sites must expect to be denied such high levels of support. Thus, the HQ purposefully creates stronger ties with prioritized sites than with deprioritized sites.

In addition to contribution, companies differentiate their sites by maturity. This also ties to the level of support given by the HQ. Contrary to the contribution-based discrimination, this maturity-based distinction is less rigid and rather serves as an indicator of how much support the HQ must expect to provide to each site. As sites progress in the rollout and mature, they typically require less support and become more independent in implementing the OPEX program. However, exceptions occur.

Two complementary forces guide the rollout, namely push and pull. Regardless of the strategy, the HQ always aims to generate pull from its sites, acknowledging that their commitment and participation are essential for an effective rollout. In the DS, the HQ initially exerts a very high push to overcome inertia. Gradually building up momentum, pull is created slowly as the rollout progresses. The strong initial push is, for example, expressed by the HQ defining the priority ranking, rollout sequences, timelines, and milestones. Initially, pull is very low but can, for example, come in the form of sites volunteering to move ahead in the rollout sequence. In the DS, push and pull pivot at a certain point during the rollout, once the built-up momentum ultimately exceeds the inertia.

In essence, the DS is characterized by an asynchronous deployment initiation, initially driven by push with limited pull, followed by a late pivot of these forces. Figure 7.1 features a conceptual depiction of the DS.

The conceptual depiction of the DS consists of three elements. The diagram on the top left shows the maturity development of the different rollout waves (here I, II, III as examples) over time. As our depiction holds true regardless of the graphs' shape, we simplify the relation to be linear. Notably, in the DS, the rate of maturity

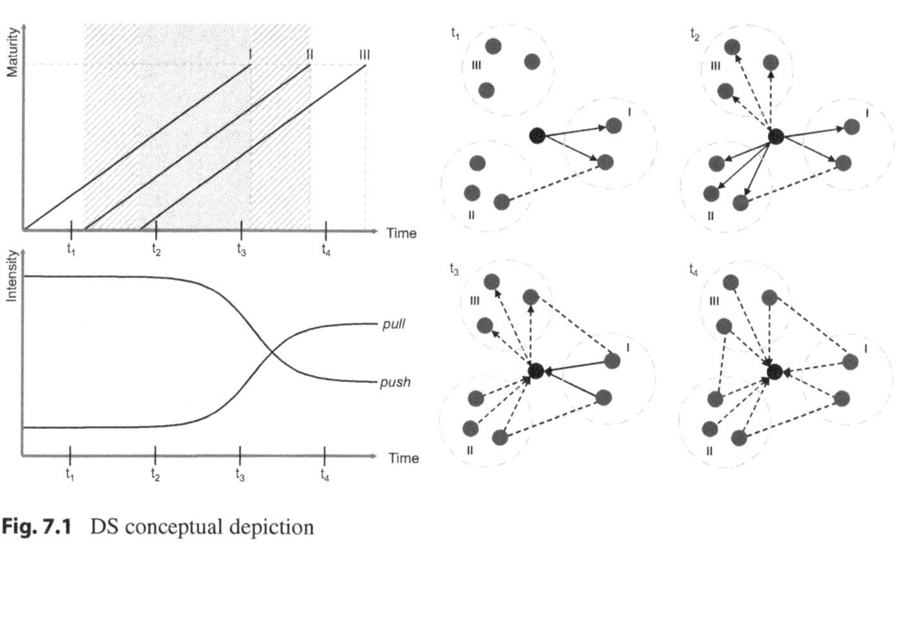

Fig. 7.1 DS conceptual depiction

development is consistent across all waves. Despite the deployment being staggered, we observed that the different waves overlap. This is captured by the shaded areas in the diagram, where darker shading indicates greater overlap.

The bottom-left diagram shows the distribution of push and pull over time. Pull is a proxy for the rate of adoption, following an S-shaped curve in relation to rollout progression. Push, being reciprocal to pull, consequently follows an inversely shaped curve. The diagram further features the late pivot point of the two curves.

The right side visualizes the DS through social network interactions amidst the rollout across four discrete points in time (here t_1–t_4), which correspond to the axes on the left. The HQ is represented by a black node, and the sites by gray nodes. Based on site differentiation by contribution, structurally equivalent entities are circled and labeled corresponding to the rollout waves. Push is reflected by edges directed from the HQ to the sites, and pull vice versa. Solid-line edges represent strong ties, while dotted-line edges indicate weak ties. Between HQ and sites, this relates to the level of HQ support. Cross-site interactions via bridges (i.e., weak ties) are portrayed by undirected edges between the site nodes. Based on our conceptual depiction, we detail the DS from initiation throughout the four discrete points in time hereafter.

Initially, the degree of push is very high. This directive is expressed by the HQ mandating the program's implementation and prescribing the rollout strategy, as well as differentiating, prioritizing, and sequencing sites. Pull is very low, as the engagement of the sites is limited to potentially requesting to pre- or postpone the deployment.

At t_1, only the top priority sites of the first wave are subject to the deployment. The rollout is dominated by push, which is evident in HQ management of both content and process (cf. Chap. 5). Accordingly, foundational practices are mandated strictly with little room for site-specific adaptation, due to limited experience (i.e., maturity). Meanwhile, the HQ steers the rollout via rigid top-down directives by imposing rollout timelines and progression milestones on the sites. Further, the HQ reserves itself the right to define the level of support each site receives. As the focus at this point lies exclusively on the first wave, the support is high, while no rollout-related interactions between the HQ and deprioritized sites occur yet. Commitment to the OPEX program from the sites is still minimal. Few lateral interactions between sites emerge at this point. Typically, these are represented by site-level OPEX personnel of later waves participating in rollout activities of the current wave to learn through observation.

At t_2, all sites are subject to the rollout, leading to overlapping waves. Push continues to be the dominant force driving the rollout. However, as first improvements from the OPEX program proliferate, bottom-up pull begins to rise. Evidently, sites begin to proactively request the deployment of additional, not necessarily mandatory, OPEX content or start to adapt practices to site-specific circumstances. With its limited resources, the HQ is forced to focus its efforts on higher-priority sites, providing them with stronger support than deprioritized sites. Cross-site interactions remain similar to t_1. Additionally, more mature sites take on an ambassador status, serving as role models for the remaining network. The HQ establishes formats that facilitate successful practice sharing between sites, such as communities of practice, social intranet platforms detailing local OPEX use cases, reward programs for improvement suggestions, or benchmarking procedures between sites.

At t_3, all sites continue to be subject to the rollout. At this point, wave I and II sites have reached a high maturity level. Consequently, they are more self-reliant in rolling out the OPEX program and benefit from greater autonomy, but still receive strong HQ support where necessary. The higher autonomy materializes in growing engagement from the sites. This trend is amplified by now noticeable performance improvements, which strengthen the commitment of previously hesitant sites. Consequently, pull continues to increase, building on the developments apparent at t_2. In return, the level of push decreases further, now only dominating wave III. As a result of the HQ's previous efforts to facilitate cross-site interactions, they have increased further. Meanwhile, the HQ shifts its activities towards elevating successful practices from highly mature sites to global standards.

At t_4, sites of wave I and II have completed most of the deployment and have begun to transition to sustainment. At this point, the sites of wave III have also reached a high level of maturity. Having surpassed the pivot point, pull now dominates the rollout. This is captured by additional OPEX adaptation at the site level and by upheld demand for nonmandatory content. The remaining push is evident in continued site prioritization and sequencing when deploying updates to the program. Cross-site interactions have increased further, spanning between all three waves.

7.2.2 Emergent Sequence

Following the ES, sites are not differentiated by default but treated alike, forming a single, structurally equivalent entity. The rollout is initiated simultaneously across the entire manufacturing network. After this initial push, the forces guiding the deployment pivot instantly, with pull becoming the dominant one. This is substantiated by high site engagement and, consequently, low HQ directive.

Sites set their own progression milestones and timelines and decide when to proceed with the deployment, thus developing their own rates of progression. As these differ across the network, the rollout gradually becomes staggered, resulting in an emergent sequence. Therefore, we label this strategy as such. The HQ sets periodic progression gates, which all sites must clear on time. Contrary to the progression milestones in the DS, these gates are fewer, not site specific, and represent lower boundaries rather than well-defined maturity targets. Through this directive, the HQ maintains control over the rollout by exerting a minimal level of push.

The high engagement is also expressed by the level of HQ support being co-defined by the sites. Corresponding claims are submitted to the HQ, which allocates its resources accordingly to fulfill the demand. As the rollout staggers, the required level of support varies across the network. The ties created between HQ and sites are diverse but tend to weaken with growing maturity. Resource constraints limit the HQ's ability to provide equally high support to all sites simultaneously. While a scenario where all sites require substantially high support at the same time is unlikely for the above reason, it is nevertheless accounted for in the ES. We observed that the HQ may reserve itself the right to overrule the sites'

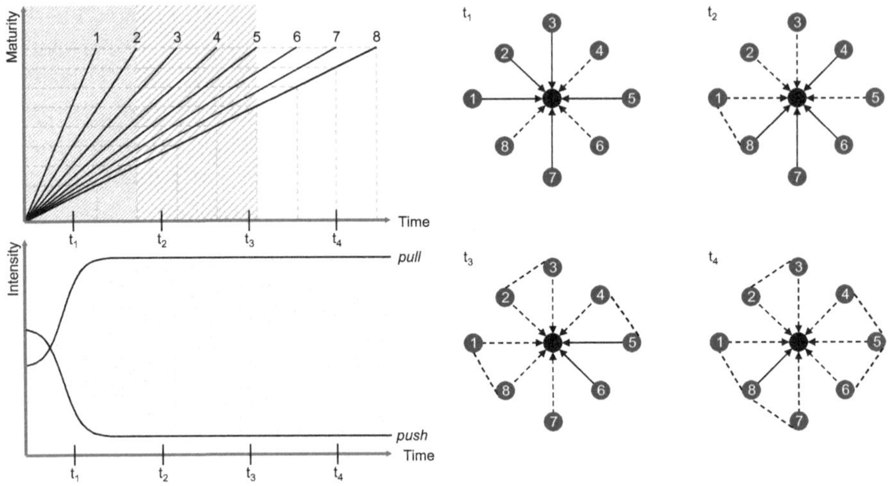

Fig. 7.2 ES conceptual depiction

support claims. By exception, the HQ then distinguishes different manageable entities and prioritizes sites based on their contribution to allocate its resources. As this occurs only momentarily, it does not represent a partition of the network into multiple structurally equivalent groupings. This directive complements the push exerted by the HQ.

Essentially, the ES is characterized by a synchronous deployment initiation and relies on pull with residual push, following an early pivot of these forces. Figure 7.2 features the conceptual depiction of the ES.

The conceptual depiction of the ES is consistent with the notation of Fig. 7.1. The top-left diagram shows the maturity development of all sites (here 1–8) over time. Again, we simplify this relation to be linear. As outlined above, progression rates differ in the ES. The graphs' varying slopes capture this. The synchronous deployment initiation inherently creates significant overlap, indicated again by the shaded areas.

The bottom-left diagram depicts the distribution of the forces guiding the rollout over time. In line with our previous arguments, push and pull follow inversely and regularly shaped S-curves. The diagram further features the instant pivot of the two. Here, we want to emphasize that our conceptual depictions are schematic and not to scale regarding the delta between the pivot point and any other point on the time axes. The same applies to the delta in push (or pull) between the DS and ES at any given time.

Finally, the social network interactions amidst the rollout are again depicted on the right. Nodes are labeled correspondingly to the sites. The four discrete points in time (again t_1–t_4) correspond to the axes on the left and Fig. 7.1. Based on our conceptual depiction, we detail the ES from initiation throughout the four discrete points in time hereafter.

Initially, push and pull are moderate. We assume the former to be slightly higher, with the HQ mandating the OPEX program's implementation, defining the rollout strategy, and initiating the deployment. Pull materializes in sites informing the definition of early progression gates and the time of network-wide initiation.

At t_1, all sites are subject to the rollout, while push and pull have already pivoted. The former is low but evident in the management of both content and process dimensions. In this sense, the HQ mandates foundational practices that all sites must implement, linking these to progression gates. If necessary, due to the significant overlap at this point, the HQ uses its directive to prioritize sites in its provision of rollout support. Apart from these instances, where the HQ exerts residual push, it relies exclusively on the sites' pull to progress the deployment. Buy-in at this point is moderate across the network, while more committed sites begin to pull ahead in the rollout. These sites already exhibit a greater maturity than their less committed counterparts and benefit from initial performance improvements. Due to limited experience with the OPEX program, most sites require significant support from the HQ. However, because of site-specific progression timelines and individual capabilities, exceptions can occur. The HQ focuses on implementing the necessary mechanisms to share successful practices laterally. These correspond to the DS but have not been established yet. Consequently, cross-site interactions are not apparent at this point.

At t_2, the first three sites have reached high maturity in the deployment, while the rest of the network is still trailing behind. Push continues to decrease further as the HQ allows for greater local adaptation to foster the emergence of successful practices. Although the previously described residual push is maintained, the HQ must less frequently mitigate resource constraints through prioritization than before. Growing performance improvements at high-maturity sites are now becoming visible to the entire manufacturing network, leading to a surge in commitment to the program and additional pull. Sites, which had previously held off on the deployment or progressed at a slower pace, now proceed with the rollout and increasingly request additional, nonmandatory content elements. Due to the HQ's prior efforts to facilitate successful practice sharing, first cross-site interactions now emerge.

At t_3, most sites have reached high maturity in the deployment. Benefiting from their gained experience, the remaining sites now tend to require less support in implementing the program's content, with only a few maintaining strong ties with the HQ. Nevertheless, it still encounters overlap, upholding the eventuality of having to account for conflicting support claims by prioritizing selected sites over others. Therefore, the degree of push remains unchanged. Performance improvements continue to arise across the manufacturing network, perpetuating the high level of pull. The HQ now shifts its efforts to elevating successful practices to global standards. Meanwhile, additional lateral interactions between sites form.

At t_4, all sites have reached high maturity in the deployment. Site 8 remains the last one to trail behind in the rollout, allowing the HQ to fully focus its deployment support here. The distribution of push and pull, as well as the HQ's efforts to diffuse successful practices, remains unchanged. Finally, all sites are now connected laterally to others in the network.

7.3 Headquarters Support

While in the DS the given support is solely defined by the HQ, in the ES, sites are more closely involved in this decision, with the HQ having the final say. From our research on rollout strategies, we know that how much support the HQ is willing to provide is, in part, influenced by its perception of a site's priority and maturity. The former relates to the site's strategic importance (i.e., size, manufactured products, served markets), revenue, and general business needs. The latter relates to the degree of OPEX integration and therefore a site's progression in implementing the program. Further, two additional parameters influence the support provision, namely capacity and accessibility.

Capacity outlines a site's ability to facilitate the rollout and encompasses three sub-elements. As the first sub-element, capability relates to the expertise of a site's OPEX resources and key personnel involved in the deployment (e.g., site leadership team (SLT)) in relation to lean, production systems, and OPEX programs. Depending on their competencies and qualifications, they can carry out rollout-related activities with more or less supervision from the HQ. Notably, capability and maturity typically correlate positively, due to progression in the program's implementation enhancing the respective know-how of site personnel, assuming that they are continuously involved in the process. While some sites may already have highly qualified OPEX managers when starting their local rollout, others draw from little prior experience when setting out on this journey. Consequently, the HQ considers this aspect when determining the necessary support. The second sub-element of capacity is a site's engagement towards the OPEX program and its rollout. Indicating how much resistance to change the HQ must overcome, engagement serves as an indicator of the ease of the deployment. In this sense, engagement also describes a site's willingness to cooperate with the HQ during the rollout. As the OPEX program is mandated by the top management, the HQ must prevent sites from avoiding its implementation (cf. Netland & Aspelund, 2014). Therefore, corporate supervision intensifies if a site appears to sidestep the program. Finally, availability represents the third sub-element of capacity, relating to the allocation of dedicated resources to the OPEX program and its rollout. Availability may be influenced by the size of a site's OPEX team, the time site personnel can designate to the rollout without impairing their other responsibilities, employee turnover, or budgetary restrictions, which prohibit the hiring of dedicated OPEX personnel.

Complementing the other parameters, accessibility refers to how well a site can be supported in person. Like capacity, accessibility builds on three sub-elements. First, proximity estimates the feasibility of traveling to a site to provide support during the deployment. This may be impacted by the geographical distance between the HQ personnel's location and the site, as well as the existing transportation infrastructure. Regionally divided corporate OPEX teams and deliberate recruiting of HQ OPEX managers close to secluded sites are measures to mitigate insufficient proximity. Second, accessibility is impacted by the travel budget of both the HQ and the sites. Respective HQ-level budgetary limitations may require the HQ to increase its remote support instead of frequently traveling to a site to be physically present during the local deployment. Meanwhile, site-level monetary restrictions may cause one of the two opposing effects. On the one hand, these may increase the on-site presence, for example, requiring the HQ to facilitate cross-site OPEX trainings at sites that do not have the financial means to travel to other locations for such activities. On the other hand, if a site cannot afford to host such activities, the HQ may revert to remote formats instead. Finally, circumstances represent the third sub-element of accessibility, addressing contingencies that limit the HQ's ability to travel to a site in person. This may include restrictions due to ongoing pandemics or occurrences with similar restraining effects (e.g., geopolitical conflicts).

While the four parameters, priority, maturity, capacity, and accessibility, embody the main components considered by the HQ, it further factors in the context of each site. Although this does not directly determine the frequency, intensity, and mode of support, it can weigh in on the final support decision. For example, the HQ may decide to still conduct a site visit, despite already having exhausted its approved annual travel budget, if contextualities at a specific site require the HQ's physical presence. In relation to context, the HQ therefore considers two sub-elements, namely a site's current needs as well as its individual and regional culture.

During the rollout, the HQ's support for its sites evolves. Over the course of the deployment, the HQ thus embraces three consecutive support roles, namely prepare, enable, and confirm. Sites are assigned roles reciprocal to those of the HQ, namely engage, learn, and execute. Together, HQ and site roles translate into distinct support phases, henceforth referred to as the combination of these roles. While the prepare/engage phase extends over a relatively short amount of time, the enable/learn and confirm/execute phases span a greater timeframe. In their respective prepare and engage roles, the HQ is, for example, tasked with introducing the OPEX program to the site, whereas the site plans the logistics of the deployment. Moreover, the HQ may even provide funding to backfill positions that become vacant as site employees are reallocated to support the deployment. Subsequently, the enable/learn phase revolves around building the corresponding capabilities of site OPEX personnel. Finally, responsibility for facilitating the rollout shifts from the HQ to the site towards the confirm/execute phase. Notably, the relation between the HQ and its sites is characterized by trust and collaboration.

The conceptual framework, depicted in Fig. 7.3, captures and visualizes our findings on HQ support. It embeds the four decisive parameters in the context, displaying HQ support at the center.

7.4 Practical Implications

Our findings present several implications for OPEX managers. As HQ resources are limited, corporate OPEX managers should abandon any attempts at rolling out their OPEX program in a de facto simultaneous manner. Instead, they should

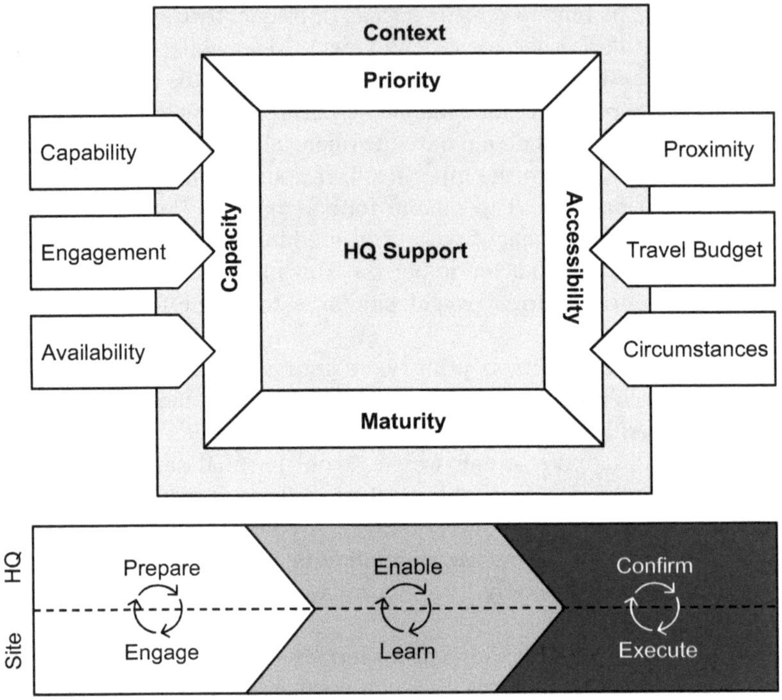

Fig. 7.3 HQ support conceptual framework

opt for a sequenced approach to avoid skin-deep or even failed implementation. This sequence can be either deliberate or emergent. We present managers with two corresponding strategies to choose from. These give guidance on the HQ's role during the deployment, providing support in identifying determinants to consider, and hints on how to steer this process. Managers should not only consider network size or manufacturing locations to select the right rollout strategy for their company. Rather, they should also evaluate whether they can expect immediate commitment to the OPEX program from their sites, whether they already possess the right capabilities to implement the program autonomously, and whether they differ in their strategic positioning within the manufacturing network. Ultimately, managers must assess whether it is necessary to coerce sites to embrace the OPEX program or whether the sites can be expected to commit to it themselves.

Our insights also suggest that HQ OPEX managers should assess their sites' priority, maturity, capacity, and accessibility, as well as the respective context, when determining how to support the sites. Further, the definition of the three support phases provides guidance for the HQ on which roles it must embrace throughout the rollout. At the same time, the results indicate which reciprocal roles the HQ should demand that the sites take on during the deployment.

References

Boscari, S., Danese, P., & Romano, P. (2016). Implementation of lean production in multinational corporations: A case study of the transfer process from headquarters to subsidiaries. *International Journal of Production Economics, 176*, 53–68. https://doi.org/10.1016/j.ijpe.2016.03.013

Danese, P., Romano, P., & Boscari, S. (2017). The transfer process of lean practices in multi-plant companies. *International Journal of Operations & Production Management, 37*(4), 468–488. https://doi.org/10.1108/IJOPM-12-2014-0571

Demeter, K., & Losonci, D. (2019). Transferring lean knowledge within multinational networks. *Production Planning & Control, 30*(2–3), 211–224. https://doi.org/10.1080/09537287.2018.1534272

Klinkner, F. (2026). *Rolling out company-specific production systems: A network perspective* [Dissertation]. University of St.Gallen.

Linderson, S. (2024). Context-driven implementation strategies: Exploring three approaches to implement a lean capability framework within a global production company. *Knowledge and Process Management*. https://doi.org/10.1002/kpm.1790

Linderson, S., Larsson, J. L. A., Birkie, S. E., & Bellgran, M. (2025). Navigating implementation strategies: Insights from XPS programmes. *Journal of Manufacturing Technology Management, 36*(9), 69–87. https://doi.org/10.1108/JMTM-05-2024-0233

Netland, T. H. (2012). Managing strategic improvement programs: The XPS program management framework. *Journal of Project, Program & Portfolio Management, 3*(1), 31–44. https://doi.org/10.5130/pppm.v3i1.2556

Netland, T. H. (2015). Critical success factors for implementing lean production: The effect of contingencies. *International Journal of Production Research, 54*(8), 2433–2448. https://doi.org/10.1080/00207543.2015.1096976

Netland, T. H., & Aspelund, A. (2014). Multi-plant improvement programmes: A literature review and research agenda. *International Journal of Operations & Production Management, 34*(3), 390–418. https://doi.org/10.1108/IJOPM-02-2012-0087

Netland, T. H., & Ferdows, K. (2016). The S-curve effect of lean implementation. *Production and Operations Management, 25*(6), 1106–1120. https://doi.org/10.1111/poms.12539

Secchi, R., & Camuffo, A. (2019). Lean implementation failures: The role of organizational ambidexterity. *International Journal of Production Economics, 210*, 145–154. https://doi.org/10.1016/j.ijpe.2019.01.007

Shah, R., & Ward, P. T. (2003). Lean manufacturing: Context, practice bundles, and performance. *Journal of Operations Management, 21*(2), 129–149. https://doi.org/10.1016/S0272-6963(02)00108-0

Yadav, O. P., Nepal, B. P., Rahaman, M. M., & Lal, V. (2017). Lean implementation and organizational transformation: A literature review. *Engineering Management Journal, 29*(1), 2–16. https://doi.org/10.1080/10429247.2016.1263914

Towards a Better Understanding of What Really Makes the Difference

8

Thomas Friedli

Abstract

In this chapter, we compare leading companies for Operational Excellence (OPEX) with what we call followers to sharpen the understanding of how to avoid superficial OPEX implementation. The main findings are based on our recent OPEX program benchmarking study, conducted in 2022.

8.1 Introduction

In the opening address of our St.Gallen Production Management Conference 2025, we emphasized the following key message for Operational Excellence (OPEX) in 2026 and beyond: "skin-deep is not enough." In this chapter, a comparison between leading companies for OPEX and what we call followers will be made to sharpen the understanding of how to go deeper and not just stay on the surface. The main findings are based on an OPEX program[1] benchmarking study,[2] conducted in 2022 by the OPEX research team at the University of St.Gallen.

[1] The terms OPEX program and production system are used interchangeably in this chapter.

[2] In other chapters, we refer to the St.Gallen OPEX Benchmarking. This benchmarking, developed back in 2004, is focused on a specific site and structured along the St.Gallen OPEX model. Throughout this chapter, we talk about benchmarking the global company-wide approach.

T. Friedli (✉)
Institute for Production and Supply Chain Management, University of St.Gallen, St.Gallen, Switzerland
e-mail: thomas.friedli@unisg.ch

T. Friedli et al. (eds.), *Striving for Operational Excellence in the Pharmaceutical Industry*, Future of Business and Finance,
https://doi.org/10.1007/978-3-032-26287-5_8

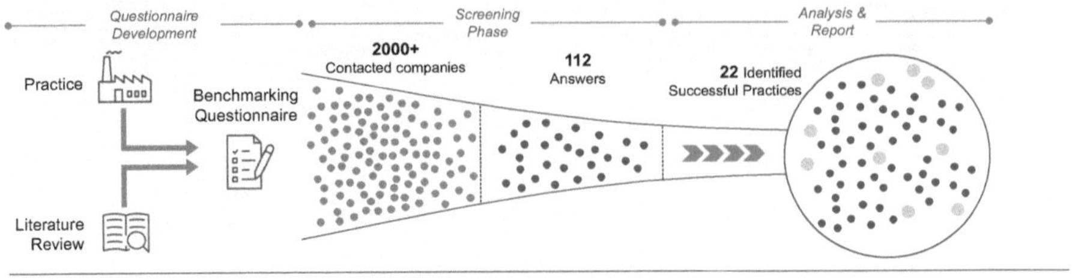

Fig. 8.1 Methodology

For a detailed description of the methodology and especially the developed constructs, see Grothkopp (2023). This research was based on a benchmarking study following the renowned St.Gallen consortia benchmarking approach (cf. Friedli et al., 2022).

Our benchmarking followed a three-step approach (see Fig. 8.1). First, we designed an initial draft questionnaire based on an extensive literature review. A consortium group with practitioners from various industries reviewed the questionnaire and gave feedback. Hence, the resulting questionnaire provided a holistic baseline to benchmark companies from different industries on how they manage their OPEX programs. Second, we reached out to more than 2000 companies contacting their OPEX program managers in positions such as "head of production systems," "COO," or similar. Over a period of 4 months, 112 responses were collected from companies completing the questionnaire (see Table 8.1). In a third step, we validated and analyzed all respondents' replies to compile the result. These results provide insights into how successful practice companies (top 20%) distinguish themselves from the rest (lower 80%). This gives us an opportunity to reveal what is needed for a sustainable and integrated implementation of OPEX, avoiding skin-deep approaches.

8.2 How We Selected the Successful Practice Companies

The successful practice companies have been selected based on extraordinary improvement performance. We constructed and validated an overall improvement performance score (see Fig. 8.2), which was used to determine the top 20% companies as successful practices. Followers refer to those companies that make up the remaining 80%.

8.3 Maturity/Implementation

Based on our OPEX understanding (cf. Chap. 1), we asked a number of questions related to the degree of implementation of the respective OPEX program. We

Table 8.1 Benchmarking sample

Panel A: Industry		Panel B: FTEs		Panel C: Sites	
Pharmaceutical	19	<1000	11	<6	28
Automotive	16	1000–5000	21	6–29	50
Machinery and plant engineering	14	5001–10,000	19	30–53	18
Electronics and electronic equipment	12	10,001–20,000	30	54–78	5
Plastics and rubber	7	>20,000	31	79–102	0
Metal production	6			103–126	3
Mechanical engineering	6			127–150	0
Chemical	5			>150	5
Food and beverages	4			No data	3
Consumer goods, consumables, and fast-moving consumer goods	4				
Aerospace and aviation	4				
Energy	4				
Paper	3				
Glass, glassware, ceramics	3				
Medical technology and instruments	2				
Defense	2				
Textile and apparel	1				
Optics and precision engineering	0				
Tobacco	0				
Total	112	Total	112	Total	112

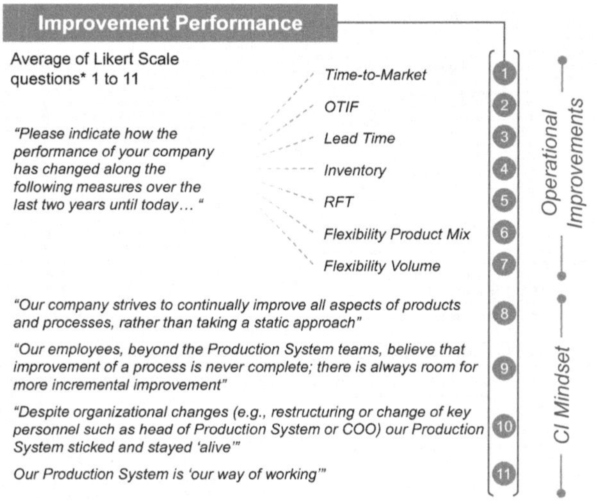

Fig. 8.2 Improvement performance score

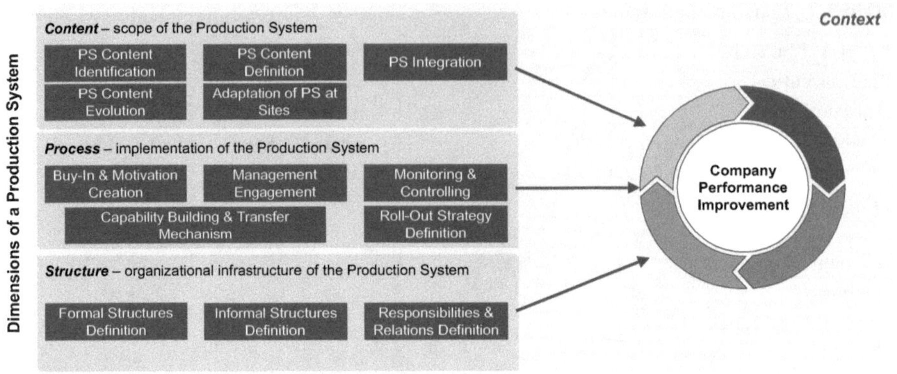

Fig. 8.3 OPEX program dimensions

followed the dimensions provided by Netland (2012) to ensure that we covered the implementation holistically (see Fig. 8.3).

8.4 Results

Three-fourths of respondents initiated or reinitiated their OPEX program in the last 10 years. The earliest initiation in this data sample refers to the year 1950.

Half of the successful practice companies operate between 5 and 29 production sites, while 5% operate less than 5 sites and 18% operate between 30 and 53 sites. Moreover, 9% of them operate with 54–78 sites, 5% with 103–126 sites, and another 9% with more than 150 sites.

Figure 8.4 shows the participants' level of performance and maturity. The level of maturity explains around 46% of the variance in performance, which is rather high for this kind of study, and demonstrates the benefits of OPEX program implementation.

When it comes to our performance construct, the successful practice companies outperform the followers in every single category we asked for.

Whereas the two groups are rather close together when it comes to the unit costs compared to competition, there are considerable gaps in the delivery category, and, interestingly, the successful practice companies do also outperform the followers in the quality conformance category. This illustrates that the conceptual model of the sand cone (cf. Chap. 6 and Ferdows and Meyer (1990)) holds another practice test and that the best performers do not have to compromise in any performance category to achieve their level.

Across the sample, OPEX programs are not seen as simple toolboxes, which demonstrates that all participating companies were already on a rather high maturity level. Nevertheless, systems thinking is considerably higher in the successful practice group (see Fig. 8.5).

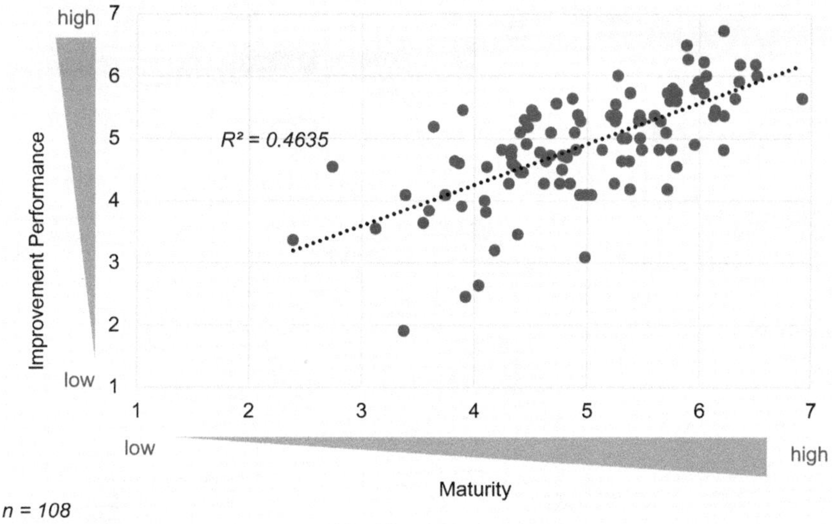

Fig. 8.4 Performance-maturity interplay

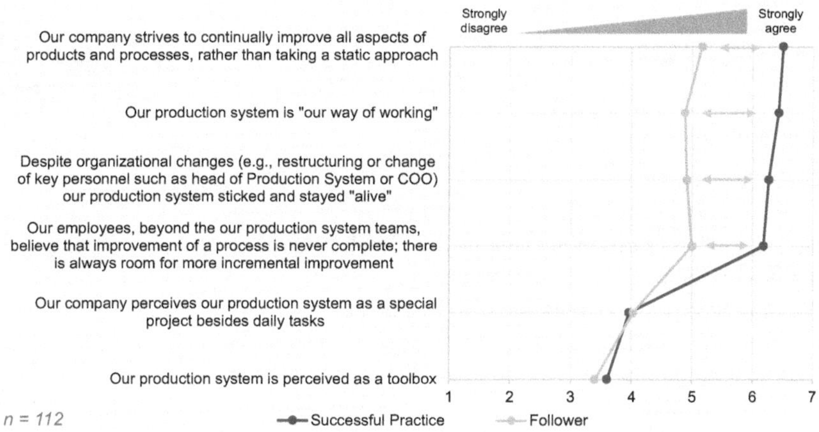

Fig. 8.5 Understanding of OPEX programs

Obviously, the successful practice companies were capable of embedding OPEX in a way that it became robust against disturbances like management changes. It is also much more than just a structure or some defined processes; it is the very way to work in the successful practice companies, and it brings a level of engagement with it that goes way beyond the OPEX experts; it embraces all other employees too.

However, in both groups, there is still a risk that the OPEX program could be seen as special and extra or just a toolbox and not as an integral part of the daily

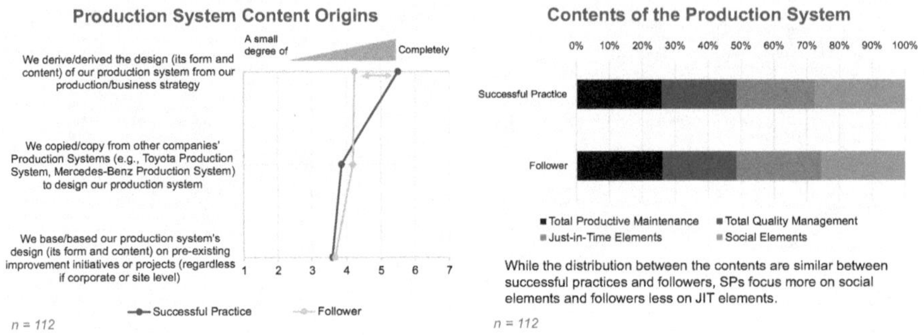

Fig. 8.6 OPEX program content

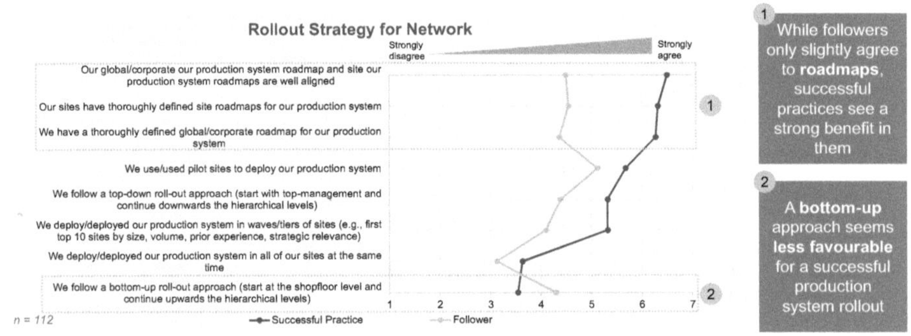

Fig. 8.7 Rollout strategy

work. This becomes obvious from the fact that both groups do not, on average, fully disagree with the respective statements.

When it comes to the question of how production systems are designed, the successful practice companies do this to a higher degree based on the production/business strategy and therefore achieve a better alignment (see Fig. 8.6).

When it comes to the rollout of the production systems, successful practice companies embrace road mapping much more, as well on-site as on the corporate level (see Fig. 8.7).

Additionally, communication seems to be key for a successful implementation (see Fig. 8.8).

We summarize the main factors to focus on to support the deep implementation of OPEX in Fig. 8.9. As leadership is crucial on the way to successfully implementing and sustaining OPEX, we will follow up on this in Chap. 9.

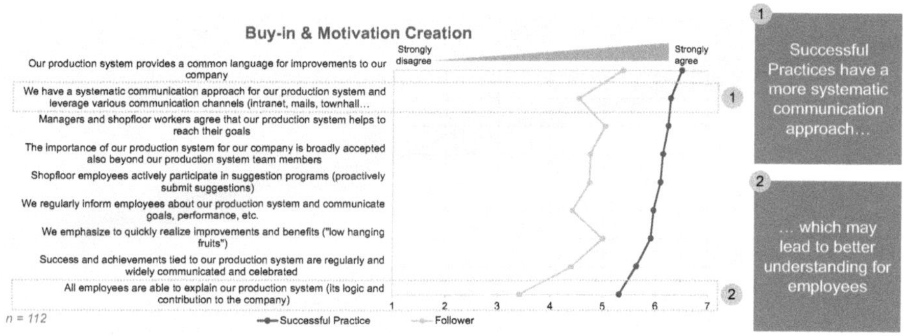

Fig. 8.8 Buy-in and motivation creation

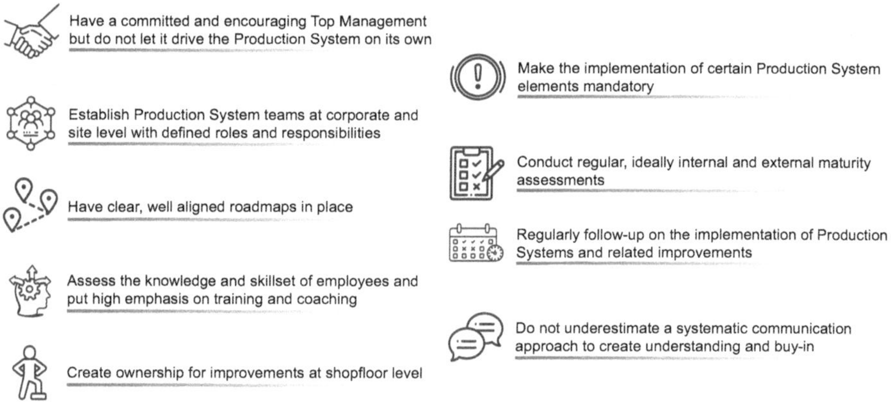

Fig. 8.9 Key factors of deep OPEX implementation

References

Ferdows, K., & Meyer, A. d. (1990). Lasting improvements in manufacturing performance: In search of a new theory. *Journal of Operations Management, 9*(2), 168–184. https://doi.org/10.1016/0272-6963(90)90094-T

Friedli, T., Grothkopp, M., Kaiser, J., Pirrone, L., & Liefde, T. d. (2022). *Benchmarking study: "Future of production systems"*. https://www.alexandria.unisg.ch/entities/publication/-c41e0450-6111-4121-b546-49446c14a812

Grothkopp, M. (2023). *Managing production systems: A corporate-wide improvement program perspective*. Dissertation. University of St.Gallen.

Netland, T. H. (2012). Managing strategic improvement programs: the XPS program management framework. *Journal of Project, Program & Portfolio Management, 3*(1), 31–44. https://doi.org/10.5130/pppm.v3i1.2556

Lead It or Leave It

9

Thomas Friedli

Abstract

In this chapter, we explore the critical role leadership plays for managing operational excellence (OPEX). To clarify the expectations when it comes to leading OPEX, we draw from a historical example from a different context but perfectly

---

Note: Leadership is instrumental for true and lasting Operational Excellence (OPEX). To clarify the expectations when it comes to leading OPEX, we will refer to a historical example from a different context but perfectly suited to demonstrate what is really needed. We will build our reasoning on the amazing story of the Berlin Airlift from 1948 and focus on the role of its main architect, General William H. Tunner. This chapter draws the information from the following works: William H. Tunner (1964), Over the Hump; Tusa, A., & Tusa, J. (1998), The Berlin Airlift: The Cold War Mission to Safe a City; and Miller, R. G. (1998): The Berlin Airlift, 1948–1949, Air Force History and Museums Program. Additionally, the following article was used for background information: Brunhaver, J. S. (1996): The Berlin Airlift, Lifeline from the Sky, The Doctrinal Implications of Supplying an Enclave from the Air, Air University Press.

We show that to build the airlift, a Toyota-like system had to be designed and sustained. To be able to serve a city with 2.5 million inhabitants in the West Sector of Berlin alone, only a takted conveyor belt-like system could do the job. Additionally, almost all elements later summarized in the Toyota Production System (TPS) are visible in the airlift. This is of interest as the example is older than the TPS and had to work in no time and there was no option for failure. We will only shortly describe some elements of the airlift in this chapter as the focus will be on the leadership part. What is however of interest is the different phases that the buildup of the airlift went through. The first phase before General Tunner took over the command was characterized by mostly operational hectic and heroism. It was only when Tunner touched ground in Berlin that professionalism took over and heroism was replaced by routine.

T. Friedli (✉)
Institute for Production and Supply Chain Management, University of St.Gallen, St.Gallen, Switzerland
e-mail: thomas.friedli@unisg.ch

T. Friedli et al. (eds.), *Striving for Operational Excellence in the Pharmaceutical Industry*, Future of Business and Finance,
https://doi.org/10.1007/978-3-032-26287-5_9

suited to demonstrate what is really needed. We build our reasoning on the story of the Berlin Airlift from 1948 and focus on the role of its main architect, General William H. Tunner. We conclude this chapter by summarizing key takeaways from Tunner's airlift and showcase how these apply and translate to (pharmaceutical) OPEX.

9.1 Introduction

After Germany surrendered in 1945, Berlin was divided into four sectors during the Potsdam Conference despite being 100 miles inside soviet-controlled Eastern Germany. In March 1948, the allied powers formed a single economic zone in Berlin (and in Western Germany) and later introduced the new "Deutsche Mark" in June of the same year. The Soviets reacted to this with a total blockade of West Berlin. Railway, road, and waterways into Berlin were closed on June 24th. The only access way remaining to Berlin agreed in writing was three 20-mile-wide air corridors between the city and West Germany for French, British, and American planes.

The challenge was formidable; the population of West Berlin alone was roughly 2.5 million inhabitants. The degree of self-sustainment was around 5%, and stocks were critically low.

Initially, the estimate was that Berlin required daily deliveries of 4000 tons to survive. However, a more realistic figure was 5600 tons daily (see Fig. 9.1). The population in whole West Germany was quite close to starving anyhow. The daily calorie available to the German population in the British Zone of Germany, e.g., was around 750. Before the blockade, the city imported between 9000 and 15,500 tons daily to meet the needs.

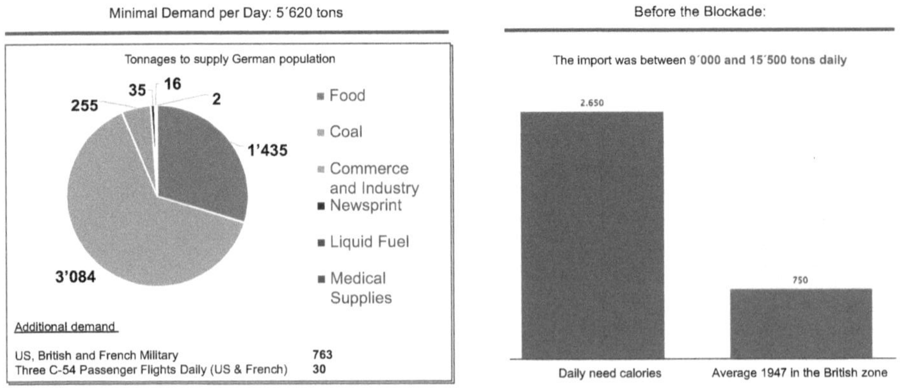

Fig. 9.1 Demand of West Berlin (numbers taken from Tunner, 1964, p. 207)

9.2 Decision Taken

There was more than one variant discussed on how to react to the blockade. Some favored sending an armed convoy to Berlin and hoped that this would not trigger the Third World War; others opted for a withdrawal as Berlin was not important from a military perspective. However, finally, Harry S. Truman, the US President, decided that Berlin will not be given up: "We are going to stay—period." As the only way open was through the air, an airlift was the answer to the challenge. There was no former experience to serve a city of this size from the air over an extended period. Additionally, there was no time for planning such an operation. The only contingency plans in place were focused on supplying the allied troops and family members stationed in Berlin. The balance of power was likewise unpromising, whereas the allied troops in Berlin numbered around 10,000 soldiers, they faced 250,000 soviet ones ready to be deployed.

9.3 Berlin Airlift Phase 1

In the first stage, British and Americans mobilized all resources they could get hold on. A lot was just improvisation. Based on a "challenges accepted and do your best" mentality, everybody was going on adrenalin. Desk-bound men who had not flown since earning their wings were called back to the cockpit. However, by the end of June, the combined US/British efforts only delivered 1404 tons per day, far away from what was necessary. It was obvious that if there would not be a drastic change in the way the airlift was managed, it would not be sustainable, especially as winter was looming.

It took a change in leadership to finally operate the airlift in a way that it could sustain. With the arrival of General William H. Tunner as commander of the airlift, everything changed. Already as an observer in faraway Washington, Tunner, back then second in command of the US Military Air Service Command (MATS), commented about the early days of the airlift:

"This was all very exciting, and loads of fun, but successful operations are not built on such methods .... Some of the features of Operation Vittles (the airlift) which were most enthusiastically reported by the press were contraindications of efficient administration" (Tunner, 1964, p. 160). Tunner, as probably the most experienced air transportation expert of his time, having, e.g., commanded "the hump," the airlift supplying Chinese troops from India crossing the Himalaya in World War II, looked for professionalism instead of adventure: "A successful airlift is about as glamorous as drops of water on a stone" (Tunner, 1964, p. 162).

The real excitement "comes from seeing a dozen lines climbing steadily on a dozen charts—tonnage delivered, utilization of aircraft—and the lines representing accidents and injuries going sharply down" (Tunner, 1964, p. 162).

Figure 9.2 summarizes just some of the challenges he encountered taking over the airlift on July 28.

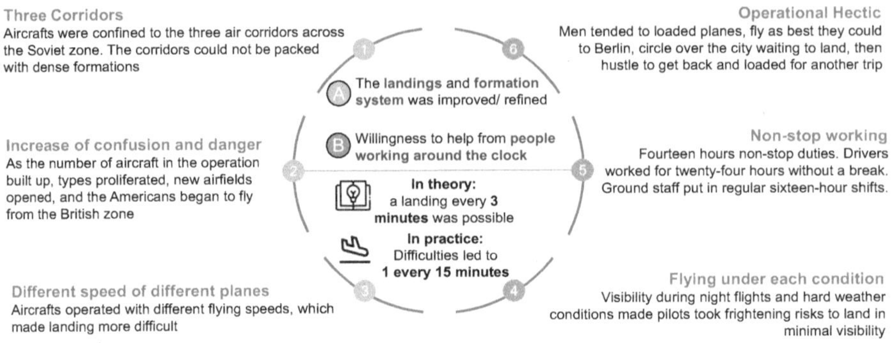

Fig. 9.2 Challenges in phase 1

9.4 Berlin Airlift Phase 2

As this chapter focuses on the leadership part of OPEX following the personality of General William H. Tunner, we will only shortly describe the technical elements Tunner introduced to the airlift. We will do so by comparing them briefly to elements of a TPS. Initially, Tunner sought order and flow, building a conveyor-like system. This initial system underwent some modifications based on severe incidents like what Tunner named "black Friday," which occurred on August 13th, 1948. In his own words: "[…] It was a day of black scudding clouds, of driving rain. Weather conditions were not too bad at Wiesbaden as we took off for Berlin, but as we gained altitude to clear the Harz Mountains we soon ran into those heavy, thick German clouds … Lieutenant Colonel Sterling P. Bettinger was piloting my C-54, good old Number 5549, which had served me so well on the Hump, and my old friend Red Forman was copilot. I sat on the jump seat behind them and helped them peer at the dark gray nothing ahead through the rain-washed windshield. We were not alone in the sky. As Bett followed the prescribed flight path to Tempelhof Field, calling out into the radio mike the exact second he passed over the Fulda beacon and swinging the nose of the plane to the exact heading of 057 degrees, we knew that some twenty C-54's were flying the same route ahead of us, each three minutes apart on the nose, each proceeding at 180 miles per hour. Ahead they stretched out like figures on a conveyor belt; behind we could hear each addition to the club as he passed over Fulda and gave us his time, loud and clear. All was well. The Berlin Airlift was seven weeks old. I had been its commander just fifteen days, and already, I felt, it was beginning to shape up into my kind of operation. But at that very moment everything was going completely to hell in Berlin. The ceiling had suddenly fallen in on Tempelhof. The clouds dropped to the tops of the apartment buildings surrounding the field, and then they suddenly gave way in a cloudburst that obscured the run-way from the tower. The radar could not penetrate the sheets of rain. Apparently both tower operators and ground-control approach operators lost control of the situation. One C-54 overshot the runway, crashed into a ditch at the end of the field, and caught fire; the crew got out alive. Another big Skymaster, coming in with a maximum load of coal, landed too far down the runway. To avoid piling into the fire ahead, the pilot had to brake with all he had; both tires blew. Another pilot, coming in over the housetops, saw what seemed to be a runway and let down. Too late he discovered that he'd picked an auxiliary runway that was still under construction, and he slithered and slipped in the rubber base for several precarious moments, then ground-looped.

With all that confusion on the ground, the traffic-control people began stacking up the planes coming in—and they were coming in at three-minute intervals. By the time we came in, the stack was packed from three thousand to twelve thousand feet. A space was saved for my plane at eight thousand feet, and we flew right on into it. Planes behind us, however, had to climb to the top of the stack—God knows why there were no collisions. As their planes bucked around like gray monsters in the murk, the pilots filled the air with chatter, calling in constantly in near-panic to find out what was going on. On the ground, a traffic jam was building up as planes came off the unloading line to climb on the homeward-bound three-minute conveyor belt, but were refused permission to take off for fear of collision with the planes milling around overhead.

'This is a hell of a way to run a railroad,' I snarled. Red and Bett wisely said nothing. They knew I had full confidence in both of them, but they were also aware that at that moment I'd have snapped my grandmother's head off" (Tunner, 1964, p. 152f.).

Tunner sends every plane with the exception of his own back to the airfields they came from and leaves his companions there to figure out a way to avoid something similar happening again: "And as for you two," I said, "I want you to stay in Berlin until you've figured out a way to eliminate any possibility of this mess ever happening again—ever/I don't care if it takes you two hours or two weeks, that's your job" (Tunner, 1964, p. 154f.).

This event triggered an even stricter regime to keep up the flow, e.g., only allowing landings with instruments even with good weather or only allowing for one try to land in Berlin; if this failed, the plane had to return to the airfield.

Figure 9.3 shows some of the characteristics of Tunner's airlift.

- **A perfect flow operation**:
  - Land a plane every three minutes on berlin airfields (1440 landings at the three airfields daily)
  - From stacking to return aircrafts that missed the approach
- **Customer pull (added value)**:
  - Careful elaborations what was needed and in which form (see optimized capacity)
- **SMED, elaborated planned maintenance**
- **Optimized capacity utilization (flour and coal vs., Bread, real coffee vs. «Ersatz-coffee», dehydrated potatoes, ...)**
- **Standardization (single conveyor belt – blocked conveyor belts)**:
  - Detailed procedures, exact duplication, and precise execution (e.g., speed of aircrafts)
- **Root cause analysis in case of events**
- **Continuous improvement**

Fig. 9.3 Characteristics of the Berlin Airlift

Fig. 9.4 Tunner's actions to improve the Berlin Airlift

Figure 9.4 shows an overview of Tunner's actions. We will discuss some of them to highlight his practices and leadership skills.

1. Personality: demanding but role model:

   Tunner was certainly a demanding commander, but at the same time, he was the very role model for everything he expected others to do and to be. To demonstrate this, a statement from him on the very first day he arrived in Germany provides evidence for this and is ample proof for his passion to get the job done. Arriving in Germany, nothing was ready for him. He gets the advice to go shopping until the rooms are ready. His reaction addressed to his staff: "Now look," I said, "we came here to work. I'm not asking you men to put in twenty-four hours a day, but dammit, if I can do eighteen hours a day, you can do fifteen. We didn't come to Germany to go shopping at the PX or the commissary, so I think we can just skip that little ceremony this afternoon" (Tunner, 1964, p. 167). At the same time, Tunner devoted considerable attention to the living conditions of the airlift personnel. He did his utmost to provide them with reasonable lodgings, food, and some distractions in the free time: "All I could do was to keep hammering away on higher headquarters to provide all the necessities and comforts of military life that were possible. Some of my base commanders thought I was going a little far when I proposed curtains for barracks windows, but I insisted. We were flying day and night. If a pilot had flown three round trips to Berlin the night before and was going to fly three more the next night, he deserved something to keep the sun out of his eyes" (Tunner, 1964, p. 179).

   Tunner also introduced proper shifts and rest for air crews (no excessive overtime: everyone engaged in the airlift now had regular shifts and periods of leave in requisitioned hotels for proper rest and real nights' sleep from time to time). Again, an element that can be found in Toyota principles: never overburden equipment nor people. Tunner believed in people and thought more in "we" as in "I." Crossing the Atlantic to take over command, he remembers: "Well, I looked around me on the plane and saw my staff on an enterprise about which we knew nothing other than that, with us on the job, it was going to be finished promptly and properly. I did know that even at this stage I was further along than I had been when I first took over the Hump. For I knew that already in Germany was a large number of aircraft crews who had learned their know-how and earned their laurels in the old ATC (Air Transport Command) and on the Hump. A pilot who had completed a tour on the Hump could go anywhere any pilot could go and return safely" (Tunner, 1964, p. 165).

Tunner also put paramount attention on safety, overcoming the initial "Cowboy Operation": Never, from the very beginning of my command until the end, had I subordinated flying safety to any other phase of operation. Despite our round-the-clock operation and the miserable weather conditions, our accident rate on the Berlin Airlift was less than the overall average for the United States Air Force. Of the total number of lives lost on the Airlift, seventy-two in all, of whom thirty-one were American, the great majority resulted from nonflying accidents. One of the many journalists who visited us, on looking at our accident figures, burst out: "Why I'm safer on the Berlin Airlift than I am flying between Washington and New York!" (Tunner, 1964, p. 218).

The experience from numerous OPEX programs clearly demonstrated the need for leaders and not for pure administrators. In the early days of OPEX in Pharma, people were chosen for OPEX jobs because of their statistical skills only to find out that to make a difference, they had to lead teams and to communicate in every direction. Besides being passionate about the task to be done, true leaders also care for the people they work with.

2. Understanding the issues/challenges:

Tunner believed in getting real information from the source: "After this new system had been in effect for several days, long enough for the pilots to become familiar with it, I set about getting their reactions. I didn't want to get this information from squadron commanders or group commanders, men who might conceivably be influenced by what they thought I would wish to hear, but directly from the lips of the men who were flying the route themselves, day in, day out. So, I took over the meeting room in the Schwarzer Bock Hotel and arranged to have a keg of good German beer and some cold cuts on hand. I sent out invitations to a group of about thirty pilots to come share the beer with me and some of my operating staff. After a stein or two, I threw out a question, and there was no hesitancy in answering it—everybody started talking at once" (Tunner, 1964, p. 175).

This helped to get unfiltered information about the realities on the ground and the most pressing issues from the very people who experienced them.

In some companies, OPEX was more visible in PowerPoint slides than on the ground, and managers often spend more time in the office than on the shop floor. The whole program became just skin-deep and never made roots to build on.

3. Doing real GEMBA:

Tunner makes quite some references to the way he led the airlift, e.g., he states: "I made constant rounds. I was available to everyone-dozens of times. Making myself available to the pilots and crew members, seeking them out and listening to their complaints, gave me an insight into the true operations of the lift that I could never have gotten sitting at my desk reading filtered and second-hand reports. Frequently I learned of an accident before the commander of the base on which it had occurred. The opportunity to report to him what was going on at his own base was extremely valuable, especially when I dragged him out of bed to do so. It tended to shake him up and keep him on his toes.

My meanderings were by no means restricted to the daytime. I would purposely pick late at night or the very early morning hours to check up on what I considered important or vital activities. It wasn't unusual for a sleepy tower operator to suddenly realize that I was standing there with him. Sometimes there would be no conversation other than a greeting or a good night; there wasn't much need for words when things were going properly. The word got around that the General had been there during the small hours. It was a twenty-four-hour operation every day, and I wanted everyone in the Airlift to remember that. The men I talked with knew, too, that I was still a long way from my bed" (Tunner, 1964, p. 177).

Such walks could lead to the identification of issue that had to be solved. To come to solutions, careful analysis in the sense of finding the root cause was done: "As I made my periodic informal visits to air bases, frequently at night and wearing an old field jacket with almost indistinguishable stars, I began hearing pilots muttering about the strange way the ships were handling. 'It seems an awful lot more sluggish than it ought to,' one airplane driver said. 'It handles like a truck. First couple of times I thought it must be the cargo, but I watched them load it, and it checked out right on the button—200 sacks of coal at one hundred pounds per sack. One day I went down to the big loading yard on the Main River, where the coal barges were brought in, and watched the unloading proceedings. The coal came rattling down out of the barges in chutes, then into sacks held by the civilian workers. The sacks were then loaded on trucks for the haul to the air base. The men were really working away, filling the sacks to the brim and hoisting them onto the big trucks with vim and vigor. They only bothered to weigh about one sack out of a hundred, and when it was overweight, as it invariably was, they simply poured a little coal out and threw it on up in the truck. I stepped forward and had fifty of the sacks already on the truck taken down and weighed. Some of those so-called one-hundred-pound sacks ran up to 125 pounds; they averaged out at 115—a 15-percent overloading that would mean a ton and a half extra weight on a C-54 full of coal. No wonder the plane seemed sluggish. The Army Transportation Corps took over the problem, and it was corrected. It had to be carefully explained to the loading crews that their overzealousness was hurting us instead of helping us'" (Tunner, 1964, p. 204).

In numerous pharma companies, GEMBA walks have been introduced. The challenge is that they often follow a standard schedule and standard plans. Everybody in the plant knows when who is showing up. Therefore, everybody also knows when nobody is showing up often leading to a situation where behavior is different before, during, and after such GEMBA walks. The real issues do not make it up to the managers to get the attention they deserve.

4. Enforce flow even if this could be unpopular:

   It is mandatory if operations should be successful to take decisions and make them happen. In some companies, it can take weeks or months or decisions are not taken at all. In some other companies, the reason for this can be found in a reluctance of the leader to take unpopular decisions. Not so Tunner: "On my trip to Berlin I had strolled around at Tempelhof, observing the planes being unloaded, the pilots checking in with Operations, and things in general. There was a lot of milling around. In the Operations room I'd seen a dozen pilots and copilots crowding around the Operations desk, waiting for clearance. From there I'd followed the crowd into an adjoining room, a kind of snack bar. Here more crew members were drinking coffee, munching on doughnuts, smoking, talking, and laughing. I'd seen a couple of planes being unloaded swiftly and efficiently by sweat-drenched German civilians, and I knew that that phase of the turnaround activity was being well taken care of. The Germans were personally involved; they had their own well-being and freedom at stake, and they were working like beavers. But the way the crews were lounging around in the Operations room and the snack bar, I wondered how in the world they'd get their planes off on time. Well, a look at the records showed they were not getting off on time. There were frequent delays. The schedule was ragged.

   There was a good, clear-cut solution to this problem: On the thirty-first of July, the third day on the job, I put out an order to the effect that no crew member would leave the site of his airplane at Tempelhof and Gatow. The order created no end of consternation and gripping when it first came out, but the bellyaching diminished quickly with each crew's first trip to Berlin. For in the meantime, we had made hasty preparations. Even as an incoming pilot was cutting off his engines after taxiing to the unloading ramp, a big truck with an unloading crew aboard was backing up to the cargo door to transfer the load. As the pilot got down from the plane, an Operations officer roared up in a jeep and handed him his clearance slip. If there was anything at all, the pilot should know—an accident at the other end of the corridor, for example—the operations officer passed along this information with the clearance slip. Then the weather officer came up in his jeep, to give the pilot the latest word in that department. But the piece de resistance was yet to come. The third jeep to arrive was fitted out like a snack bar, stocked with such items as hot coffee, hot dogs and doughnuts, and equipped with a canopy that could be extended in case of rain.

   The system clicked efficiently from the very first. Thanks to it, the second the plane was unloaded, the pilot was in his seat, ready to go. And the turn-around time came down to thirty minutes flat" (Tunner, 1964, p. 170f.).

   Decision was taken, consequences as far as possible were mitigated, and robustness of the airlift improved. Tunner could take such a decision based on his standing and his credibility. We often experience leaders who either do not take necessary decisions at all or others who take unpopular decisions but lack credibility.

5. Standardization and order:

   Tunner brought order into the organization from the very start: One of his first impressions: "My first over-all impression was that the situation was just what I had anticipated—a real cowboy operation. Few people knew what they would be doing the next day. Neither flight crews nor ground crews knew how long they'd be there, or the schedules they were working. Everything was temporary. I went out to the Wiesbaden air base, looked around, then hopped on a plane to Berlin. Confusion everywhere" (Tunner, 1964, p. 167). This was not what Tunner believed in: "The hustle and bustle and excitement of Operation Vittles in the early days all comprised a case in point. The last place you should find this type of activity is in a successful airlift. The actual operation of a successful airlift is about as glamorous as drops of water on stone. There's no frenzy, no flap, just the inexorable process of getting the job done. In a successful airlift you don't see planes parked all over the place; they're either in the air, on loading or unloading ramps, or being worked on. You don't see personnel milling around; flying crews are either flying, or resting up so that they can fly again tomorrow. Ground crews are either working on their assigned planes, or resting up so that they can work on them again tomorrow. Everyone else is also on the job, going about his work quietly and efficiently" (Tunner, 1964, p. 162).

   So, Tunner made sure that only people who had to be at a specific place were really there and he improved the overall organization. Bringing order into the airlift was kind of similar to a 5S program in a lean environment.

   Standards were not only defined, but also enforced: "You can fly by instruments in clear weather, but you sure cannot fly by visual rules in the North German fog. I thus decided that all planes under my command would fly a never-changing flight pattern by instrument rules at all times, good weather or bad, night or day. It was simply this: If a pilot should happen to miss his landing for any reason whatsoever, he would continue straight out on course and return the two hundred to four hundred miles to his home base. If the ceiling was over four hundred feet and visibility a mile or better, he would come in. If the ceiling was less than four hundred feet and visibility less than a mile, he would simply shove forward his throttles, breathe a sigh of regret at missing the hot coffee and doughnuts and pretty girls in the Red Cross truck, and proceed for home base. I stated publicly that I would reduce to copilot status any pilot who failed to land with ceiling and visibility greater than four hundred feet and a mile and that I would court-martial any pilot who did land with ceiling and visibility less than four hundred feet and one mile. I never did court-martial any pilot or reduce anyone to copilot status on these counts—I never had any intention of doing so in the first place—but the message got across" (Tunner, 1964, p. 173). Without clear standards, there would not have been a chance to get the airlift going sustainably. The same is visible in a lot of companies that either do not define precise enough standards at all or do it in a way that there is still room for doing it differently.

   Takt was everything, and to keep the takt time was key: "Why the emphasis on three minutes? Because it provided an ideal cadence of operations with the control equipment available at that time. There are 1440 minutes in a day. At three-minute intervals, this meant 480 landings at, say, Tempelhof, in a twenty-four-hour period. The planes that came in had to go out again, of course, and with the take-off interval also set at three minutes, this meant that a plane either landed or took off every 90 seconds. There was little time wasted sitting at the ends of the runways. It is this beat, this precise rhythmical cadence, which determines the success of an airlift. This steady rhythm, constant as the jungle drums, became the trademark of the Berlin Airlift, or any airlift I have operated" (Tunner, 1964, p. 174).

Just prior to takeoff, the pilot was given the numbers of the three planes ahead of him and the two which would take off after him. He took off at a specific second—the exact time was given frequently for synchronization—and proceeded to follow a climbing, circuitous route over radio beacons placed for that specific purpose. Not only were two air bases feeding planes into the southern corridor, but also planes were returning to both of these bases along separate flight plans from the central corridor. The path had to be exact. At the exact moment, the pilot crossed the Fulda range station at the entrance to the corridor; he broadcast in the clear the number of his plane. This was extremely important. Pilots of the planes ahead and behind him could thus check their watches and tell exactly the intervals between planes. From there on, it was a straight flight up the corridor, chugging along at 170 miles per hour. (Though the normal cruising speed of the C-54 is 200 miles per hour, we held it down on the Berlin Airlift to first 180 miles per hour, and then 170, because of the heavy loads we carried.) Over Berlin, the pilot turned left at a beacon a few miles on the right side of Tempelhof, proceeded directly across the airport at right angles to the runway, and started making his descent. He made three right turns, flying a box flight pattern, simultaneously lowering down to fifteen hundred feet. He would now be lined up with the runway and would come in at 120 miles per hour, lowering slowly until he was at four hundred feet (Tunner, 1964, p. 173).

6. Keeping up the momentum:

   Tunner made sure that there were always new ambitions and objectives set to avoid negligence and complacency. He motivated the different bases by adding a competitive spirit and set from time to time new record targets at special days: "Though we never became complacent, yet the steadily mounting tonnage in February and March tended to give all personnel the feeling that we now had the problem licked, that they could rest on their laurels. We even had all the planes we needed: 154 assorted British types, 225 C-54's, with an extra 75 in the maintenance pipeline and at the Great Falls school, and 200 of the 225 in daily service in the corridors. In the air room at Airlift headquarters, no less than fifty analytical charts were kept up to the minute around the clock; a quick look, any hour of the day, would give me a clear overall picture of the entire complex. Things were going too well. It was necessary, I felt, to do something to shake up the command. But what?

   The answer was competition, competition, competition" (Tunner, 1964, p. 219). "Give Americans good healthy competition and they can accomplish just about anything" (Tunner, 1964, p. 179).

One way to do this was to set up a daily quota for each unit, announce it in advance, and then report at the end of the twenty-four-hour period how each unit had done in comparison with all the others. One way of getting the quota to the men was by posting it on the howgozit boards; another was to announce it each day in the daily newspaper. And so was born the *Task Force Times*[1] (Tunner, 1964, p. 179f.).

We experienced more than one company that after some initial OPEX experience started to look for something else because they got a feeling that they had done what had been possible. Others also after decades stay focused on always improving further. Besides a culture of continuous improvement, it is leadership and some central resources that ensure that the momentum can be kept.

9.5 Summary

I am ready to try an airlift. I can't guarantee it will work. I am sure that even at its best, people are going to be cold, and people are going to be hungry. And if the people of Berlin won't stand that, it will fail. And I don't want to go into this unless I have your assurance that the people will be heavily in approval.

— General Lucius D. Clay, US Army Commander in Chief, European Command, June 1948

In the end, the Berlin Airlift was an astonishing success. It lasted from June 26th, 1948, to August 1st, 1949. The crews flew 266,600 sorties and delivered over 2,223,000 tons. Key was to plan the entire system: the flow system, the logistics infrastructure, the maintenance concept, etc. Like Brunhaver (Brunhaver, 1996, p. 28f.) states: "If any single part of the system had broken down, the whole system would have failed."

General William H. Tunner was instrumental for this success. We provided some insights not only into the design of the airlift but also into his personality and leadership capabilities. Leading is key in operations, and it means to bring in passion, understanding, and absolute willingness to succeed together with the people. In December 1950, William H. Tunner made it to the cover of Time Magazine, featured as the moving man and compared to a Detroit executive making mass production tick (see, Time Magazine, December 18, 1950), a telling achievement.

References

Brunhaver, J. S. (1996). *The Berlin Airlift, lifeline from the sky, the doctrinal implications of supplying an enclave from the air*. Air University Press.
Miller, R. G. (1998). *The Berlin Airlift, 1948–1949*. Air Force History and Museums Program.
Tunner, W. H. (1964). *Over the hump*.
Tusa, A., & Tusa, J. (1998). *The Berlin Airlift: The Cold War mission to safe a city*.

[1] For an impression, see https://www.airlift-berlin.com/newspaper-task-force-times.html.

OPEX and Quality: One Path to Excellence

10

Jessica Helbling, Matteo Bernasconi, and Thomas Friedli

Abstract

This chapter explores the relationship between Operational Excellence (OPEX) and quality organizations in pharmaceutical companies, where regulatory compliance and efficiency are critical. Traditionally, quality units operate independently to ensure compliance, while OPEX teams concentrate on continuous improvement and efficiency. This separation often leads to siloed thinking, duplicated efforts, and missed opportunities for synergy. The chapter highlights how management trends now favor cross-functional collaboration, hybrid organizational structures, and shared strategies as ways to bridge the gap between OPEX and quality. Empirical evidence illustrates various alignment models, ranging from strict separation to embedding OPEX within quality units for closer

Note: Parts of this chapter are based on the work of Helbling (forthcoming), who conducted her research as a member of the St.Gallen OPEX research team.

Funding for some of the research contained in this report was made possible, in part, by the Food and Drug Administration through BAA contracts {#75F40121C00161, #75F40119C10121} and grants {#5U01FD005675-02, #1U01FD005675-01}. The views expressed in this report do not necessarily reflect the official policies of the Department of Health and Human Services, nor does any mention of trade names, commercial practices, or organization imply endorsement by the United States Government.

J. Helbling (✉) · T. Friedli
Institute for Production and Supply Chain Management, University of St.Gallen, St.Gallen, Switzerland
e-mail: jessicarebecca.helbling@unisg.ch; thomas.friedli@unisg.ch

M. Bernasconi
Abbvie, Cham, Switzerland
e-mail: matteo.bernasconi@abbvie.com

T. Friedli et al. (eds.), *Striving for Operational Excellence in the Pharmaceutical Industry*, Future of Business and Finance,
https://doi.org/10.1007/978-3-032-26287-5_10

collaboration. Despite regulatory requirements for quality unit independence, companies that successfully align OPEX and quality benefit from enhanced culture, improved operational performance, and increased agility.

10.1 Introduction

Pharmaceutical companies operate in highly regulated environments where efficiency and compliance must coexist. Yet, organizational structures often reflect a fragmented reality. Quality units are traditionally positioned as independent guardians of compliance, while Operational Excellence (OPEX) functions focus on efficiency, lean implementation, and continuous improvement (CI). This separation can foster silo thinking, where quality is perceived as a policing function and OPEX as a parallel initiative. The absence of a shared strategy risks duplication of efforts, unclear responsibilities, and missed opportunities for synergy (Anand & Daft, What is the Right Organization Design?, 2007).

The independence of the quality unit is essential to guarantee credibility with customers and is required by regulators (FDA, 2006). The quality unit consists of two functions: quality control (QC) ensures that products meet specifications, while quality assurance (QA) validates processes and compliance. However, this independence can create distance from other functions, reinforcing the perception that product quality is the sole responsibility of the quality department. In reality, product quality is the outcome of the entire manufacturing organization. Studies show that high levels of quality behavior among employees are strongly correlated with a robust quality culture, which in turn drives operational performance (Friedli et al., 2018; Patel et al., 2015).

OPEX, by contrast, has emerged as a driver of process stability, waste reduction, and cultural change. Its focus lies in embedding standardized practices, enabling learning across sites, and fostering employee engagement. When OPEX and quality remain disconnected, organizations struggle to achieve excellence that is both efficient and compliant. The challenge is not whether these functions should collaborate, but how they can be structurally and culturally integrated to reinforce each other.

Recent developments in management practice suggest that boundaries between functions are becoming more permeable. Cross-functional collaboration, project-based work, and integrated improvement frameworks are increasingly replacing rigid departmental silos. In this context, the convergence of OPEX and quality is not only desirable but also necessary. Both functions ultimately pursue the same objective: delivering reliable, high-quality products to patients and customers while ensuring sustainable performance. The alignment of OPEX and quality thus represents a strategic opportunity to align compliance with efficiency, creating organizations that are resilient, trusted, and continuously improving (O'Callaghan et al., 2023).

This chapter therefore examines how companies design their OPEX structures, how quality units are organized, and most importantly how the two functions can be integrated. By drawing on empirical insights from pharmaceutical companies and selected academic studies, it highlights the mechanisms through which OPEX and quality can reinforce each other, ensuring that excellence is not pursued in parallel tracks but through a unified path to performance and compliance.

10.2 Organizational Design

"Organization structure defines how tasks are divided, resources are deployed, and departments are coordinated" (Daft, 2010). Traditionally, organizations have maintained clear boundaries between internal departments and external stakeholders (Anand & Daft, 2007). In recent years, however, these boundaries have begun to fade. The evolution of management practice has shifted from strictly function-based work toward team- and project-based collaboration. Cross-functional initiatives, project portfolios, and integrated improvement frameworks increasingly complement traditional hierarchies. Management relies on organizational structures to fulfill the organizational goals and cascade down competitive priorities. The development of organizational structures reflects the growing need for agility to guarantee the organizational capability of adapting and reacting to changes. This is particularly relevant for pharmaceutical companies due to their regulatory environment. Numerous health authorities define and enforce standards that drug manufacturers must follow to participate in the market. Therefore, the industry must simultaneously ensure compliance and drive efficiency.

The organizational setup provides the foundation for how improvement and compliance activities are coordinated. Structures are not static charts but strategic instruments that balance efficiency, flexibility, and control. Classical organizational theory distinguishes between functional, divisional, and matrix arrangements (Galbraith, 2002; Mintzberg, 1979). Pharmaceutical companies have historically relied on functional setups, but hybrid forms are becoming more common. These hybrids combine global governance with local adaptation, and formal independence with cross-functional collaboration.

Against the backdrop, the design of OPEX and quality units must be understood as part of a broader organizational trajectory. The way companies anchor these functions, whether as separate departments, embedded units, or cross-functional teams, reflects their attempt to reconcile regulatory independence with OPEX. The following sections therefore examine in detail how OPEX structures are designed, how quality units are organized, and how both can be integrated to reinforce each other.

10.2.1 OPEX and Quality Unit Structures

OPEX structures typically evolve along two structural axes: the vertical distribution of roles between global, divisional, and local levels and the logic of interaction between these levels, expressed in top-down, bottom-up, or hybrid mechanisms (Bessant & Caffyn, 1997; Gronauer, 2012).

At the global level, OPEX functions are primarily described as the bearers of methodological competence, standardization, and strategic orientation. They define company-wide programs such as production systems, lean standards, role profiles, or competence models (Liker, 2004; Shah & Ward, 2007). This global standardization creates comparability, promotes process stability, and enables organization-wide learning (Spear & Bower, 1999). At the same time, the global level ensures that improvement activities are closely aligned with corporate goals, thereby creating strategic coherence (Morgan & Liker, 2006).

Local OPEX units focus on operational implementation and contextualization of global principles. The literature emphasizes that local units serve as "reality filters" by adapting global standards to specific requirements (Anand et al., 2017). Moreover, the local level generates most of the inputs for improvement, as employees identify problems at an early stage and propose quick solutions (Bessant & Francis, 1999). In practice, this means that shop floor teams and site-level experts play a critical role in translating abstract frameworks into tangible improvements.

Three basic forms of interaction between global and local levels can be distinguished. Top-down models enable strong strategic control, clear priorities, and rapid, coordinated implementation. Bottom-up models promote initiative, learning-oriented work, cultural anchoring, and employee engagement (Bessant & Caffyn, 1997). Hybrid models combine both approaches and are considered particularly effective because they balance strategic consistency with operational flexibility (Hines et al., 2004; Netland, 2015).

While the underlying principles of OPEX structures are broadly applicable across industries, their concrete expression varies. In fact, OPEX structures are designed to address each company's strategy, business goals, and operational challenges. Insights from an interview study with 14 pharmaceutical companies reveal distinct patterns (Helbling, Forthcoming). Research-driven firms typically establish dual-CI structures that integrate global and local OPEX units and combine top-down direction with bottom-up engagement. By contrast, CDMOs and generics manufacturers tend to favor leaner, centralized arrangements, with CDMOs in particular relying heavily on top-down mechanisms. Larger pharmaceutical companies usually operate within formalized CI frameworks, supported by global OPEX teams and standardized production systems to ensure consistency across multiple sites. Smaller firms, however, adopt site-specific, demand-driven structures that are more responsive to local cultures. Interestingly, the geographical location of a company's headquarters appears to have little bearing on its chosen OPEX setup.

Pharmaceutical companies tend to adopt dual-CI approaches that blend top-down ("push") direction with bottom-up ("pull") engagement. In practice, this means that strategic guidance from global OPEX teams is complemented by site-level initiatives driven by employees, creating a balance between consistency and adaptability. As shown in Fig. 10.1, the majority of firms maintain dedicated OPEX functions at both the global and local levels, ensuring that standards are aligned while still allowing for contextual implementation.

A handful of exceptions illustrate the diversity of practice. Companies M, I, and E operate exclusively with a global setup, and company C relies primarily on local

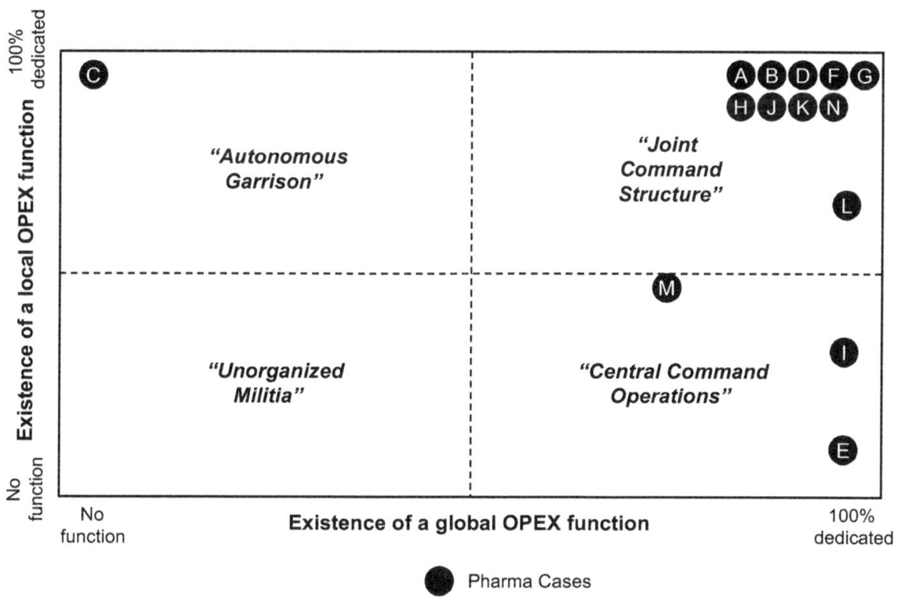

Fig. 10.1 Global and local OPEX functions (Helbling, Forthcoming)

structures. These cases are typically small to medium-sized firms, where limited resources make fully formalized frameworks difficult to sustain.

For practitioners, the key lesson is that OPEX structures must balance strategic coherence with local adaptability. Global teams provide the methodological backbone and ensure alignment with corporate goals, while local units contextualize these frameworks and generate the majority of improvement ideas. The most effective setups combine both perspectives through hybrid interaction models. Company differences highlight that there is no "one-size-fits-all" solution. The success of OPEX structures depends less on their formal design and more on how effectively they integrate global standards with local realities, ensuring both consistency and responsiveness.

Quality unit follows similar organizational structures as in the case of OPEX functions. However, the organization of the quality unit is generally more standardized across the industry to comply with regulations (see, for example, FDA, 2006). The corporate quality organization is responsible for defining the procedures and guidelines applicable to the entire manufacturing network. The local quality function is in charge of the operative tasks related to their work and eventually of implementing local relevant standards. OPEX and quality functions often work towards the same objective, namely, delivering high-quality products that function as intended for the customers. A strong "silo" thinking might risk the loss of the sense of responsibility of certain functions since they might tend to interpret it as quality unit role. However, multiple studies show that the quality unit and OPEX departments should not work separately but as a team. Patel et al. (2015) and Friedli et al. (2018) argue that high level of quality behaviors from manufacturing employees is correlated with an establishment's quality culture. Focusing on the creation of robust quality practices will result in a CI culture for the entire manufacturing site, which is also supported by the regulators (Bernasconi et al., 2025). Fellows et al. (2022) show that improving quality practices associated with production help to increase delivery performance of the manufacturing facilities.

10.2.2 Alignment of OPEX and Quality

The structural alignment of OPEX and quality requires more than functional alignment. It demands a deliberate orchestration of roles across organizational levels and interaction mechanisms. Both functions pursue the same goal: delivering stable, compliant, and continuously improving processes. Yet, they operate with distinct institutional logics. Institutional logics refer to the socially constructed belief systems and practices that guide organizational behavior, and multiple logics can coexist and shape how functions interpret their roles (Thornton & Ocasio, 1999). This implies that OPEX focuses on efficiency, learning, and standardization, whereas quality ensures regulatory adherence, risk control, and reliability. Alignment succeeds when these logics are structurally connected across the organization.

Insights from a multiple case study (Helbling, Forthcoming) show that companies differ significantly in how they anchor these functions. The sample includes R&D-driven pharmaceutical companies, generics manufacturers, CDMOs/CMOs, and animal health companies headquartered in the United States, Europe, and Asia, covering a broad range of organizational sizes in terms of FTE and turnover. Figure 10.2 summarizes these findings in a 2 × 2 matrix, with the axes representing "existence of a local OPEX function" and "existence of an OPEX function within the local quality unit." The distribution of companies across the matrix highlights three dominant patterns:

(I) **Lower-right quadrant:** Most companies fall into this category. OPEX and quality are structurally separated, each with its own mandate.
(II) **Upper-right quadrant:** A smaller group of companies integrates OPEX responsibilities directly into the quality unit. This setup creates close alignment and tight collaboration, as improvement and compliance activities are handled within one organizational structure.
(III) **Lower-left quadrant:** In these cases, the focus on OPEX is generally weaker as companies lack dedicated teams for OPEX.

The upper-left quadrant was not observed in the study.

Taken together, the matrix illustrates that structural alignment of OPEX and quality remains uneven across the companies. In most companies, the two functions continue to operate in parallel, with clear boundaries and separate mandates. This reflects the persistence of silo structures, even when both units pursue the same

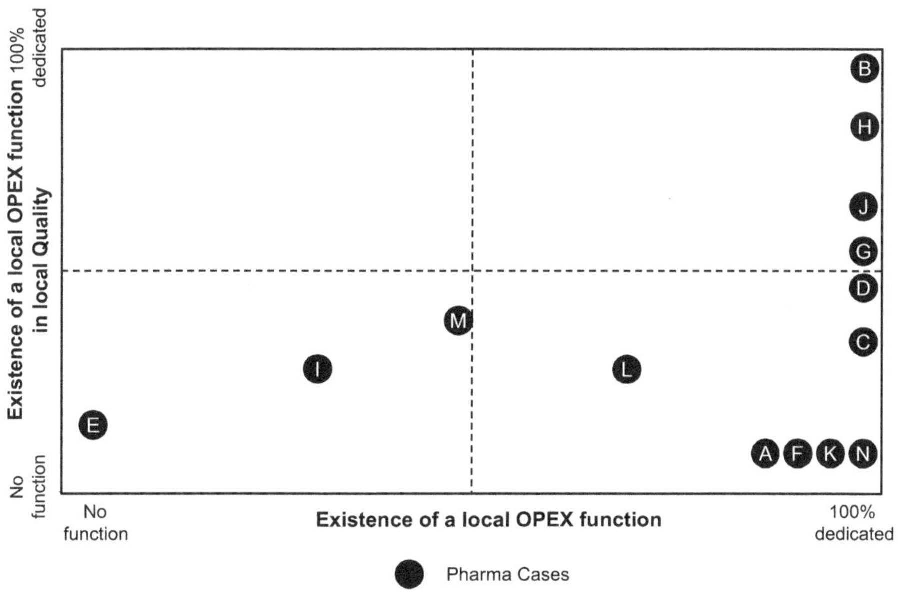

Fig. 10.2 Local quality and OPEX functions (Helbling, Forthcoming)

overarching objective. At the same time, a smaller number of cases demonstrate that closer structural alignment is possible. Where OPEX responsibilities are embedded within quality, improvement and compliance activities are handled within one organizational framework, reducing fragmentation and fostering collaboration. The findings thus highlight both the prevalence of separation and the emerging examples of structural alignment.

10.3 Key Challenges and Opportunities

The alignment of OPEX and quality is not only a structural question but also a matter of culture, governance, and strategic alignment. As already mentioned, both functions pursue the same overarching objective, but their logics differ. Bringing these logics together creates both challenges and opportunities for pharmaceutical companies.

10.3.1 Key Challenges

A first challenge lies in the cultural gap between OPEX and quality. OPEX teams typically promote empowerment, experimentation, and continuous learning, while quality units emphasize discipline, documentation, and compliance. These differences can create friction when joint initiatives are launched. For example, OPEX may encourage rapid problem-solving cycles, whereas quality insists on formal validation and documentation before changes are implemented. This tension can slow down improvement projects or lead to frustration among employees. Research confirms that a strong quality culture requires active involvement from all employees, not only the quality unit (Friedli et al., 2018). Without shared ownership, alignment risks reinforcing silos instead of breaking them down.

Another challenge is role clarity and accountability. In many firms, quality is perceived as the sole owner of product integrity, while OPEX is seen as a driver of efficiency and cost reduction. This division can result in duplicated initiatives or conflicting priorities. For instance, quality may launch a compliance-driven improvement program, while OPEX simultaneously runs a lean initiative targeting the same process. Without clear governance, employees face overlapping demands and confusion about which standards to follow. Studies on OPEX in pharma highlight that unclear governance often leads to inefficiencies and diluted impact (O'Callaghan et al., 2023).

A third challenge relates to regulatory independence. Regulators expect quality units to remain organizationally independent from manufacturing (FDA, 2006). This independence is essential for credibility but makes alignment with OPEX difficult. OPEX initiatives often require cross-functional collaboration, yet quality must demonstrate autonomy in oversight. Balancing independence with collaboration is therefore a delicate task. Companies that fail to manage this balance risk either undermining regulatory trust or limiting the effectiveness of OPEX. For example, when streamlining the deviation-handling workflow, OPEX teams may push for faster cycle times, while quality must preserve its independent assessment of each case. Even small changes in responsibilities or handovers can quickly raise concerns about whether quality's oversight remains sufficiently autonomous.

Finally, governance complexity poses challenges. Global OPEX teams may define lean standards, while quality ensures compliance. If these standards are not harmonized, sites risk confusion or conflicting directives. In multinational firms, differences in local regulatory environments further complicate alignment. The result can be fragmented practices that undermine both efficiency and compliance.

10.3.2 Key Opportunities

Despite these challenges, alignment offers significant opportunities. When OPEX and quality collaborate, process stability and compliance reinforce each other. OPEX reduces variability, which directly supports quality objectives. Quality ensures that improvements are embedded in a culture of reliability. Industry reports emphasize that integrated approaches streamline operations, enhance compliance, and support innovation. This aligns with the Center for Drug Evaluation and Research's efforts to establish a Quality Management Maturity (QMM) Program. The goal of the QMM Program is to incentivize and recognize mature quality practices that promote CI (FDA, 2022). As a result, drug manufacturers are incentivized to go beyond compliance and thus embed CI, innovation, and quality in every process to strive for excellence. The collaboration of OPEX and quality, eventually under the concept of the QMM, can lead to multiple opportunities.

One major opportunity is efficiency gains without compromising compliance. Joint governance and shared key performance indicators can reduce duplication, clarify responsibilities, and ensure that efficiency gains do not undermine regulatory requirements. For example, when OPEX and quality jointly define metrics, employees receive consistent signals about priorities, reducing confusion and increasing accountability.

Another opportunity lies in innovation and organizational learning. By building on the concept of knowledge management, organizations should manage their knowledge on the processes and products (ICH, 2008). The combination of OPEX's problem-solving methods with quality's risk-based perspective offers to companies the possibility to experiment within safe boundaries. This balance enables faster improvement cycles without jeopardizing compliance. Companies with integrated OPEX and quality functions are more agile, drawing on operator involvement and external expertise to supplement internal capabilities.

Alignment also strengthens the quality culture across the organization. Employees no longer see quality as a policing function but as a partner in improvement. This shift enhances engagement and ownership, which are critical for sustaining CI. Friedli et al. (2018) show that high levels of quality behavior among employees correlate directly with robust quality culture and improved operational performance.

Finally, alignment creates strategic resilience. Companies that align OPEX and quality are better positioned to respond to regulatory changes, market pressures, and technological disruptions. By embedding CI into compliance routines, they ensure that excellence is achieved not in parallel tracks but through a unified path.

10.4 Conclusion and Solution Approaches

The analysis of OPEX and quality structures demonstrates that alignment remains uneven. While both functions pursue the same overarching objective, their institutional logics differ. OPEX emphasizes efficiency, learning, and standardization, whereas quality focuses on regulatory adherence, risk control, and reliability. The case study evidence shows that most companies continue to operate inside clear boundaries and separate mandates, reflecting the persistence of silo structures. At the same time, a smaller number of firms have begun to structurally embed OPEX responsibilities within quality, thereby reducing fragmentation and fostering closer collaboration.

This uneven landscape highlights a central tension. Independence of the quality unit is essential for regulatory credibility, yet separation can limit the effectiveness of improvement initiatives. Conversely, alignment promises efficiency gains and cultural alignment, but risks undermining regulatory autonomy if not carefully designed. The challenge for practitioners is therefore not whether OPEX and quality should collaborate, but how they can be structurally and culturally aligned without compromising their distinct mandates.

Establishing Joint Governance Mechanisms

One promising solution is the creation of joint governance structures where OPEX and quality jointly define priorities, metrics, and improvement programs. Shared boards or steering committees can ensure that efficiency initiatives are aligned with compliance requirements. Literature on OPEX in pharma emphasizes that governance clarity is critical to avoid duplication and conflicting directives (O'Callaghan et al., 2023). By embedding quality representatives into OPEX decision-making, companies can balance independence with collaboration.

Embedding Cross-Functional Routines

Another solution lies in operational routines that deliberately connect OPEX and quality. Examples include joint audits, integrated corrective and preventive action (CAPA) processes, and shared key performance indicator reviews. These routines create touchpoints where improvement and compliance activities intersect, fostering mutual understanding and reducing silo dynamics. Empirical studies show that high levels of quality behavior across the workforce correlate directly with robust quality culture and improved operational performance (Friedli et al., 2018). Embedding OPEX practices into quality routines strengthens this culture and ensures that improvements are sustained.

Developing Hybrid Competence Models

A third solution is to design competence models that combine OPEX and quality skills. Training programs can equip quality staff with lean and problem-solving methods, while OPEX experts gain familiarity with regulatory requirements. Companies which empower engineers and operators to take active roles in improvement achieve greater agility and resilience. Hybrid competence models thus create a workforce capable of bridging efficiency and compliance, reducing dependency on single functions.

Taken together, these solution approaches suggest that alignment is not about merging OPEX and quality into a single unit but about designing interfaces that enable collaboration while preserving independence. Joint governance, cross-functional routines, and hybrid competence models provide practical pathways to achieve this balance. They allow companies to reduce duplication, enhance compliance, and foster a culture of improvement without undermining regulatory credibility.

For practitioners, the key lesson is that alignment must be deliberate and visible. It requires structural anchoring of both functions, clear governance mechanisms, and cultural reinforcement through shared routines. Companies that succeed in this integration are better positioned to respond to regulatory changes, market pressures, and technological disruptions. Ultimately, excellence is achieved not in parallel tracks but through a unified path where OPEX and quality reinforce each other.

References

Anand, N., & Daft, R. L. (2007). What is the right organization design? *Organizational Dynamics, 36*(4), 329–344. https://doi.org/10.1016/j.orgdyn.2007.06.001

Anand, N., Gardner, H. K., & Morris, T. (2017). Knowledge-based innovation: Emergence and embedding of new practice areas in management consulting firms. *Academy of Management Journal, 50*(2), 406–428. https://doi.org/10.5465/AMJ.2007.24634457

Bernasconi, M., Friedli, T., & von Dzengelevski, O. (2025). The impact of current good manufacturing practices inspections on continuous improvement mindset in the pharmaceutical industry. *Journal of Pharmaceutical Innovation.* https://doi.org/10.1007/s12247-025-10200-7

Bessant, J., & Caffyn, S. (1997). High-involvement innovation through continuous improvement. *International Journal of Technology Management, 14*(1), 7–28. https://doi.org/10.1504/IJTM.1997.001705

Bessant, J., & Francis, D. (1999). Developing strategic continuous improvement capability. *International Journal of Operations & Production Management, 19*(11), 1106–1119. https://doi.org/10.1108/01443579910291032

Daft, R. L. (2010). *Organization theory and design.* South-Western Cengage Learning.

FDA. (2006). *Quality systems approach to pharmaceutical current good manufacturing practice regulations*. FDA.

FDA. (2022). *Quality management maturity: Essential for stable U.S. supply chains of quality pharmaceuticals*. Food and Drug Administration.

Fellows, M., Friedli, T., Li, Y., Maguire, J., Rakala, N., Ritz, M., Bernasconi, M., Seiss, M., Stiber, N., Swatek, M., & Viehmann, A. (2022). Benchmarking the quality practices of global pharmaceutical manufacturing to advance supply chain resilience. *AAPS Journal.* https://doi.org/10.1208/s12248-022-00761-7

Friedli, T., Buess, P., Köhler, S., Chen, C., Mendivil, S., & Baker, D. (2018). The impact of quality culture on operational performance—An empirical study from the pharmaceutical industry. *PDA Journal of Pharmaceutical Science and Technology, 72*(5), 531–542. https://doi.org/10.5731/pdajpst.2018.008771

Galbraith, J. R. (2002). *Designing organizations: An executive guide to strategy, structure, and process*. Jossey-Bass.

Gronauer, T. (2012). *Einfluss von internen Kontext-und Prozesselementen auf die Etablierung eines Operational Excellence Program: eine vergleichende Fallstudieanalyse aus der pharmazeutischen Industrie*. University of St.Gallen.

Helbling, J. (Forthcoming). *Fostering continuous improvement in the life sciences industry.* University of St.Gallen.

Hines, P., Holweg, M., & Rich, N. (2004). Learning to evolve: A review of contemporary lean thinking. *International Journal of Operations & Production Management, 24*(10), 994–1011. https://doi.org/10.1108/01443570410558049

ICH. (2008). *Pharmaceutical quality system Q10*. ICH Tripartite Guideline.

Liker, J. K. (2004). *The Toyota way: 14 management principles from the world's greatest manufacturer*. McGraw Hill.

Mintzberg, H. (1979). *The structuring of organizations: A synthesis of the research.* Prentice-Hall.

Morgan, J. M., & Liker, J. K. (2006). *The Toyota product development system: Integrating people, process, and technology*. Productivity Press. https://doi.org/10.4324/9781482293746

Netland, T. H. (2015). Critical success factors for implementing lean production: The effect of contingencies. *International Journal of Production Research, 54*(8), 2015–2031. https://doi.org/10.1080/00207543.2015.1096976

O'Callaghan, E., McDermott, O., Walker, G., & Southern, M. (2023). Critical success factors for operational excellence in the pharmaceutical industry. *Lean, Green and Sustainability.* https://doi.org/10.1007/978-3-031-25741-4_17

Patel, P., Baker, D., Burdick, R., Chen, C., Hill, J., Holland, M., & Sawant, A. (2015). Quality culture survey report. *PDA Journal of Pharmaceutical Science and Technology, 69*(5), 631–642. https://doi.org/10.5731/pdajpst.2015.01078

Shah, R., & Ward, P. (2007). Defining and developing measures of lean production. *Journal of Operations Management, 25*(4), 785–805. https://doi.org/10.1016/j.jom.2007.01.019

Spear, S., & Bower, H. K. (1999). Decoding the DNA of the Toyota production system. *Harvard Business Review, 77*(5), 96–106. Retrieved from https://hbr.org/1999/09/decoding-the-dna-of-the-toyota-production-system

Thornton, P., & Ocasio, W. (1999). Institutional logics and the historical contingency of power in organizations. *American Journal of Sociology, 105*(3), 801–843.

Reconciling OPEX and Digital

11

Lorenzo Pirrone and Thomas Friedli

Abstract

Manufacturing companies typically strive for process improvements to keep a competitive edge in the market. Especially OPEX programs are well known for providing such advantages. With the emergence of advanced digital technologies, firms seek the next improvement frontier, implementing a wide variety of manufacturing digitalization projects. Emerging digitalization programs typically aim to enhance existing OPEX programs, although many companies still struggle to reap expected benefits while juggling OPEX and digitalization at the same time. This chapter takes a closer look at why and where challenges occur as well as how OPEX and digitalization can be reconciled.

This chapter is based on the main author's dissertation: ***Managing OPEX and Digitalization Programs in Manufacturing Companies*** published at the University of St.Gallen.

L. Pirrone (✉)
Deloitte Consulting GmbH, Munich, Germany
e-mail: lpirrone@deloitte.de

T. Friedli
Institute for Production and Supply Chain Management, University of St.Gallen, St.Gallen, Switzerland
e-mail: thomas.friedli@unisg.ch

T. Friedli et al. (eds.), *Striving for Operational Excellence in the Pharmaceutical Industry*, Future of Business and Finance,
https://doi.org/10.1007/978-3-032-26287-5_11

11.1 The Promise and the Pitfalls: What Companies Really Face

Manufacturing companies typically strive for process improvements to keep a competitive edge in the market. Especially OPEX programs are well known for providing such advantages (Netland & Aspelund, 2013). Now, with the emergence of advanced digital technologies, firms seek the next improvement frontier (Calabrese et al., 2020). They started implementing a wide variety of manufacturing digitalization projects (cf. Buer, Semini, et al., 2021; Buer, Strandhagen, et al., 2021) which eventually begin to merge into digitalization programs (Baudin & Netland, 2023). These programs typically aim to enhance existing OPEX programs, although many companies still struggle to reap expected benefits while juggling OPEX and digitalization at the same time (Baudin & Netland, 2023). They experience that it requires a fundamental transformation of "organizational decision-making, processes, and architectures" (Holmström et al., 2019, p. 728), and that is why some companies reduced their effort in their OPEX implementations. Apparently, this does not seem to be the right path moving forward (Buer, Strandhagen, et al., 2021). In fact, research largely proliferates a complementary relationship presenting both digitalization as an enabler of OPEX and OPEX as an important catalyst for digitalization (e.g., Buer et al., 2018; Lorenz et al., 2019). Consulting firms, in particular, consistently assert that the simultaneous integration of OPEX practices and digitalization will result in significant boosts in efficiency (e.g., Küpper et al., 2017; Rao et al., 2020). For example, it is assumed that the concurrent implementation of OPEX and digitalization will lead to a cost reduction of 40% within 5–10 years of implementation (Küpper et al., 2017). Other studies such as the study OPEX 2020: The Future of Operational Excellence, conducted by the Institute of Technology Management (ITEM-HSG, St.Gallen), approximately corroborates that finding: 56% of 73 participants state that digitalization will support OPEX, 38% perceive OPEX as the foundation to successfully implement digital solutions, and only 4% do not see any impact from one to another (Friedli et al., 2018). Simply put, both OPEX and digitalization are needed for successful transformation.

However, recent studies introduce more nuanced perspectives to this realization, rendering the debate increasingly complex and inconclusive (Frank et al., 2024; Yilmaz et al., 2022). In particular, at the organizational level, the interplay between OPEX and digitalization implies significant challenges for companies and their employees (Holmemo & Korsen, 2023; Johansson et al., 2024). For instance, digital tools may create paradoxes when they lag behind shop floor improvements (Holmemo & Korsen, 2023); tensions arise due to managers' and workers' contrasting interpretations of technologies (Margherita & Braccini, 2024); and OPEX and digitalization programs as a whole compete for resources and priorities (Pirrone et al., 2024). Consequently, rather than fostering synergies, the OPEX and digitalization relationship is regularly marked by friction and incompatibility. Moreover, it is well known that individual implementation of OPEX or digitalization is already challenging (Yilmaz et al., 2022). For example, only 14% of all digitalization programs (cf. Petit et al., 2019) and only 10–40% of all OPEX programs are successfully implemented (Pearce et al., 2018). Lastly, synergies at the program management level often go unnoticed. For example, the study Manufacturing Data Analytics, recently conducted by ITEM-HSG and the Laboratory for Machine Tools and Production Engineering (WZL, Aachen), highlighted misalignment between OPEX and digitalization teams. Thereby, two-thirds of participants do not have structures in place to support coordination, resulting in only one-third of participants who assess their OPEX maturity before digitalizing processes (Beckschulte et al., 2023).

Thought Experiment: Millions Spent. Gains Stalled

You've invested millions in digital tools. Dashboards light up, AI models hum in the background ... but a year later, performance gains have stalled. The OPEX team says that it is a process discipline issue. The digital team says that it is a tech adoption issue. Both are right—and wrong.

This is the moment many companies find themselves in—ambitious technology programs under-deliver, process improvement initiatives lose momentum, and the two camps quietly blame each other. The result: stalled transformation, wasted investment, and competitors who pull ahead. This chapter is about fixing that. We will show how to reconcile OPEX and digital into one integrated transformation engine—so that technology amplifies process excellence, and process discipline ensures that digital solutions stick and scale.

11.2 The Clash Zone: Why OPEX and Digital Collide

Although both OPEX and digitalization programs share numerous similarities and in essence seek the same, namely "to reduce costs and to increase productivity within companies" (Bittencourt et al., 2021, p. 1506), they are inherently different as well. At its core, OPEX puts customer value at the center and reduces waste through incremental improvements (Imai, 1986). Digitalization in the context of manufacturing, in contrast, relies on advanced developments in information technologies, often summarized by the term Industry 4.0 (Buer et al., 2018). Whereas OPEX programs are process oriented (they start with process and people who manage these processes), digitalization programs are technology oriented, usually starting with possibilities of technology. OPEX and digitalization programs are further based on different motivations. When OPEX programs are considered problem driven, centered around efficiency and effectiveness, digitalization programs are use case driven, often coming with uncertain customer and economic values. Moreover, the scope of OPEX and digitalization programs differs. As OPEX programs mainly span over a narrow scope of manufacturing and indirect areas, digitalization programs can span into various areas beyond manufacturing. They can range from manufacturing transformation to wide-ranging business model innovations. Lastly, the required expertise in OPEX programs is comparably low; OPEX principles can be easily taught to the workforce, while digital technologies require specialized knowledge. Ultimately, the effort that is needed for manufacturers to engage in technology experimentation, screening, learning, engaging with research institutions and vendors, etc. can help to improve productivity, but it can also drain resources and energy away from value added at the shop floor (Baudin & Netland, 2023).

Inherent in such tensions is the lack of coordination and alignment during the implementation process (e.g., Pozzi et al., 2023), a typical challenge related to poor program management (Martinsuo & Ahola, 2022). Therefore, it can be assumed that reconciling OPEX and digital becomes an increasingly relevant topic in the field of program management. The Project Management Institute defines program management as "a centralized coördinated management to achieve a program's strategic objectives and benefits" (Ross & Shaltry, 2006). A key characteristic of program management is the challenge of integrating and coordinating a complex network of resources, which simultaneously represents the primary driver of benefits that would not be realized if projects were managed independently (Netland, 2012; Ross & Shaltry, 2006).

Signs you are in the clash zone

- **Strategic Misalignment:** Strategic objectives are not effectively translated into operational execution, resulting in fragmented and inconsistent use case development.
- **Conflicting Program Priorities:** A lack of balance between OPEX's emphasis on stability and efficiency and digitalization's focus on innovation and speed creates tension and misalignment between departments, managers, and workers.
- **Limited Measurable Impact:** Digital initiatives often fail to deliver clear, quantifiable improvements, making it difficult to justify investments.
- **Mutual Literacy Gaps:** Digital program managers lack deep operational understanding, while OPEX managers face challenges with data analytics and digital technologies.
- **Misaligned Expectations:** Differing views on timing, scalability, and problem-solving approaches lead to frustration and reduced collaboration effectiveness.
- **Lack of Joint Engagement:** Digital teams are perceived as reactive IT service providers, whereas OPEX teams expect proactive support aligned with business needs.
- **Power Struggles and Self-Assertion:** Control over systems and restricted access foster mistrust and hinder open collaboration.
- **Unclear Ownership:** Ambiguity in system, process, and life cycle ownership results in overlapping responsibilities and inefficient workflows.
- **Weak Governance and Transparency:** Inadequate communication and coordination, especially in global environments, limit synergy and timely decision-making.
- **Inefficient Decision-Making:** Decisions are often driven by resource constraints rather than strategic value, with no standardized criteria guiding prioritization.

- **Fragmented Collaboration Structures:** Absence of shared frameworks and methodologies undermines joint execution and alignment.
- **Rigid Project Structures:** Fixed and inflexible requirements conflict with the need for agile, iterative development approaches.
- **Data Silos and Access Barriers:** Limited access to critical systems and data restricts experimentation, validation, and cross-functional learning.
- **Underutilized Cross-Functional Expertise:** Lack of mechanisms to leverage complementary skills across teams leads to missed opportunities for holistic problem-solving.
- **Inconsistent Knowledge Sharing:** Poor documentation and limited sharing of lessons learned reduce scalability and repeatability of successful initiatives.
- **Low Strategic Visibility:** Teams operate without a clear understanding of broader organizational goals, weakening alignment and impact.

11.3 The Playbook: How to Reconcile OPEX and Digital

Based on our experience with numerous pharmaceutical companies and their evolving OPEX and digital programs, we have identified six principles for successful reconciliation. These principles provide a structured approach to align strategy, governance, capabilities, collaboration, and technology, ensuring that digital initiatives reinforce OPEX programs rather than compete with them.

1. **Shared Direction**
   OPEX and digital must operate under one strategy with joint KPIs such as on-time delivery, first-pass yield, and OEE. Combined road maps link digital frameworks with OPEX road maps. Business cases are only valid if they solve real manufacturing problems and show tangible ROI beyond cost, including scalability, energy efficiency, and decision support.
2. **Common Language and Narratives**
   Digital use cases need to be mapped to OPEX pillars (e.g., predictive maintenance → TPM). A digital-OPEX playbook provides shared references, while cross-training ensures mutual understanding. OPEX practitioners learn agile and digital concepts; digital teams learn OPEX principles. Shared success stories and aligned narratives help bridge tensions between incremental OPEX improvements and MVP-driven digital approaches.
3. **Integrated Governance and Structures**
   Joint steering committees, program management offices, and translator roles provide structural alignment. Expertise clusters mirrored across OPEX, IT, and OT functions enable coordinated execution. A combined funding model ensures that process change and technology investments are inseparable.
4. **Change Architecture**
   Rewards should be based on cross-functional team outcomes rather than departmental KPIs. Change management is embedded jointly—adoption by end users, not tool delivery, is the real success measure.

5. **Scaling Through Proof Points**
   Pilots should demonstrate benefits for both OPEX and digital objectives. Scaling is achieved through structured replication—using templates, playbooks, and networks of champions. Rather than ad hoc expansion, successful scaling begins with those sites that have the most appetite, strategic relevance, and capability for adoption. This ensures that local wins translate into sustainable global impact and empowerment. Adoption of strategic agendas happens through top-down, bottom-up alignment with site-specific incorporation of needs.
6. **Trust and Social Capital**[1]
   Mutual trust and respect are the foundation for reconciliation. OPEX validates process value, and digital contributes with advanced technological solutions. Shared norms, such as "OPEX first, digital second" and "simplify rather than complicate," guide decision-making. Informal exchanges like shop floor huddles, digital days, and communities of practice reinforce collaboration and accelerate adoption.
   By using a social capital lens, we can map these principles to the concrete alignment strategies OPEX and digital program managers (O&DPMs) should follow to effectively reconcile their programs. These strategies have been gathered from 22 interviews and workshop interactions with O&DPMs from 10 pharmaceutical companies. Alignment strategies are described as first-order codes and aggregated into second-order themes as well as three related dimensions of social capital (see Fig. 11.1).

In addition to the principles and alignment strategies identified above, we observed a phenomenon in how OPEX and digital programs reconcile. Since program managers largely drive alignment through social capital (principle 6), we found that their efforts follow a bell curve (see Fig. 11.2). This curve reflects three stages of alignment—loose, evolving, and close—defined by O&DPM's social capital and the challenges they encounter.

In the *loose alignment stage*, minimal interaction between O&DPMs results in fewer challenges, as the programs operate independently. At this stage, social capital among O&DPMs is underdeveloped. Structural capital remains low due to limited formal connections or structured exchanges. Cognitive capital is weak, as shared narratives and strategic alignment are either missing or inconsistently developed. Relational capital is similarly scarce, with O&DPMs engaging in temporary, project-based interactions rather than forming systematic collaboration networks. As a result, alignment challenges remain minimal, not because alignment is

[1] Social capital (for an extensive review, see Adler & Kwon, 2002) acts as a facilitator of competitive advantage (Nahapiet & Ghoshal, 1998) and enfolds in relationships which are developed through exchange (Bourdieu, 1986). A distinct characteristic of social capital is that, once it is built, it reinforces further establishment (Nahapiet & Ghoshal, 1998). Social capital research typically proposes three dimensions: cognitive, relational, and structural (Nahapiet & Ghoshal, 1998). Following the theory of the firm paradigm of Kogut and Zander (1996, p. 503) who have proposed "that a firm [can] be understood as a social community […]," social capital theory can shed light on those institutional dynamics (Nahapiet & Ghoshal, 1998) that determine the alignment of OPEX and digitalization programs.

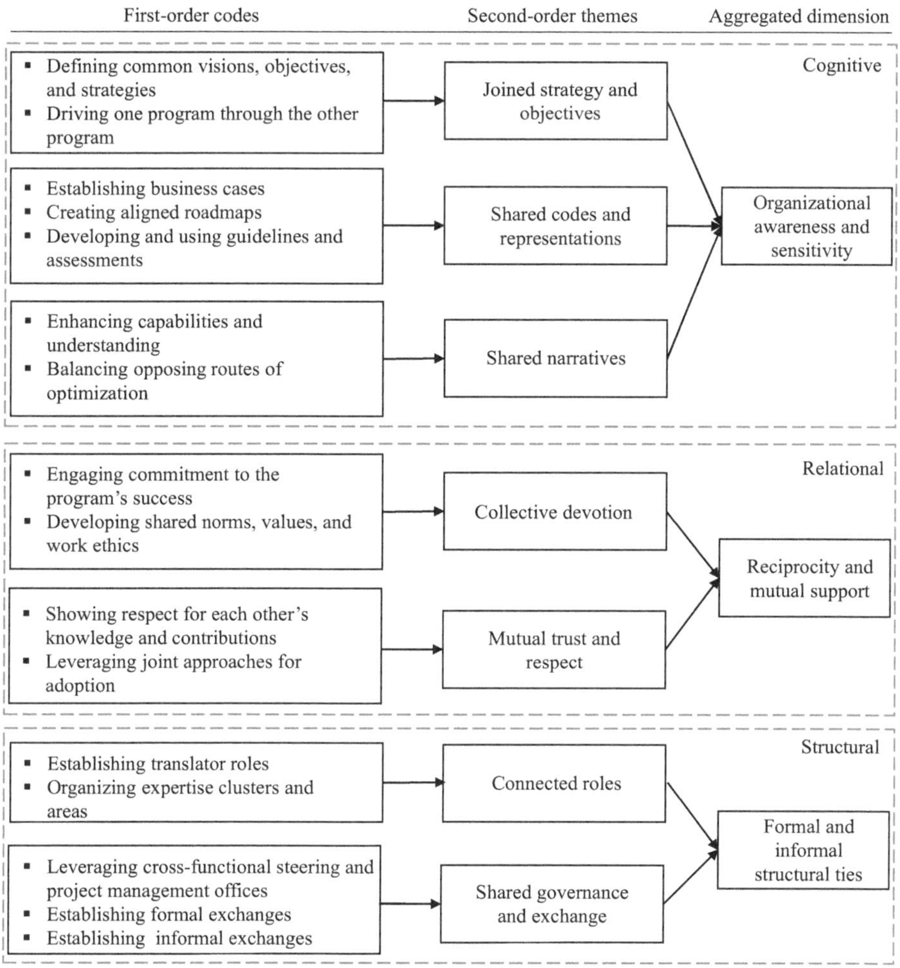

Fig. 11.1 Alignment strategies of OPEX and digital managers (Pirrone et al., 2026)

effective, but because interaction is infrequent. Simply put, O&DPMs operate in separate domains, minimizing friction by avoiding integration. This works if O&DPMs can identify which aspects of their programs can be independently managed from each other without having a mutual impact.

As O&DPMs recognize the potential synergies between their programs, they enter the *evolving alignment stage*. This stage is marked by increasing but unevenly distributed social capital. While structural capital begins to take shape through emerging governance structures, it remains insufficient to facilitate seamless collaboration. Cognitive capital improves as discussions on shared business cases and implementation strategies emerge, but fundamental misalignments persist. A key challenge at this stage is the paradoxical effect of increased interaction. Misaligned objectives, unclear roles, and inconsistent strategic narratives lead to heightened alignment challenges where O&DPMs have to balance various conflicting areas.

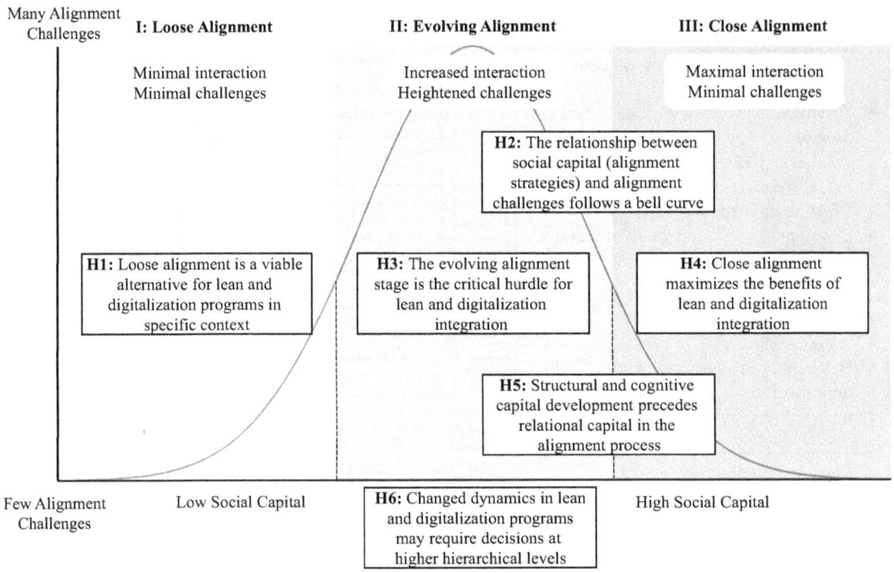

Fig. 11.2 Bell curve of OPEX and digital program alignment (Pirrone et al., 2026)

This stage also exhibits a vicious reinforcement behavior. In some instances, structural interventions, such as cross-functional teams, successfully support cognitive capital development. However, when these interventions lack clarity, they exacerbate relational tensions.

With continued effort, O&DPMs can resolve evolving alignment challenges and reach *close alignment.* This stage is characterized by high social capital across all three social capital dimensions. Structural capital is reinforced through dedicated governance bodies, formalized roles, and established cross-functional exchanges. Cognitive capital strengthens as shared narratives become institutionalized via aligned business cases, common road maps, and clearly defined implementation frameworks. Relational capital also reaches maturity, as O&DPMs develop trust, shared work ethics, and mutual commitment to joint success. In this stage, reinforcing behaviors dominate. Structural alignment mechanisms, such as steering committees and integrated workflows, facilitate knowledge exchange and problem-solving, reinforcing cognitive and relational capital.

Key Takeaway

Alignment is not a one-off decision but a continuous process. Friction rises as interaction grows, but with deliberate effort, OPEX and digital leaders can build governance, shared strategy, and trust to reach close alignment. Managers play an active role in shaping this journey—by fostering collaboration, advocating for structures, and ensuring that digital initiatives and OPEX objectives reinforce each other.

11.4 The Assessment: How Mature Is Your Integration?

"There is undoubtedly a need to create an index that offers a real-world and usable charter for organizations to decide how to start and scale and sustain the initiatives of OPEX 4.0" (Narula et al., 2023, p. 134). Following this statement, we want to take a closer look at Toyota. The founder of OPEX is often seen as a slow adopter of new technology; for example, Toyota's principles, such as "Use only reliable, thoroughly tested technology that serves your people and processes" (Liker & Meier, 2006, p. 11), are commonly misunderstood. However, opposite to what it is commonly associated with, Toyota advocates bringing new technology into their processes. That way, Toyota invented the famous "autonomation" concept (Holmemo & Korsen, 2023) and implemented cloud and software-as-a-service platform technologies or strategically driven advanced analytics, artificial intelligence, or autonomous driving (Betancourt et al., 2018). The main difference to other companies, however, is that technology implementation, first, always follows a clear need accompanied by an in-depth technology study (Liker & Meier, 2006) and, second, typically prioritizes the human element over technology (Holmemo & Korsen, 2023). Understanding the TPS from a holistic program management lens (e.g., Pirrone et al., 2024), not the OPEX practice and tool level (e.g., 5S, value stream mapping), requires a comprehensive system thinking of social and technical components. The most adequate and elaborated interpretation as such can be found in The Toyota Way published by Liker and Meier (2006). They present 14 underlying management principles of the successful car manufacturer relying on in-depth experiences at the shop floor.

The essence of their work is built on the 4P model, which is comprised of philosophy, processes, people and partners, and problem-solving. **Philosophy** entails having a strong customer focus and responsibility for creating value for the organization and society. **Processes** include creating continuous flow, leveraging pull approaches, leveling out of workloads, fixing problems right away, standardization of tasks and processes, visual controls, and implementation of reliable technology. The ultimate goal of the process dimension is to link people and processes, so that transparency is achieved, and problems are surfaced. **People and partners** involve having leaders that thoroughly understand the work and live or teach the philosophy. In that sense, people lead and follow and respect the extended network of partners and suppliers. Lastly, **problem-solving** requires a thorough understanding of problems at the location where they occur, making well-founded decisions by building consensus, and an ongoing effort to become a learning organization. Liker and Meier (2006) affirm that all OPEX principles must be present across the organization and utilized by the whole staff to achieve set goals.

We follow this understanding and argue that the degree to which this is achieved can be referred to as OPEX program maturity. The level of digitalization program maturity, and hence digital technology implementation, is assessed following similar studies (Lorenz et al., 2019; Tortorella & Fettermann, 2018). By operationalizing this understanding, we surveyed 74 companies from various industries through

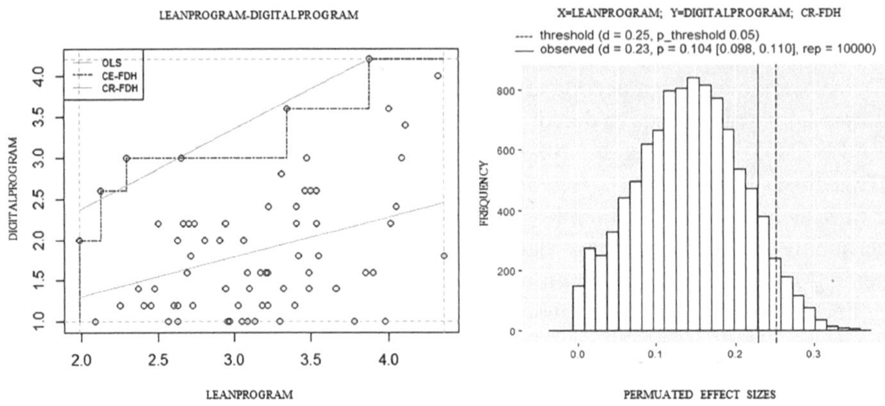

Fig. 11.3 NCA plot of OPEX and digitalization program maturity and ceiling line significance

necessary condition analysis (NCA) and subsequently conducted an interview study with 19 program managers from 11 pharmaceutical companies.

11.4.1 Quantitative Assessment

The final NCA analysis investigated the discussed OPEX program maturity dimensions as an aggregated value together with the digitalization program maturity. The NCA plot (left side in Fig. 11.3) shows an empty space and a ceiling zone with 1.733 at a scope of 7.581. The CR-FDH has a moderate effect size with 0.229 and an accuracy of 97.3% with two observations falling into the ceiling zone. To validate the significance level of this ceiling line, the data sample was simulated 10,000 times (right side in Fig. 11.3). The result shows that observed and expected effect sizes are very close to each other. Moreover, the significance level of the CR-FDH represents a p-value of 0.104, which is, due to the exploratory nature of this research phase, acceptable.

The bottleneck table (see Table 11.1) shows how OPEX maturity becomes essential at different stages of digitalization maturity. At the early stage (0–20%), no OPEX element acts as a bottleneck, meaning companies can start with digital initiatives even without strong OPEX foundations. Once programs reach around 30%, however, customer value orientation, strategic supplier integration, and a basic understanding of OPEX in manufacturing become critical, linking digital progress to both customer needs and supply chain alignment. As maturity grows further (40–70%), process discipline and operational partnerships become decisive, with supplier collaboration emerging as the single most important factor at advanced levels. Beyond 70%, sustaining transformation requires systematic problem-solving, continuous improvement, and, eventually, extending OPEX into support functions. This progression highlights that while early digitalization can advance independently, long-term success depends on embedding OPEX principles more deeply into operations, partnerships, and supporting functions. In summary, a

Table 11.1 Bottleneck table for OPEX - Digital program maturity constructs

Digital program	Philosophy	Process	People and partner (MAN)	People and partner (SUP)	People and partner (STR)	People and PARTNER (OPE)	Problem-solving
0	NN	NN	NN	NN	NN	NN	NN
10	NN	NN	NN	NN	NN	NN	NN
20	NN	NN	NN	NN	NN	NN	NN
30	2.6	NN	8.4	NN	3.0	NN	NN
40	11.8	7.9	18.9	NN	13.9	7.3	NN
50	20.9	18.6	29.4	NN	24.7	22.3	2.2
60	30.1	29.3	40.0	NN	35.5	37.2	14.6
70	39.3	40.0	50.5	NN	46.4	52.2	27.0
80	48.5	50.7	61.0	NN	57.2	67.1	39.4
90	57.6	61.4	71.6	21.0	68.0	82.1	51.9
100	66.8	72.1	82.1	45.0	78.9	97.0	64.3

distinct maturity progression path can be identified, beginning with philosophy, followed by people and partners, then process, and finally problem-solving. Furthermore, the results indicate that the maturity of OPEX programs accumulates toward the maturity of digitalization programs. This is evident in the increasing percentage of values across each dimension of OPEX program maturity. This suggests that development in OPEX program maturity must be followed to achieve higher digitalization program maturity.

11.4.2 Qualitative Assessment

Philosophy Interview results suggest that companies with a lower OPEX program maturity lack a strong establishment of OPEX philosophy. They tend to focus on acquiring new tools, instead of using tools, either OPEX or digital, to enhance customer value. Thereby, the likelihood of short-term prioritization of digitalization initiatives that are not solving problems is typically increased. With growing maturity, OPEX and digitalization are defined as prominent and concurrent pillars of a company's strategy. However, concepts are not yet fully understood and embraced. With higher maturity, integrating OPEX with digital technologies is seen as an effective approach to prioritizing customer value.

Process Standardization, automation, and visualization were the most prominent themes occurring in the second program dimension. Companies with low maturity in standardization have not yet established common approaches to designing, displaying, and analyzing processes and their respective data. This starts to change when processes are, to some extent, globally standardized, databases are centralized, and data catalogs provide helpful guidelines to drive further standardization.

In that regard, companies understand that simplification and standardization, typical elements of the OPEX program, are necessary to drive digitalization program maturity. Cumulating in the highest form of maturity, companies have perfected the adoption of scalable, digital tools. Thereby, they fully leverage standardized processes, reuse digital tools, and seamlessly fit global tools to local needs. While few companies represent low levels of automation, evidence for very mature companies was also found. Specifically, the physical flow, e.g., in packaging lines, represents typical applications for automation. The maturity level of visualization and the representation of information significantly vary across companies. Visualization tools were named by various applications such as deviation monitoring, batch tracking, OEE reporting, preventive maintenance applications, or guided changeovers. The implementation degree of such tools is mainly determined by manual, semiautomated, or automated data-gathering processes.

People and Partners Leader, people, and partner are distinct themes related to the third program dimension. In low-maturity companies, leaders rarely act as OPEX-digital role models, lacking critical future skills, visionary thinking, and an understanding of digital business needs. As maturity grows, leaders recognize that successful digital initiatives require advocating OPEX to sustain a culture of excellence. Regarding people, all companies report certain OPEX program maturity within the manufacturing area. By starting to expand into adjacent areas such as supply chain, R&D, finance, or marketing, they grow their OPEX program maturity beyond functional silos. Thereby, they shift the understanding of operational processes to the full view of business processes. To foster digitalization, one company launched a low-code/no-code initiative, enabling knowledge sharing across sites. Very much aligned with Toyota's tenth principle "Develop exceptional people and teams who follow your company's philosophy" (Liker & Meier, 2006, p. 12), higher maturity levels show that the interaction between people, the exchange of knowledge, and the perceived value of solutions drive that further engagement. As a critical factor of this, all companies highlighted the role of OPEX and digital upskilling. Whereas OPEX practitioners need to be aware of digital opportunities to further improve processes, digital personnel need to be trained to have an OPEX mindset. However, training approaches remain immature, often limited to specific applications or transactional methods like classroom training and e-learning. Personalized, role-specific learning journeys are still in conceptual stages. Lastly, regarding the partner dimension and its corresponding principle "Respect your extended network of partners and suppliers by challenging them and helping them improve" (Liker & Meier, 2006, p. 12), most cases exhibited relatively low maturity as well. Although some companies stated to work closely together with universities or suppliers, the lack of technical interoperability of machines, immaturity of processes, or regulatory requirements are seen as the main drivers of reduced interaction. Hence, supplier development programs are rarely executed.

Problem-Solving Understanding, assessment, and improvement have been identified in the fourth program dimension. Understanding refers to the principle of "Go and see yourself to thoroughly understand the situation" (Liker & Meier, 2006,

p. 13). This principle is embedded to varying degrees. Some companies have no aligned problem-solving method between OPEX and digitalization personnel. They barely identify problems where they happen and base their decision to digitalize on assumptions, instead of taking the customer's point of view. Other companies directly access their systems relying on proprietary data from the processes. They combine forces and bring them to the shop floor to really understand the problem where it occurs. With the principle "Make decisions slowly by consensus, thoroughly considering all options, and implement decisions rapidly" (Liker & Meier, 2006, p. 13), companies encourage comprehensive assessments. In mature companies, these assessments are characterized by systematic approaches. Just like Toyota, who "Use(s) only reliable, thoroughly tested technology that serves (its) people and process" (Liker & Meier, 2006, p. 11), companies implement emerging technologies based on joint, in-depth assessments, cross-functional technology transfer workshops, and detailed process analysis. Not only are OPEX and digital personnel typically involved, but finance and the customer are also engaged to assess the value of digitalization projects. Lastly, with regard to improvement and the idea of becoming a learning organization (Liker & Meier, 2006, p. 13), it was found that companies still must increase their capabilities. Maturity spans from manual problem-solving to guided, digitized root cause analysis with operator support. However, across various cases, it was found that the ability to act on data and perform proper problem-solving has not advanced in the same way as the possibilities that come with those technologies.

Finally, all companies rated their OPEX and digitalization maturity on a scale from one to ten. One represented no OPEX foundations and no formally established OPEX program, while ten represented a well-accepted OPEX program and an organization-wide establishment of a continuous improvement mindset. Likewise, one, with respect to digitalization program maturity, stood for isolated digitalization efforts and a large share of manual processes. In contrast, ten could be interpreted as a state where cutting-edge digital technologies regularly find their ways into the manufacturing landscape, including partners' and customers' seamless integration, as well as data-driven decision-making. Figure 11.4 shows the result of that assessment and reveals three insights. First, pharmaceutical companies typically exhibit lower to medium maturity levels for OPEX and digitalization programs, with particularly lower maturity levels observed in digitalization efforts. Hence, many cases are in area I (low OPEX maturity, low digitalization maturity) and area II (high OPEX maturity, low digitalization maturity). Area III maturity (high OPEX maturity, high digitalization maturity) has only been achieved by a minority of companies. Second, no case can be identified which has a low OPEX maturity on the one hand and a high digitalization maturity on the other hand. Third, the delta of OPEX maturity to digitalization maturity seems to be at least equal (case I), if not positive, representing the general existence of higher OPEX maturity compared to digitalization maturity. These findings suggest that it is hardly possible to achieve high digitalization maturity without having a strong OPEX foundation established.

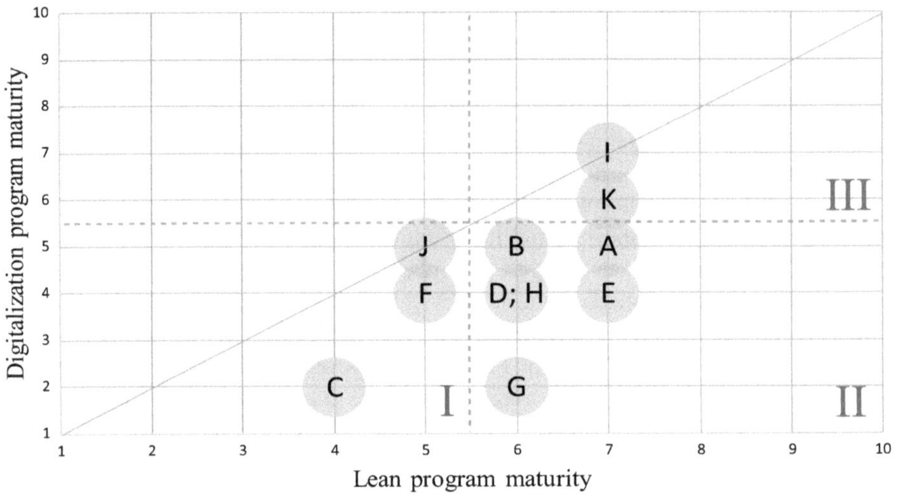

Fig. 11.4 Self-assessed maturity

11.4.3 Synthesis

A combined OPEX and digitalization program maturity model is derived from the study's quantitative and qualitative results. Reflecting on the study's findings, a two-dimensional diagram is created, with OPEX program maturity represented on the *x*-axis and digitalization program maturity on the *y*-axis. Final NCA as well as qualitative results indicate that the diagram needs to be separated into two spaces: an unworkable and a feasible space. In the unworkable space, relationships represent combinations of OPEX and digitalization program maturities for which no evidence has been found. Hence, it is concluded that a relationship where OPEX program maturity is low and digitalization program maturity is high is not, or is hardly, possible to achieve. Considering ceiling zone and effect sizes, hypotheses such as "without OPEX program maturity there cannot be high digitalization program maturity" can be formulated. Hypotheses such as "higher digitalization program maturity only exists if higher OPEX program maturity is present" are representative of the feasible space.

Three distinct levels describe how the OPEX-digital maturity relationship within the model evolves. At the first maturity level, companies begin to establish strategic relevance of the OPEX and digitalization program, emphasizing customer value as the primary goal. Vision formation and early implementation efforts typically start in manufacturing areas with initial attempts of strategic supplier integration. At this stage, only two OPEX program dimensions, philosophy and people and partners, start to take shape. So far, companies often lack a solid foundation in OPEX practices and tend to initiate digitalization projects as ad hoc solutions. Hence, this maturity level is mainly driven by foundation building and the focus on layering the groundwork for subsequent higher maturity levels.

With the second maturity level, labeled execution and deployment, the two OPEX program dimensions progress toward higher implementation (e.g., relationships with suppliers on operational level). Additionally, companies are now developing OPEX and digitalization program maturity within the process dimension. Thereby, they focus on strengthening the implementation of OPEX principles and the operational translation of their visions and strategies. Now companies are starting to see OPEX and digitalization as complementary elements of their strategy, but full organizational implementation remains challenging. Nevertheless, the OPEX program takes on a formalized approach and expands beyond manufacturing areas. By shifting the view beyond functional silos, processes are increasingly simplified and standardized. Digitalization is now viewed as a useful tool to support standardization and optimization.

At the most advanced maturity level, and with the continued development of the program dimensions, companies finally address the problem-solving dimension. In other words, they only transition to implementing a continuous improvement system after successfully developing sufficient capabilities within the first two maturity levels. Hence, only when the necessary OPEX and digital skill sets are established can the most advanced maturity stage be reached. Organizations now have a clear understanding of not competing but synergetic approaches. They build on strong OPEX and digital capabilities and acknowledge the critical role of leadership in always solidifying this organizational belief. Moreover, they realize that partnerships are crucial and deliberately advance their suppliers. This maturity stage is named optimization and refinement. Overall, the exploratory-quantitative together with the explanatory-qualitative study finds that the relationship is progressive and cumulative. This demonstrates that, as digitalization program maturity increases, specific OPEX program dimensions must also advance. Simultaneously, already established OPEX dimensions require further development to deepen their impact and enhance digitalization maturity.

The final model is depicted in Fig. 11.5. The four program dimensions, ten maturity dimensions, three maturity levels, as well as unworkable and feasible spaces are named as such. The progressive characteristic is symbolized by the filled quadrants within each maturity level. The cumulative property is represented by the increasing diameters of the three maturity circles.

11.5 The Game Plan: What to Do Monday Morning

Key Takeaway
High digitalization maturity is not possible without OPEX maturity. During any assessment, both dimensions should be respected. OPEX foundations need to be continuously improved following a progressive and cumulative pattern. Hence, only when investments into OPEX programs are continued can advanced stages for digitalization be achieved. The combined maturity model highlights this dependency, with progress unfolding in three stages: foundation building around customer value and basic OPEX principles, execution and deployment through standardized processes and supplier collaboration, and finally optimization and refinement with continuous improvement and leadership commitment.

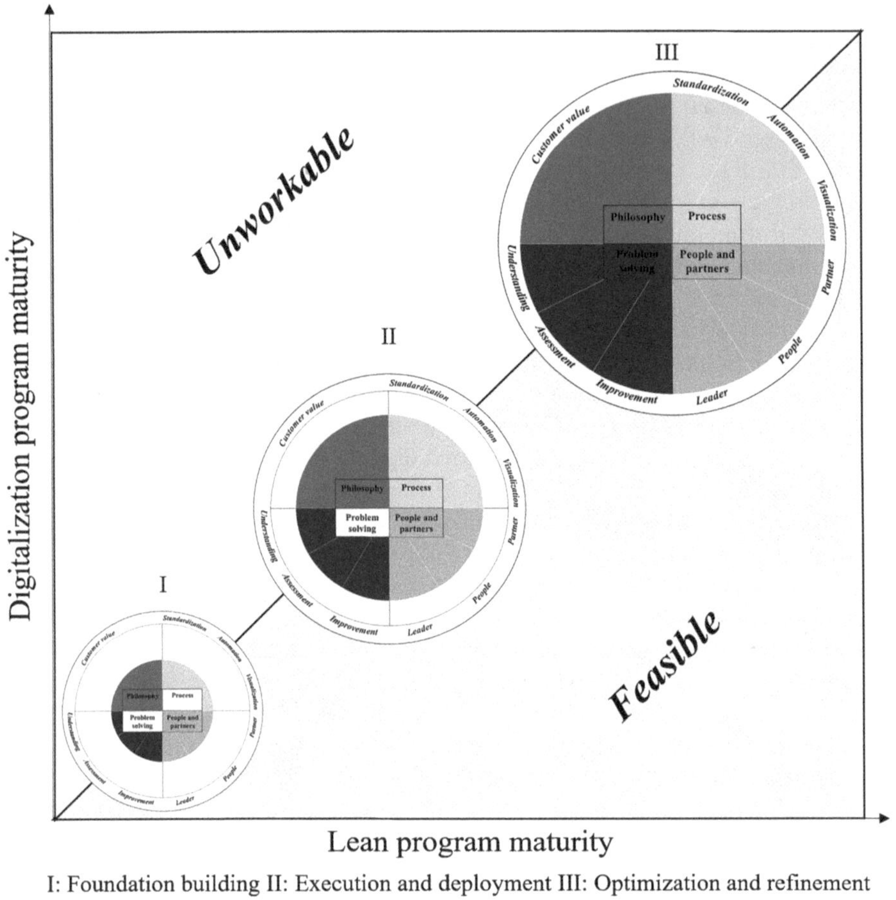

Fig. 11.5 OPEX and digitalization program maturity model

In the future, there will not be OPEX programs and digital programs—there will only be transformation programs that deliver results. Our research shows that OPEX and digital are not competing transformation agendas—they are complementary engines which, when aligned, turn incremental improvements into exponential performance gains. We formulate a ten-point action plan designed to help you in that undertaking:

1. **Audit your current portfolio:** Identify which digital initiatives are process orphaned and which OPEX initiatives could be tech-enabled.
2. **Assess your maturity:** Use the OPEX-digital maturity model to determine where you stand and where the biggest integration gaps lie.
3. **Create your first joint priority list:** Within the next 30 days, align OPEX and digital leaders on the top 3–5 initiatives with joint ownership.

4. **Build a shared language session:** Within the next 90 days, run a workshop where both teams map digital use cases to OPEX pillars and agree on a common vocabulary.
5. **Integrate KPIs:** Update scorecards so that digital adoption metrics link directly to operational performance metrics.
6. **Establish joint governance:** Form a steering group with both OPEX and digital leads; align budgets, timelines, and reporting.
7. **Pilot for impact, not tech:** Start with one high-value use case that delivers both process improvement and digital adoption ROI within 6–12 months.
8. **Train cross-functional champions:** Develop talents who understand both OPEX/OPEX methods and digital solution delivery.
9. **Plan the scale-up path:** Define how successful pilots will be replicated across sites or functions without losing momentum.
10. **Celebrate integrated wins**: Publicly recognize initiatives where OPEX and digital teams achieved results together to reinforce the cultural shift.

References

Adler, P. S., & Kwon, S.-W. (2002). Social capital: Prospects for a new concept. *The Academy of Management Review, 27*(1), 17–40.

Baudin, M., & Netland, T. H. (2023). *Introduction to manufacturing; An industrial engineering and management perspective* (1st ed.). Routledge.

Beckschulte, S., Hinrichs, M. P., Pirrone, L., Grothkopp, M., Sohnius, F., Schmitt, R., & Friedli, T. (2023). *Manufacturing data analytics study 2023*. Apprimus.

Betancourt, P., Mooney, J., & Ross, J. W. (2018, February). *Digital innovation at Toyota Motor North America: Revamping the role of IT*. Massachusetts Institute of Technology Sloan School of Management, Center for Information Systems Research.

Bittencourt, V. L., Alves, A. C., & Leão, C. P. (2021). Industry 4.0 triggered by lean thinking: Insights from a systematic literature review. *International Journal of Production Research, 59*(5), 1496–1510. https://doi.org/10.1080/00207543.2020.1832274

Bourdieu, P. (1986). The forms of capital. In J. In Richardson (Ed.), *Handbook of theory and research for the sociology of education* (pp. 241–258).

Buer, S.-V., Semini, M., Strandhagen, J. O., & Sgarbossa, F. (2021). The complementary effect of lean manufacturing and digitalisation on operational performance. *International Journal of Production Research, 59*(7), 1976–1992. https://doi.org/10.1080/00207543.2020.1790684

Buer, S.-V., Strandhagen, J. O., & Chan, F. T. S. (2018). The link between Industry 4.0 and lean manufacturing: mapping current research and establishing a research agenda. *International Journal of Production Research, 56*(8), 2924–2940. https://doi.org/10.1080/00207543.2018.1442945

Buer, S.-V., Strandhagen, J. W., Semini, M., & Strandhagen, J. O. (2021). The digitalization of manufacturing: investigating the impact of production environment and company size. *Journal of Manufacturing Technology Management, 32*(3), 621–645. https://doi.org/10.1108/JMTM-05-2019-0174

Calabrese, A., Dora, M., Levialdi Ghiron, N., & Tiburzi, L. (2020). Industry's 4.0 transformation process: how to start, where to aim, what to be aware of. *Production Planning & Control, 33*(5), 492–512.

Frank, A. G., Thürer, M., Filho, M. G., & Marodin, G. A. (2024). Beyond Industry 4.0—Integrating lean, digital technologies and people. *International Journal of Operations & Production Management, 44*(6), 1109–1126. https://doi.org/10.1108/IJOPM-01-2024-0069

Friedli, T., Macuvele, J., & Buess, P. (2018). *Benchmarking—Lean 2020—The future of operational excellence.*
Holmemo, M. D. Q., & Korsen, E. B. H. (2023). The growing gap between lean production and digital lean tools. *International Journal of Lean Six Sigma, 14*(6), 1188–1206. https://doi.org/10.1108/ijlss-05-2022-0119
Holmström, J., Holweg, M., Lawson, B., Pil, F. K., & Wagner, S. M. (2019). The Digitalization of operations and supply chain management: Theoretical and methodological implications. *Journal of Operations Management, 65*, 728–734. https://doi.org/10.1002/joom.1073
Imai, M. (1986). *Kaizen: The key to Japan's competitive success*. McGraw-Hill.
Johansson, P. J., Bruch, J., Chirumalla, K., Osterman, C., & Stalberg, L. (2024). Integrating advanced digital technologies in existing lean-based production systems: analysis of paradoxes, imbalances and management strategies. *International Journal of Operations & Production Management, 44*(6), 1158–1191. https://doi.org/10.1108/IJOPM-05-2023-0434
Kogut, B., & Zander, U. (1996). What firms do? Coordination, identity, and learning. *Organization Science, 7*(5), 502–518.
Küpper, D., Heidemann, A., Ströhle, J., Spindelndreier, D., & Knizek, C. (2017). *When lean meets Industry 4.0: The next level of operational excellence* (pp. 1–15). Boston Consulting Group.
Liker, J. K., & Meier, D. (2006). *The Toyota way fieldbook: A practical guide for implementing Toyota's 4Ps*. McGraw-Hill.
Lorenz, R., Buess, P., Macuvele, J., Friedli, T., & Netland, T. H. (2019). Lean and digitalization—Contradictions or complements? In F. Ameri, K. E. Stecke, G. von Cieminski, & D. Kiritsis (Eds.), *IFIP advances in information and communication technology. Advances in production management systems. Production management for the factory of the future* (Vol. 566, pp. 77–84). Springer. https://doi.org/10.1007/978-3-030-30000-5_10
Margherita, E. G., & Braccini, A. M. (2024). Exploring tensions of Industry 4.0 adoption in lean production systems from a dialectical perspective. *International Journal of Operations & Production Management, 44*(6), 1127–1157.
Martinsuo, M., & Ahola, T. (2022). Multi-project management in inter-organizational contexts. *International Journal of Project Management, 40*(7), 813–826. https://doi.org/10.1016/j.ijproman.2022.09.003
Nahapiet, J., & Ghoshal, S. (1998). Social capital, intellectual capital, and the organizational advantage. *The Academy of Management Review, 23*(2), 242–266.
Narula, A., Puppala, H., Kumar, A., Luthra, S., Dwivedy, M., Prakash, S., & Talwar, V. (2023). Are Industry 4.0 technologies enablers of lean? Evidence from manufacturing industries. *International Journal of Lean Six Sigma, 14*(1), 115–138.
Netland, T. H. (2012). Managing strategic improvement programs: the XPS program management framework. *Journal of Project, Program & Portfolio Management, 3*(1), 31–44. https://doi.org/10.5130/pppm.v3i1.2556
Netland, T. H., & Aspelund, A. (2013). Company-specific production systems and competitive advantage: A resource-based view on the Volvo production system. *International Journal of Operations & Production Management, 33*(11/12), 1511–1531.
Pearce, A., Pons, D., & Neitzert, T. (2018). Implementing lean—Outcomes from SME case studies. *Operations Research Perspectives, 5*, 94–104. https://doi.org/10.1016/j.orp.2018.02.002
Petit, J. P., Brosset, P., & Bagnon, P. (2019). *Smart factories at scale*. Capgemini.
Pirrone, L., Bernasconi, M., Grothkopp, M., & Friedli, T. (2026). Riding the bell curve of lean and digitalisation programme alignment—A social capital perspective. *International Journal of Production Research, 64*(3), 819–838. https://doi.org/10.1080/00207543.2025.2558133
Pirrone, L., Grothkopp, M., Budde, L., & Friedli, T. (2024). Integrate or separate: lean and digitalization programs through the lens of boundary work management. *Manufacturing Technology Management*. Advance online publication. https://doi.org/10.1108/JMTM-11-2023-0518
Pozzi, R., Rossi, T., & Secchi, R. (2023). Industry 4.0 technologies: critical success factors for implementation and improvements in manufacturing companies. *Production Planning & Control, 34*(2), 139–158. https://doi.org/10.1080/09537287.2021.1891481

Rao, A., Prusty, A., Bhat, R., & Khan, A. (2020). *Digital lean manufacturing: Industry 4.0 technologies transform lean processes to advance the enterprise* (pp. 1–17).

Ross, D. W., & Shaltry, P. E. (2006). *The new PMI standard for program management.* Paper presented at PMI® Global Congress 2006—North America, Seattle, WA. Project Management Institute.

Tortorella, G. L., & Fettermann, D. C. (2018). Implementation of Industry 4.0 and lean production in Brazilian manufacturing companies. *International Journal of Production Research, 56*(8), 2975–2987. https://doi.org/10.1080/00207543.2017.1391420

Yilmaz, A., Dora, M., Hezarkhani, B., & Kumar, M. (2022). Lean and industry 4.0: Mapping determinants and barriers from a social, environmental, and operational perspective. *Technological Forecasting and Social Change, 175*, 1–16. https://doi.org/10.1016/j.techfore.2021.121320

Continuous Improvement in the Life Sciences Industry

12

Jessica Helbling and Thomas Friedli

Abstract

This chapter examines the current state of continuous improvement (CI) in the life sciences industry, focusing on pharmaceutical and MedTech companies. Based on 66 interviews across 23 firms, the analysis highlights both the barriers that hinder CI and the enablers that support its implementation. While pharma tends to pursue CI in a structured, top-down, and compliance-driven manner, MedTech favors pragmatic, locally adaptable approaches. Despite these differences, both sectors face common challenges such as initiative overload, lack of empowerment, resistance to change, and difficulty quantifying benefits.

Twelve overarching enabler categories were identified. These enablers interact dynamically with barriers, forming a barrier–enabler matrix that illustrates which mechanisms are most effective in overcoming specific obstacles. A cluster

---

Note: Parts of this chapter are based on the work of Helbling (forthcoming), who conducted her research as a member of the St.Gallen OPEX research team.

Funding for some of the research contained in this report was made possible, in part, by the Food and Drug Administration through BAA contracts {#75F40121C00161, #75F40119C10121} and grants {#5U01FD005675-02, #1U01FD005675-01}. The views expressed in this report do not necessarily reflect the official policies of the Department of Health and Human Services, nor does any mention of trade names, commercial practices, or organization imply endorsement by the United States Government.

J. Helbling (✉) · T. Friedli
Institute for Production and Supply Chain Management, University of St.Gallen, St.Gallen, Switzerland
e-mail: jessicarebecca.helbling@unisg.ch; thomas.friedli@unisg.ch

T. Friedli et al. (eds.), *Striving for Operational Excellence in the Pharmaceutical Industry*, Future of Business and Finance,
https://doi.org/10.1007/978-3-032-26287-5_12

analysis further distinguishes three organizational profiles—overburdened, underpowered, and inertia bound—each requiring tailored enabler constellations.

Cross-industry insights from automotive, electronics, and consumer goods reinforce the importance of systemic integration, leadership alignment, and capability building. The chapter concludes with five key learnings that emphasize the need to embed CI systemically, evolve from technical to natural practice, balance top-down guidance with bottom-up engagement, broaden the value perspective, and reconcile regulatory constraints with CI enablement. Together, these findings provide a practical framework for embedding CI as a strategic capability in pharma and MedTech.

12.1 Introduction

This chapter explores continuous improvement (CI) in the life sciences industry, focusing on the pharmaceutical and MedTech sectors. The analysis is grounded in an empirical study involving 14 pharmaceutical and 9 MedTech companies. In total, 66 interviews were conducted with representatives from Operational Excellence (OPEX) and quality departments, providing the empirical foundation for the subsequent discussion.

12.1.1 Life Sciences Industry

The life sciences industry encompasses a wide range of fields dedicated to improving health and quality of life. It includes biotechnology, pharmaceuticals, MedTech, nutraceuticals, cosmetics, food processing, and environmental sciences (Hofstetter & Lichtenberger, 2017; Wohlfarth, 2024; Wolf, 2023). MedTech covers medical devices, in vitro diagnostics, and digital health (MedTech Europe, 2021). Even before the SARS-CoV-2 pandemic, demographic change and aging populations had already underscored the industry's societal and economic relevance (Wolf, 2023).

A notable trend is the convergence of pharmaceuticals and MedTech, visible in products that combine hardware and active ingredients (Hofstetter & Lichtenberger, 2017). In this chapter, life sciences therefore refers primarily to these two sectors. Both operate under strict regulation, including cGMP requirements and routine inspections (Califf, 2024; FDA, 2023). Despite the strict regulations, shortages and recalls remain frequent. Manufacturing issues caused 62% of drug shortages between 2013 and 2017 and 40% in 2022/2023 (FDA, 2019, 2024). MedTech recalls rose by 20% in Europe and 7% in the United States in 2023, driven by device failures, software errors, and quality defects (Sedgwick, 2024a, 2024b). Rising costs, shrinking margins, global disruptions, and evolving regulation intensify these challenges (Kaltenbach et al., 2023; Thomas, 2023; Wettstein & Huber, 2024). To ensure patient safety and competitiveness, firms should increasingly rely on CI (ICH, 2008; ISO, 2016; Maguire et al., 2023).

12.1.2 Continuous Improvement

CI, often associated with the Japanese concept of *Kaizen*, is widely defined as a management philosophy, a set of improvement methodologies, and a cultural orientation toward ongoing enhancement of productivity and quality (Anand et al., 2009; Berger, 1997). At its core, CI rests on two principles: continuity—the sustained application of improvement processes—and improvement—targeted, data-driven modifications of existing systems. Unlike disruptive innovation, CI emphasizes incremental, low-risk changes that accumulate over time to strengthen efficiency, quality, and customer satisfaction (Imai, 2012; ISO, 2015).

The historical trajectory of CI illustrates its dual roots. Early suggestion schemes in Japan, such as the meyasubako in the eighteenth century, laid the groundwork for systematic idea generation. The modern form of CI emerged after World War II with the Toyota Production System, profoundly shaped by Deming and Juran's quality theories. From the 1980s onward, Kaizen gained global recognition, diffusing into Western industries and later into European manufacturing, where it was adapted to local organizational contexts (Holweg, 2007; Schroeder & Robinson, 1991). This cross-fertilization established CI both as a Japanese practice rooted in employee involvement and as a Western approach emphasizing structured systems and planned initiatives.

Today, CI is closely intertwined with OPEX. While OPEX provides the overarching vision of excellence and integrates diverse management concepts, CI delivers the methodological and cultural infrastructure for its realization. Without CI, OPEX risks becoming static and detached from daily practice; without OPEX, CI risks fragmentation and lack of strategic anchoring. Together, they form a dynamic interplay that enables organizations to adapt continuously to evolving market and regulatory demands (Ahlström et al., 2021; Found et al., 2018).

12.2 Status Quo CI

CI in the life sciences industry remains at a relatively early stage of maturity. While most companies have introduced structures and practices, CI is still evolving from fragmented initiatives toward more systematic approaches. The following sections highlight how pharma and MedTech companies currently handle idea generation, prioritization, and sustainment of CI activities.

Idea Gathering

Pharmaceutical companies typically rely on formalized, standardized systems for idea generation. Large firms employ Web-based platforms such as SharePoint, ensuring consistency across sites and enabling systematic tracking of suggestions. Idea generation is often embedded in structured processes, including formal suggestion schemes, daily huddles, and Gemba walks. Research-driven companies emphasize broad employee involvement but within a highly regulated framework, ensuring that ideas align with compliance requirements. Smaller pharmaceutical firms, by contrast, tend to use simpler, site-specific methods such as idea walls or suggestion boxes, reflecting their more localized cultures.

MedTech companies also encourage bottom-up contributions but often rely on more flexible and informal channels. Additionally, firm size can make a difference, whereas larger firms use centralized digital platforms; smaller firms emphasize direct employee involvement through local tools, mobile apps, or management-led Gemba walks. European MedTech firms lean toward formal tools, whereas US firms favor mobile solutions and management-driven practices that facilitate rapid submission of ideas.

Across both industries, employees at all levels are empowered to contribute ideas. Digital platforms are often combined with informal practices such as huddles and Gemba walks. The key difference lies in the degree of formalization. Pharma emphasizes structured, company-wide systems, while MedTech favors less formalized, locally adaptable solutions.

Prioritization

Pharmaceutical companies prioritize CI initiatives based on strategic alignment, regulatory compliance, business impact, and financial return. Larger firms employ formalized, quantitative evaluation methods. Research-driven companies focus on initiatives that reduce operational losses and ensure compliance with cGMP standards. CDMOs, due to their service-oriented business model, emphasize quick wins and initiatives that prevent delays in customer projects.

MedTech firms also prioritize initiatives by business impact, financial return, and regulatory requirements, but their approaches are more pragmatic and locally driven. Smaller firms rely on flexible tools such as four-field matrices or escalation mechanisms, enabling rapid decision-making. Larger MedTech companies employ multistage evaluation processes, often with strong regulatory focus, but still allow local prioritization before escalation to global teams.

Both industries use business impact, financial benefits, and compliance as key criteria, where compliance refers both to the continuous strengthening of regulatory adherence and to the assurance that established requirements are never undermined. The difference lies in methodology and scope. Pharma relies on formalized, quantitative tools and global evaluation structures, while MedTech employs more flexible, locally adaptable mechanisms. Pharma's prioritization is more hierarchical and method driven, and MedTech's is more adaptive and responsive to local needs.

Sustainment

Pharmaceutical companies sustain CI initiatives by embedding changes into standardized documents such as SOPs, work instructions, and role descriptions. Large firms employ multilevel monitoring systems, audits, KPIs, and maturity assessments to track progress. Research-driven companies often use structured maturity models to evaluate CI development, while CDMOs involve controlling departments to measure actual savings after implementation. Sustaining improvements is closely tied to integration into the quality management system (QMS), with strong emphasis on statistical validation and compliance.

MedTech firms also formalize improvements through documentation, but their sustainment practices are more flexible. European companies emphasize employee accountability and visible benefits, while US firms rely on continuous follow-up cycles after implementation. Larger MedTech companies combine data-driven monitoring with leadership involvement and systematic reviews, whereas smaller firms depend on direct observation, short monitoring periods, and active employee participation. ISO 13485 compliance plays a central role in ensuring sustainability.

Across both industries, CI progress is embedded into standardized documents and supported by monitoring mechanisms. The difference lies in approach and emphasis. Pharma sustains improvements through formal, multilevel systems integrated into the QMS, while MedTech emphasizes employee responsibility, follow-up cycles, and ISO 13485 compliance. Pharma's sustainment is more formalized and statistically validated, and MedTech's is more practice grounded and people driven.

Summary
Across both industries, CI is recognized as essential but remains at an early stage of maturity. Pharma tends to pursue CI in a structured, hierarchical, and compliance-oriented manner, reflecting its process-driven production and regulatory environment. MedTech, by contrast, adopts a flexible, employee-driven approach, reflecting its discrete manufacturing and product diversity. Commonalities include the dual push–pull structure, reliance on digital platforms and suggestion systems, prioritization by business impact and compliance, and embedding improvements into documentation. Differences emerge in the degree of formalization, methodology, and sustainment practices, with pharma leaning toward global, standardized frameworks and MedTech toward local, adaptive solutions.

12.3 Barriers and Enablers

12.3.1 Barriers

Research on CI has long emphasized the obstacles organizations face when embedding improvement practices. Zhang et al. (2017) conducted one of the most comprehensive literature reviews, identifying 44 distinct barriers. Among the most frequently cited were insufficient management time and employee resistance to change, followed by lack of training, insufficient investment, limited internal funding, weak understanding of benefits, cultural issues, and backsliding due to past experiences. Other barriers, such as unwillingness to learn, technology-related challenges, or customer pressure, appeared only sporadically. Building on this overview, Macuvele (2021) consolidated the 44 barriers into five overarching categories: knowledge, conflicts, employee and culture, resources, and financial outcome.

For the life sciences industry, 77 barriers were identified, of which 26 are specific to pharmaceutical companies and 13 to MedTech firms. From this broad set, 11 barriers emerge as particularly salient across the industry: lack of knowledge, too many initiatives, prioritization of running business, resistance to change, lack of top management commitment, lack of empowerment, missing overarching strategy, missing alignment / silo thinking, fear of regulations, efforts linked to changes, and hard to quantify savings. These barriers can be grouped into six overarching categories: knowledge and capabilities, organization and structure, leadership and strategy, culture and behavior, resources and capacity, and external factors. Compared with earlier studies, the relevance of financial outcome has diminished, while aspects such as organization and structure and leadership and strategy have gained importance.

To better understand the patterns within this complex landscape, a k-means clustering algorithm (MacQueen, 1967) was applied to the aggregated barrier data. From this analysis, three distinct barrier clusters can be distinguished:

(I) **Overburdened (mixed cluster):** 36% of pharmaceutical and 33% of MedTech companies fall into this cluster. They struggle primarily with initiative overload and the tendency to prioritize ongoing operations over CI efforts.

(II) **Underpowered (MedTech cluster):** 7% of pharmaceutical and 56% of MedTech companies belong here. Their defining challenges include lack of knowledge, insufficient top management commitment, lack of empowerment, and difficulty quantifying savings.

(III) **Inertia bound (pharma cluster):** 57% of pharmaceutical and 11% of MedTech companies fall into this cluster. They are characterized by resistance to change, missing overarching strategy, silo thinking, fear of regulations, and high effort linked to changes.

Further investigation into company headquarters, size, and business model did not reveal systematic differences, suggesting that industry context and organizational mindset play a greater role than structural factors. Small, medium, and large firms are represented across all clusters, indicating that CI barriers are shaped more by governance and internal attitudes than by scale. A notable exception emerges in the case of CDMOs and generics manufacturers, all of which fall into the inertia-bound cluster. This pattern suggests that certain business model characteristics can reinforce structural resistance to change.

Overall, the analysis demonstrates that barriers to CI vary widely across the life sciences industry—from initiative overload to skill gaps to strategic misalignment—underscoring the need for tailored approaches that account for both organizational context and industry-specific dynamics.

12.3.2 Enablers

CI does not unfold in isolation—it depends on a constellation of enabling mechanisms that interact dynamically with organizational barriers. Based on the interview data, 12 overarching enabler categories were identified, each addressing specific challenges and shaping the conditions under which CI can thrive. These include strategy and structure (including prioritization), leadership, dedicated resources and timing, standardization and simplification, training and coaching, two-way communication, collaboration and exchange, involvement, recognition, showcasing of benefits (including data access), performance management (including reviews), and risk management.

Rather than acting as isolated levers, these enablers form an interdependent system. Strategic clarity and prioritization mechanisms help reduce initiative overload and align CI efforts with long-term goals. Leadership commitment signals organizational intent and ensures that CI is not sidelined by operational pressures. Dedicated resources and protected time enable continuity, while standardization lowers entry barriers and fosters consistency across sites.

Training and coaching build both technical competence and cultural readiness, while two-way communication and collaboration platforms foster transparency, trust, and cross-functional learning. Involvement and recognition mechanisms strengthen ownership and motivation, and showcasing tangible benefits—especially through data—reinforces credibility and supports executive buy-in. Performance management systems provide structure and accountability, while risk management enables calculated experimentation even in regulated environments.

Importantly, these enablers are not unique to the life sciences sector. Comparative insights from industries such as automotive, electronics, and consumer goods reveal additional practices that can inform CI design in pharma and MedTech:

(I) **Centralized CI functions:** Successful organizations often establish a central CI team that provides strategic guidance, training, and alignment, while local teams retain autonomy in execution. This model balances consistency with site-level flexibility.

(II) **Flexible, pull-based support:** Rather than enforcing top-down mandates, CI support is offered on demand, allowing teams to request expertise when needed. This fosters ownership and responsiveness.

(III) **Multi-site knowledge sharing:** Structured formats such as CI summits, exchange programs, and shared digital platforms promote cross-site learning and reduce duplication.

(IV) **Blended CI approaches:** Combining Lean, Six Sigma, and Agile elements enables organizations to tailor CI to different contexts, from manufacturing to R&D.

(V) **Leadership alignment:** CI is most effective when senior leaders visibly support and participate in improvement efforts, ensuring strategic coherence and cultural reinforcement.

(VI) **Capability building:** Long-term success depends on systematic skill development, from onboarding to advanced CI certifications, often embedded in career progression frameworks.

These cross-industry insights underscore that CI is not merely a technical toolkit but a strategic capability. Its success hinges on the interplay between structural enabler, cultural reinforcement, and leadership commitment. The 12 categories identified in this study provide a robust framework for diagnosing organizational readiness and designing targeted interventions that move CI from aspiration to embedded practice.

12.3.3 Barriers and Enablers

The barrier–enabler matrix (Fig. 12.1) highlights which mechanisms are most effective in addressing specific obstacles to CI. It shows that barriers are not abstract problems but concrete situations that require targeted responses. For example, when organizations struggle with missing overarching strategy, silo thinking, initiative overload, or difficulty quantifying savings, the most decisive lever is strategy and structure, including clear prioritization. Without strategic clarity, CI risks fragmentation, duplication of efforts, and loss of focus.

In contrast, when the challenge is lack of top management commitment, success depends on a dual approach: embedding CI into corporate strategy and showcasing benefits in a way that resonates with executives. Data-driven evidence, visible quick wins, and alignment with strategic priorities are crucial to secure leadership buy-in.

Where running the business is prioritized over CI, the allocation of dedicated resources and structured time windows becomes central. Without protected capacity, improvement activities are repeatedly postponed or overshadowed by operational demands. Similarly, lack of empowerment requires both involvement and leadership to foster ownership and autonomy. Lack of knowledge is best addressed through training and coaching, ensuring that employees acquire both technical skills and the CI mindset.

Barriers linked to high effort of change demand simplified processes and resource allocation to reduce perceived complexity. Resistance to change calls for a broader set of enablers—leadership, involvement, recognition, and two-way communication—to build trust, reduce defensiveness, and foster engagement. Finally, when fear of regulations dominates, risk management provides the necessary framework to balance compliance with innovation, enabling calculated experimentation within safe boundaries.

Taken together, the matrix illustrates that barriers and enablers are tightly interwoven. Each barrier requires a specific configuration of enablers, and the absence of one can undermine the effectiveness of others. For practitioners, this means that CI cannot be advanced by generic measures alone—it requires a tailored approach that matches the organization's barrier profile.

	Strategy & structure (incl. prioritization)	Leadership	Dedicated resources & timing	Standardization & simplification	Training & Coaching	2-way-communication	Collaboration & exchange	Involvement	Recognition	Showcasing benefits (incl. data access)	Performance management (incl. review)	Risk Management
Missing overarching strategy	54%	11%	0%	0%	0%	7%	14%	7%	0%	4%	4%	0%
Missing alignment / silo thinking	34%	6%	2%	2%	4%	18%	20%	4%	0%	6%	4%	0%
Lack of top management commitment	17%	10%	2%	2%	8%	10%	4%	10%	8%	21%	6%	0%
Too many initiatives	42%	8%	15%	0%	3%	13%	8%	3%	0%	0%	8%	0%
Prioritization of running business	20%	16%	34%	2%	2%	7%	5%	9%	0%	5%	0%	0%
Lack of empowerment	10%	15%	4%	0%	10%	11%	7%	19%	11%	11%	0%	1%
Lack of knowledge	20%	9%	2%	4%	38%	4%	13%	3%	1%	5%	0%	0%
Effort linked to changes	21%	5%	37%	11%	0%	5%	0%	5%	0%	11%	5%	0%
Resistance to change	7%	17%	3%	3%	8%	13%	5%	15%	14%	11%	2%	2%
Fear of regulations	9%	0%	0%	9%	6%	9%	9%	0%	0%	3%	0%	55%
Hard to quantify savings	39%	2%	0%	2%	2%	7%	23%	2%	2%	2%	18%	0%

Fig. 12.1 Barrier–enabler matrix

	Strategy & structure (incl. prioritization)	Leadership	Dedicated resources & timing	Standardization & simplification	Training & Coaching	2-way-communication	Collaboration & exchange	Involvement	Recognition	Showcasing benefits (incl. data access)	Performance management (incl. review)	Risk Management
Cluster 1	1.128962538	0.421038521	0.453405548	0.135653598	0.471610607	0.378725797	0.414591737	0.290808726	0.10512632	0.279802659	0.153433634	0.125840315
Cluster 2	1.195719371	0.473134632	0.217699922	0.13081425	0.56759107	0.445367513	0.56302293	0.396635169	0.230424208	0.396128369	0.25666726	0.126795306
Cluster 3	1.334489874	0.452690649	0.40704554	0.191077015	0.418707626	0.509562126	0.5758165	0.341215879	0.166503428	0.335496013	0.24167595	0.3577194

Fig. 12.2 Cluster–enabler matrix

Combining these insights with the cluster analysis results in the cluster–enabler matrix (Fig. 12.2). Distinct profiles emerge that highlight how different organizational contexts demand different enabler constellations:

(I) **Cluster I (overburdened):** Organizations in this cluster must emphasize strategy and structure, supported by training, resources, leadership, collaboration, and communication. The challenge here is not a lack of ideas but an overload of initiatives. Clear prioritization and disciplined governance are therefore critical to prevent dilution of effort.

(II) **Cluster II (underpowered):** These companies depend on strategic alignment, but also require training, collaboration, leadership, involvement, and benefit presentation to build capabilities. The focus is on strengthening knowledge, empowering employees, and securing management commitment. Without these enablers, CI remains fragile and easily sidelined.

(III) **Cluster III (inertia bound):** This cluster exhibits the most diverse enabler set. Organizations must combine strategy and structure with collaboration, communication, leadership, training, resources, risk management, involvement, and benefit showcasing. Here, the challenge is cultural and structural rigidity. Overcoming inertia requires a broad portfolio of enablers that simultaneously address mindset, governance, and compliance concerns.

Across all clusters, strategy and structure consistently emerge as the central enabler, underscoring the importance of prioritization and governance. In contrast, standardization, recognition, and performance management appear less critical in overcoming the dominant barriers. This does not mean that they are irrelevant, but rather that they play a supporting role compared to the strategic and cultural levers.

For practitioners, the analysis offers two key lessons. First, barriers are not evenly distributed. Some organizations are overloaded with initiatives, others lack capabilities, and still others are trapped in inertia. Recognizing which cluster a company belongs to is the first step toward designing effective interventions. Second, enablers must be applied selectively. Overburdened firms benefit most from prioritization and governance, while underpowered firms require capability building and leadership support. Inertia-bound organizations demand the broadest set of enablers, as they must simultaneously tackle cultural resistance, regulatory fears, and structural misalignment.

12.4 Key Learnings

The empirical findings and cross-industry comparisons reveal five strategic insights that are essential for embedding continuous improvement (CI) as a sustainable capability within pharma and MedTech organizations. These learnings go beyond operational optimization and point toward a more systematic, culturally anchored approach to improvement. They provide guidance not only on what needs to be done, but also on how organizations can practically move forward.

Strengthen Systematic Integration
CI must be understood not merely as a collection of tools or isolated projects, but as an integrated organizational capability. This requires embedding CI into strategic planning, leadership routines, and cultural norms. Organizations that treat CI as a peripheral activity—detached from core business objectives—struggle to sustain momentum. In contrast, those that link CI to overarching goals, foster cross-functional collaboration, and institutionalize improvement routines across levels are better positioned to achieve long-term impact. Integration also means that CI is not confined to production or quality departments but permeates R&D, support functions, and leadership agendas. For practitioners, this implies that CI should be designed as part of the organization's deep structures rather than as a temporary program. When CI is embedded in this way, OPEX is reinforced. CI provides the routines and cultural practices that give substance to OPEX's vision, ensuring that OPEX remains dynamic, strategically anchored, and capable of evolving with market and regulatory demand.

Shift from Technical to Natural CI
Maturity assessments indicate that CI in pharma and MedTech remains predominantly technical, relying on formal structures, designated teams, and predefined tools. While these elements are necessary, they are not sufficient. In high-performing organizations, CI becomes a natural part of daily work. Employees continuously identify and solve problems without waiting for formal initiatives. This cultural shift requires more than training. It demands a redefinition of roles, expectations, and leadership behavior. CI must evolve from a scheduled activity to an embedded mindset, where improvement is not an exception but the norm. For practitioners, this means creating environments where CI is experienced as part of everyday work rather than as an additional burden. Unless this works, there is a big risk of ending up in only skin-deep implementations.

Balance Top-Down Guidance and Bottom-Up Engagement
Effective CI systems balance strategic direction with local ownership. Leaders must provide clear priorities, allocate resources, and manage workload to prevent initiative overload. At the same time, employees need autonomy, access to relevant data, and authority to act. Organizations that rely solely on top-down mandates risk disengagement, while purely bottom-up approaches may lack coherence. The most resilient CI systems combine both. Central guidance ensures alignment, while local

engagement fosters relevance and adaptability. This balance is particularly important in regulated environments, where flexibility must coexist with compliance. For practitioners, the lesson is clear. CI thrives when leadership sets the frame, but employees fill it with content.

Broaden the Value Perspective of CI
CI is often framed narrowly as a tool for cost reduction or process efficiency. However, its true value lies in its contribution to strategic resilience, risk mitigation, and long-term performance. By linking CI initiatives to broader organizational objectives such as patient safety, regulatory robustness, or innovation capacity, companies can reposition CI as a strategic enabler. This reframing enhances executive buy-in, strengthens employee motivation, and ensures that improvement efforts are not only operationally sound but also strategically meaningful. It also helps justify CI investments in areas where financial returns are indirect or long-term. For practitioners, this means communicating CI not only in terms of savings but also in terms of resilience, compliance, and innovation.

Balance Regulatory Constraints with CI Enablement
Regulations are frequently perceived as barriers to CI especially when minor improvements are dismissed due to compliance concerns or documentation burdens. Yet, regulatory frameworks can also serve as catalysts for improvement by promoting transparency, standardization, and accountability. Organizations must learn to navigate this tension, enabling CI within the boundaries of compliance, rather than treating regulations as immovable constraints. This requires risk-based thinking, proactive quality leadership, and a culture that encourages experimentation within safe limits. When regulatory rigor is paired with CI enablement, organizations can achieve both robustness and agility. For practitioners, the key is to see regulation not only as a constraint but also as an opportunity to strengthen CI discipline.

References

Ahlström, P., Danese, P., Hines, P., Netland, T. H., Powell, D., Shah, R., Thürer, M., & van Dun, D. H. (2021). Is lean a theory? Viewpoints and outlook. *International Journal of Operations & Production Management, 41*(12), 1852–1878. https://doi.org/10.1108/IJOPM-06-2021-0408

Anand, G., Ward, P. T., Tatikonda, M. V., & Schilling, D. A. (2009). Dynamic capabilities through continuous improvement infrastructure. *Journal of Operations Management, 27*(6), 444–461. https://doi.org/10.1016/j.jom.2009.02.002

Berger, A. (1997). Continuous improvement and kaizen: standardisation and organisational design. *Integrated Manufacturing Systems, 8*(2), 110–117. https://doi.org/10.1108/09576069710165792

Califf, R. M. (2024). Rules and regulations. *Federal Register, 89*(23), 7496–7525. Retrieved from https://www.govinfo.gov/content/pkg/FR-2024-02-02/pdf/2024-01709.pdf

FDA. (2019). *Drug shortages: Root causes and potential solutions.* Retrieved from https://www.fda.gov/media/131130/download?attachment. Updated February 21, 2020.

FDA. (2023). *Current good manufacturing practice (CGMP) regulations*. Retrieved from https://www.fda.gov/drugs/pharmaceutical-quality-resources/current-good-manufacturing-practice-cgmp-regulations
FDA. (2024). *Fiscal year 2023 report on the state of pharmaceutical quality*. Retrieved from https://www.fda.gov/media/179254/download?attachment.
Found, P., Lahy, A., Williams, S., Hu, Q., & Mason, R. (2018). Towards a theory of operational excellence. *Total Quality Management & Business Excellence, 29*(9), 1012–1024. https://doi.org/10.1080/14783363.2018.1486544
Hofstetter, E., & Lichtenberger, R. (2017). Spezifika der Life-Science Industrie. In J. Becker & T. R. Villinger (Eds.), *Life Science Venturing: Herausforderung—Spezifika—Prozess* (pp. 3–19). Springer Fachmedien.
Holweg, M. (2007). The genealogy of lean production. *Journal of Operations Management, 25*(2), 420–437. https://doi.org/10.1016/j.jom.2006.04.001
ICH. (2008). *Pharmaceutical quality system Q10*. Retrieved from https://database.ich.org/sites/default/files/Q10%20Guideline.pdf
Imai, M. (2012). *Gemba Kaizen: A common sense approach to a continuous improvement strategy* (2nd ed.). McGraw-Hill Professional.
ISO. (2015). *ISO 9000:2015—Quality management systems—Fundamentals and vocabulary*. Retrieved from https://www.iso.org/standard/45481.html
ISO. (2016). *ISO 13485:2016—Medical devices—Quality management systems—Requirements for regulatory purposes*. Retrieved from https://www.iso.org/standard/59752.html
Kaltenbach, T., Magunia, P., Nyctelius, H., Grotelüschen, J., Bühren, M., Fath, S., Schmid, C., & Schenck, A. (2023). *Global MedTech 2023—Stem the tide*. Retrieved from https://content.rolandberger.com/hubfs/07_presse/Global_MedTech_2023.pdf
MacQueen, J. (1967). Some methods for classification and an analysis of multivariate observations. In *Proceedings of the fifth Berkeley symposium on mathematical statistics and probability, volume 1: Statistics*. Retrieved from https://projecteuclid.org/ebook/Download?urlid=bsmsp%2F1200512992&isFullBook=False
Macuvele, L. -J. (2021). *Impacting shop floor behavior in the context of pharmaceutical operational excellence programs—A management control perspective*. Dissertation. University of St. Gallen.
Maguire, J., Fisher, A., Harouaka, D., Rakala, N., Lundi, C., Yambot, M., Viehmann, A., Stiber, N., Gonzalez, K., Canida, L., Buhse, L., & Kopcha, M. (2023). Lessons from CDER's quality management maturity pilot programs. *The AAPS Journal, 15*(14), 1–8. https://doi.org/10.1208/s12248-022-00777-z
MedTech Europe. (2021). *The European medical technology industry in figures*. Retrieved from https://www.medtecheurope.org/wp-content/uploads/2021/06/medtech-europe-facts-and-figures-2021.pdf
Schroeder, D. M., & Robinson, A. G. (1991). America's most successful export to Japan: Continuous improvement programs. *Sloan Management Review, 9*(5), 67–81. Retrieved from https://sloanreview.mit.edu/article/americas-most-successful-export-to-japan-continuous-improvement-programs/
Sedgwick. (2024a). *State of the nation 2024: Product safety and recall—European edition*. Retrieved from https://marketing.sedgwick.com/acton/attachment/4952/f-88e8032f-1f1f-4575-933e-1dcbb229927c/1/-/-/-/-/Sedgwick%20Brand%20Protection%20-%20State%20of%20the%20Nation%202024%20Recall%20Index%20Report%20-%20European%20Edition.pdf?utm_term=Download%20Report&utm_campaign=2023%20Recall%20index%20report%20-%20edition%204&utm_content=landing+page&utm_source=Act-On+Software&utm_medium=landing+page&cm_mmc=Act-On%20Software-_-Landing%20Page-_-2023%20Recall%20index%20report%20-%20edition%204-_-Download%20Report&sid=TV2:mbiiMQRUv
Sedgwick. (2024b). *State of the nation 2024: Product safety and recall—United States edition*. Retrieved from https://marketing.sedgwick.com/acton/attachment/4952/f-f9fce568-6f90-40a4-9757-4c46ad16b3d4/1/-/-/-/-/Sedgwick%20Brand%20Protection%20-%20

State%20of%20the%20Nation%202024%20Recall%20Index%20Report%20-%20US%20Edition.pdf?utm_term=Download%20Report&utm_campaign=2023%20Recall%20index%20report%20-%20edition%204&utm_content=landing+page&utm_source=Act-On+Software&utm_medium=landing+page&cm_mmc=Act-On%20Software-_-Landing%20Page-_-2023%20Recall%20index%20report%20-%20edition%204-_-Download%20Report&sid=TV2:d4fDuVorL

Thomas, F. (2023). Bridging gaps and investing in change. *Pharmaceutical Technology, 48*(1), 10–12. Retrieved from https://www.pharmtech.com/view/bridging-gaps-and-investing-in-change

Wettstein, E., & Huber, C. (2024). *Die Schweizer Medizintechnikindustrie—Branchenstudie 2024.* Retrieved from https://www.swiss-medtech.ch/sites/default/files/2024-09/SMTI_2024_Bericht_240904_DE.pdf

Wohlfarth, A. (2024). *Life science industry explained: A comprehensive guide.* Retrieved from https://www.ourcrowd.com/learn/life-science-industry-explained-a-comprehensive-guide

Wolf, K. (2023). *Die Bedeutung der Life Science Branche für unsere Gesellschaft.* Retrieved from https://www.aristo-group.com/de-ch/die-bedeutung-der-life-science-branche-fuer-unsere-gesellschaft-ch/

Zhang, L., Narkhede, B. E., & Chaple, A. P. (2017). Evaluating lean manufacturing barriers: An interpretive process. *Journal of Manufacturing Technology Management, 28*(8), 1086–1114. https://doi.org/10.1108/JMTM-04-2017-0071

Quality First: Regulatory and Industry Collaborations in Pharmaceutical Manufacturing

13

Matteo Bernasconi

Abstract

This chapter examines how collaborations between regulatory authorities and pharmaceutical manufacturers drive quality in drug production. Regulatory agencies set standards, conduct inspections, and support continuous improvement, while manufacturers implement company-wide quality management systems. Dynamic, two-way collaborations help align expectations, reduce information asymmetry, and ensure that regulations are practical and relevant. Using interviews with regulators and industry representatives, the study identifies five key areas for quality maturity: management commitment, business continuity, advanced pharmaceutical quality systems, technical excellence, and employee engagement. Most technical aspects show alignment, but some social

This chapter is based on part of the author's dissertation: *Advancing Quality in Pharmaceutical Manufacturing: Analyzing Regulatory-Industry Interactions*, published at the University of St. Gallen. For more details, see Chap. 5 of Bernasconi (2024). The research was conducted as a member of the St.Gallen OPEX research team.

Funding for some of the research contained in this chapter was made possible, in part, by the Food and Drug Administration through BAA contract [#75F40121C00161]. The views expressed in this chapter do not necessarily reflect the official policies of the Department of Health and Human Services, nor does any mention of trade names, commercial practices, or organization imply endorsement by the United States Government.

All quotations and citations are reproduced verbatim from interviews. Grammatical errors and imprecise language in the original speech have been preserved to maintain accuracy and integrity of the interviewees' statements.

M. Bernasconi (✉)
Abbvie, Cham, Switzerland
e-mail: matteo.bernasconi@abbvie.com

T. Friedli et al. (eds.), *Striving for Operational Excellence in the Pharmaceutical Industry*, Future of Business and Finance,
https://doi.org/10.1007/978-3-032-26287-5_13

components—such as leadership's role in shaping culture and employee quality mentality—are interpreted differently by regulators and industry. The study finds that effective collaboration enhances transparency, mutual learning, and operational stability while highlighting that regulatory relationships are often more cooperative than punitive. Continuous improvement, risk management, and resilient manufacturing—enabled by both technical and cultural investments—are vital for enduring quality and compliance in the industry.

13.1 Introduction

Consumers expect to buy safe, effective, and high-standard quality products. For certain products, customers can easily assess the quality and decide whether to purchase the goods or not by observing the product; however, this is not always possible. In the case of pharmaceutical products, patients cannot easily assess product quality, and their well-being depends directly on drug quality. Therefore, regulatory agencies oversee pharmaceutical manufacturing by setting standards and guidelines that manufacturers must follow, such as current good manufacturing practices (cGMPs). Drug manufacturers implement these requirements and institutionalize manufacturing quality through company-wide quality management programs aimed at improving and sustaining quality, involving everyone from top management to shop floor operators, as well as suppliers and customers (Dean & Bowen, 1994). Quality management programs and associated practices facilitate early identification of product deviations and process inaccuracies, leading to cost savings, reduced remanufacturing, increased reactivity, and better lead times (Weckenmann et al., 2015).

Health authorities, such as the U.S. Food and Drug Administration (FDA), conduct regular inspections to ensure regulatory compliance and assess quality management programs. The role of health authorities is not solely punitive, but rather synergistic, working with manufacturers to ensure that products on the market meet high-quality standards. The FDA supports drug manufacturers in establishing continuous improvement programs (FDA, 2022a), and interactions during inspections can further inform the development of quality management practices (Bernascon et al., 2025). Nevertheless, improvements must be scientifically substantiated and validated to ensure regulatory alignment and maintain product quality. The highly regulated environment may lead to a focus on rigid routines that can contribute to organizational inertia, potentially resulting in resistance to change and diminished innovation (Van de Ven, 1986). Collaborations between health authorities and drug manufacturers are focused on enhancing quality in the pharmaceutical industry and promoting ongoing improvement. Despite ongoing efforts towards quality enhancement, the true nature and impact of collaborations between regulators and industry remain not fully understood, particularly in terms of their influence on evolving quality management practices. This uncertainty highlights the importance of investigating how such partnerships may foster organizational adaptability and innovation in a highly regulated setting.

13.2 Regulatory Authorities and Interactions with Industry

Collaborations between regulatory authorities are common practice for establishing standardized norms for drug products globally. Notable examples include the International Council for Harmonisation of Technical Requirements for Pharmaceuticals for Human Use (ICH) and the Pharmaceutical Inspection Co-operation Scheme (PIC/S). ICH harmonizes technical guidelines to ensure the development of safe, effective, and high-quality drugs (ICH, 2025). PIC/S is a cooperative arrangement among numerous regulatory authorities focusing on the harmonization and standardization of cGMP inspections (PIC/S, 2025). While collaborations among regulators play an essential role in establishing standards and expectations, effective implementation of quality standards and principles relies on close collaboration with drug manufacturers.

Collaborations between regulatory authorities and drug manufacturers are not intended to be a one-way process, where regulators set requirements and manufacturers merely comply. Instead, these collaborations are dynamic and mutually beneficial. Drug manufacturers provide practical insights, technical expertise, and real-world feedback that can shape regulatory guidelines and improve their implementation; for example, manufacturers can comment on draft guidelines before they are officially implemented. Regulatory authorities, in turn, provide oversight, guidance, and frameworks to ensure safe, effective, and high-quality drugs. This two-way collaboration ensures that regulations remain feasible, relevant, and adaptable to evolving scientific and industry needs.

Collaboration can occur through informal exchanges as well as during enforcement activities and sanctions. Regulatory authorities can utilize responsive regulation, ensuring that regulatory reactions are proportional to the situation, ranging from cooperative to punitive approaches (Nielsen, 2006). cGMP investigators may participate in collaboration by highlighting the agency's current rationale and, when necessary, increasing pressure on the industry. Manufacturers may also attempt to use collaboration to influence regulatory decisions (Edelman et al., 1999). Regulatory–industry collaboration has multiple positive effects (Desai, 2016): (1) The industry's overall compliance level increases over time as manufacturers better understand requirements and regulators' expectations. (2) In the event of punitive actions, organizations can recover more quickly, minimizing negative impacts. Nevertheless, organizations that lack transparency are generally less inclined to initiate voluntary regulatory collaborations (Desai, 2016).

However, to be effective, interactions between regulators and manufacturers must overcome a naturally embedded information asymmetry. On the one hand, regulatory authorities lack information on organizations to tailor requirements and inspections to their operating context (Haunschild & Rhee, 2004). On the other hand, industry leaders suffer from reduced information and knowledge about regulators' expectations, making it harder to respond to changes in requirements effectively and in a timely manner (Gray & Silbey, 2014). Both sides actively seek to reduce this information asymmetry and address the challenges that result. Drug manufacturers search for internal (e.g., subject matter experts, corporate programs) and external information (e.g., consultants) to reduce the gap and generate responses to the challenges based on these limited insights (Chen, 2008; Knudsen & Levinthal, 2007). Nevertheless, the most effective measure is to establish close collaboration between regulators and manufacturers to maintain information exchange and reduce asymmetry (Crilly & Zollo, 2012). Such collaboration is especially achieved when regulators engage with manufacturers at industry association conferences, share successful practices and failures, and maintain regular stakeholder dialogue (Short & Toffel, 2008).

The FDA's Center for Drug Evaluation and Research (CDER) Quality Management Maturity (QMM) program is a clear example of regulatory–industry collaboration. The origin of QMM can be traced to a report by the drug shortage task force that identified one of the root causes of drug shortages as the drug market's inability to reward manufacturers for mature quality management programs (FDA, 2019). Following the report, the FDA began socializing QMM with industry through multiple conferences, publishing white papers (FDA, 2022a, 2023a), scientific articles (Fellows et al., 2022), sharing results of pilot projects (Maguire et al., 2023), requesting industry comments (FDA, 2023b), and opening the program for voluntary participation (FDA, 2024). Despite the FDA's efforts to share information on the QMM program, drug manufacturers have sought additional information internally and externally to supplement the limited details from the agency and to prepare for a potential program launch. Manufacturers have begun developing their own quality management programs by mimicking available FDA information and relying on drug manufacturers' association quality management guides and assessment tools, such as the Parenteral Drug Association's Quality Culture Assessment model (Baker et al., 2021) and the International Society for Pharmaceutical Engineering's Advancing Pharmaceutical Quality program (Frederick, 2020).

13.3 Regulatory Authorities: Industry Interactions on QMM

Numerous efforts have been invested in the development and definition of the QMM program in practice by both regulators and drug manufacturers. However, the process takes time, and the amount of information disclosed to the public has not always been sufficient to answer all questions or address criticisms surrounding QMM. Industry representatives have raised four main concerns against a regulatory-supported QMM program (Jaworski & Eaton, 2023): (1) the additional resources required from manufacturers to manage a QMM program; (2) variability and subjectivity in audit programs to assess QMM, leading to low comparability across the industry; (3) the program's current sponsorship by the FDA, resulting in the absence of a coordinated global recognition effort; and (4) the FDA's ability to interpret the collected data, which may lead to questions and potential manipulation by less scrupulous manufacturers.

The integration of QMM within cGMP inspections or as a standalone initiative has been a prominent topic of discussion within the industry. According to the FDA (2023a), quality management practice programs such as QMM are not incorporated into current inspection protocols and are not planned to be included in the future. Nonetheless, official FDA cGMP investigator guidelines provide examples of indicators of an advanced quality system that encompass topics related to QMM (FDA, 2022b). cGMP investigators are recognized for evaluating more than mere compliance with cGMP (Gray et al., 2015). As a result, investigators may already, indirectly and unconsciously, assess behaviors, practices, and quality culture within a manufacturing plant as part of their pharmaceutical quality system (PQS) assessment. While the technical elements form the basis of a robust PQS and are addressed through regulatory assessments, PQS extends beyond technical components. A key element of an effective PQS is a quality-oriented culture (Wu, 2015) and the maturity practices that foster expected quality behaviors (Patel et al., 2015; Friedli et al., 2018).

Given the clear interest of regulatory authorities in further developing quality management practices, ongoing efforts by drug manufacturers and industry associations to establish their own programs, international initiatives to strengthen PQS (ICH, 2008), and industry criticism of regulatory programs (Jaworski & Eaton, 2023), it is essential to examine the impact of current regulatory collaborations on quality management practice programs. Specifically, this analysis investigates cGMP inspections as a form of regulatory collaboration and evaluates requirements for "beyond compliance quality management practices" (see Chapter OPEX 2025: A Definition). It addresses both the expectations communicated by investigators regarding satisfactory quality management practices and how the industry currently perceives regulatory expectations. The goal is to clearly define the information asymmetry gap between regulators (current expectations) and drug manufacturers (current understanding), in order to inform future research aimed at closing this gap.

13.4 Research Methodology

The aim of this research is to investigate the effect of regulatory collaboration in reducing information asymmetry on the development of mature quality management programs. The analysis encompasses both regulatory authorities and drug manufacturers' perspectives. The selection of interview participants followed two distinct approaches. On the one hand, regulatory authority interviewees were chosen based on the prerequisite that their organization is part of the PIC/S, ensuring comparable cGMP inspection systems. These interviewees were contacted via email and invited to participate in the study. On the other hand, industry representatives were selected using purposive sampling, controlled with predefined selection criteria (Bernard, 2013). First, interviewees were required to represent both corporate and site quality organizations to balance perspectives and minimize potential corporate bias regarding site operations. Second, participating companies were chosen to represent different product areas and company sizes, increasing the generalizability of results.

Semi-structured interviews were conducted to collect data. The interview guideline was developed by the author and reviewed by subject matter experts. Pre-reading materials were provided to each interviewee in preparation for the interview.

Table 13.1 Interview sample overview (Bernasconi, 2024)

Nr	Type	Duration (min)	Role (Nr. of interviewees
1	Regulatory	56:24	Senior Compliance Officer (1)
2	Regulatory	51:02	Senior Compliance Officer (1)
3	Regulatory	52:17	Compliance Officer (1)
4	Regulatory	44:24	Compliance Officer (1)
5	Regulatory	57:11	Senior Compliance Officer (1)
6	Regulatory	50:11	Senior Compliance Officer (1)
7	Regulatory	42:12	Compliance Officer (1)
8	Regulatory	55:21	Senior Compliance Officer (1)
9	Regulatory	51:54	Senior Compliance Officer (1)
10	Regulatory	51:32	Compliance Officer (1)
11	Regulatory	50:27	Senior Compliance Officer (1)
12	Regulatory	56:22	Compliance Officer (1)
13	Regulatory	50:40	Compliance Officer (1)
14	Regulatory	56:35	Senior Compliance Officer (1)
15	Regulatory	53:44	Senior Compliance Officer (2)
16	Regulatory	49:54	Senior Compliance Officer (1)
17	Industry	59:42	Policy Quality Intelligence and Inspection Leader (3)
18	Industry	41:37	Site Quality Head, Europe (2)
19	Industry	50:52	Site Quality Head, North America (2)
20	Industry	44:53	Site Quality Head, Europe (2)
21	Industry	53:37	Director of Data Analytics and Director of QA (2)
22	Industry	43:15	Site Quality Head, Asia (1)
23	Industry	49:43	Site Quality Head, North America (1)
24	Industry	56:37	Director of Quality Systems (1)
25	Industry	49:15	Site Quality Head, North America (1)
26	Industry	58:54	Head of Pharmaceutical Business Unit (1)
27	Industry	56:26	Head of Excellence Quality Control (1)
28	Industry	59:37	Site Quality Head, North America (1)
29	Industry	46:58	Head of Governance Quality Systems (2)
30	Industry	22:14	Director of Quality and Compliance (1)
		25 h: 23 min	38 Interviewees

In total, 30 cases were collected, comprising 38 interviewees: 16 from health authorities and 14 industry representatives across 7 different companies. Table 13.1 provides an overview of the interviewees.

The interviews were recorded and transcribed for analysis. To describe the information asymmetry in quality management programs, we conducted two rounds of coding. We began with open coding and focused coding (Saldaña, 2013) to identify elements that differentiate compliant sites from noncompliant sites. The first-order codes were validated by an additional researcher and during a workshop with the health authority. Subsequently, second-order codes were generated using axial coding, and the refined codes were mapped to the FDA's QMM practice areas (FDA, 2023a): management commitment to quality, business continuity, advanced PQS, technical excellence, and employee engagement and empowerment.

13.5 Results [1]

13.5.1 Synthesis

Figure 13.1 illustrates the results from two coding cycles and the respective allocation to the QMM practical areas. A total of 20 elements were identified and assigned to the five areas. The allocation of these codes to practical areas provided new insights into the elements defining each area. Moreover, we argue that the areas are not independent; rather, they are interconnected and dependent on each other. The

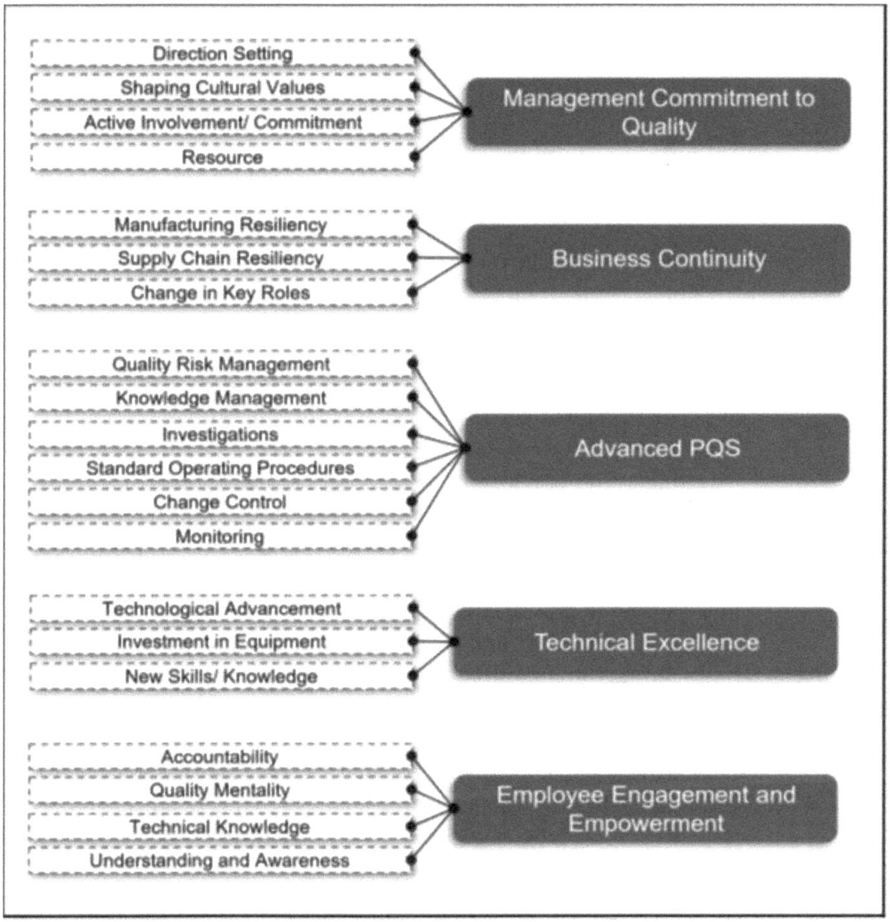

Fig. 13.1 Focused and axial coding results (Bernasconi, 2024)

[1] All quotations and citations are reproduced verbatim from interviews. Grammatical errors and imprecise language in the original speech have been preserved to maintain accuracy and integrity of the interviewees' statements.

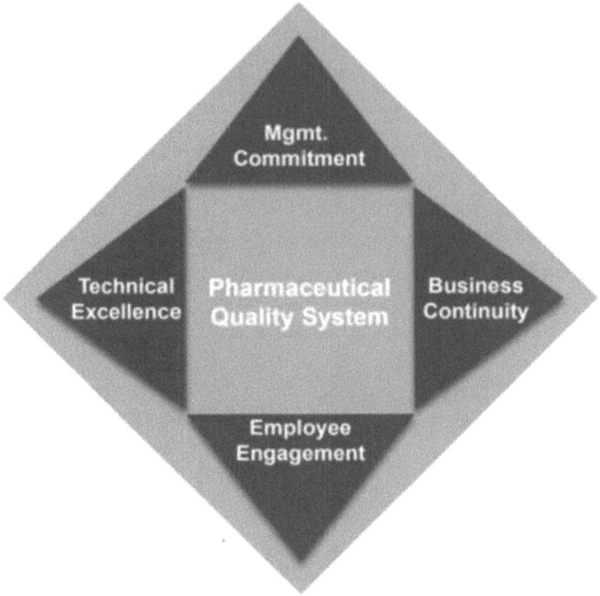

Fig. 13.2 Pharmaceutical regulatory collaboration framework (Bernasconi, 2024)

practical area serving as the baseline for all areas is the advanced PQS, serving as a regulatory technical component that sets the boundaries and minimal requirements for the system to operate. The remaining areas enhance the regulatory basis and summarize the technical and social components of regulatory collaborations on quality management practices. The five areas are further conceptualized into a guiding framework shown in Fig. 13.2. The framework is structured to depict the components of quality excellence from a regulatory and manufacturing point of view without being tailor-suited to one specific health authority. The advanced PQS area summarizes all elements necessary for an effective quality management program, from its enablers to the facilitation of continuous improvement. The other practice areas must respect the boundaries of the PQS to stay compliant.

The framework is structured in two segments, one vertical and one horizontal. The vertical segment represents the social-oriented focus area of the framework. This depicts the hierarchical scheme of organizations by having the leadership commitment on top and employees' engagement at the bottom. Both leaders and employees must operate under the boundaries set by the PQS. The leadership defines the direction that the organization will follow and cascades it down in the form of goals for the employees. The employees are engaged to achieve these goals under the PQS. However, employees are not only working towards their goals. Rather, they critically challenge the system and escalate problems and solutions to the leaders. The vertical segment depicts this combination of top-down and bottom-up approaches that are necessary to manage, run, and improve PQS. The horizontal segment represents the technical components to keep the system working. This segment portrays a time perspective. In fact, the technical components (in terms of knowledge and equipment) evolve and change over the years. The state-of-the-art PQS might be obsolete in some years if they fail to stay updated with the new technologies and necessary skills to operate them. In addition, the continuous improvement generated by employees and leaders will create new challenges that might be tested for resiliency to unexpected events. In the following, the element and the different focus areas and respective information asymmetry are described in detail.

13.5.2 Management Commitment to Quality

The first area targets the leadership commitment to ensure quality, establish a culture oriented towards quality, and set the organizational direction (FDA, 2023a). We identified four elements belonging to this focus area.

The first element is *direction setting*, which is a very established managerial responsibility in quality management practices. Both regulators and industry recognize the role of leaders in defining the direction and thus setting the expectations for the rest of the organization. The regulators highlight the necessity of having a clear strategy derived from the leadership and cascaded down to the shop floor. However, the strategy alone is not enough; it should be made measurable to guarantee that all hierarchical levels can continuously measure the adherence to it and to facilitate the steering from leadership in case of misalignment. The necessity of a clear direction setting has been stressed by the authorities: "*This is what makes the difference between good and bad firms. So a bad firm is one that does not have a plan in place and does not have good program management.*" As a result, the manufacturing facility will act very inconsistently without a clear plan to follow. The element of direction setting shows no information asymmetry in terms of understanding between the two groups.

The second element of management commitment to quality is *shaping cultural values*. The health authorities label senior leadership as the individuals in charge of defining the cultural values and resulting behaviors for the rest of the organization. There is no formal assessment protocol for these norms. Nevertheless, they are indirectly assessed during the inspection process: "*The cultural factors which are less tangible to measure, but perhaps more important, how the site manages its decision-making, how cultural norms are lived, and how people make decisions based on that.*" However, regulators expect senior leadership to establish quality values. According to them, there are three cultural values supported by committed leaders: trust, communication, and transparency. The management trusts employees and allows them to act freely since they are aware of their boundaries. According to one compliance officer interview: "*You have to trust the employees for what they do.*" As a result of the trust generated, the managers will benefit from improved bottom-up communication. A compliance officer provided the example that in a bad facility, "*the problems were not escalated to senior management, so senior management had no awareness while they were responsible.*" The limited communication, and thus missing escalation of problems, reduces the capability of leaders to make decisions based on the actual situation and might result in complacency. Transparency refers to the leader's openness to allow people to discuss and speak up about problems, issues, and processes. One way to quickly assess the degree of transparency during an inspection is to observe the following: "*Is management trying to answer over the employees to prevent them from making any verbal mistakes or just saying something wrong to the agency or do they let their employees sort of speak freely giving them the empowerment and ownership of their processes?*" In addition, investigators get a sense of the degree of transparency by screening the openness of the site leadership to communicate issues and deviations during the inspections. Interestingly, industry representatives have only mentioned transparency in their description of cultural norms but not mentioned trust and communication. The industry reinforces the relevance of a culture of transparency, supporting the fact that mature facilities perform a transparent exchange with regulators. One site quality manager stated: "*The FDA was just here, and we cannot be afraid to have a conversation with the FDA about a risk-based decision that we made [...]. Because the FDA is going to come to my site every two years. Any decisions we make during the days between inspections are all going to be based on it [risk assessment].*" During the inspection, sites are confronted with an assessment of their daily routines and practices. They should be confident of their approaches and be able to openly show them to the regulators to support their actions and decisions. The fear of sharing elements with the investigators is a sign that even the people working at the site are not confident with their own operations. We observe a misalignment between health authorities and industry in the element of shaping cultural values: Both agree on the fact that leadership is responsible for defining the cultural norms for the organization. However, the regulators have clear expectations on the three pillars of building the culture of an organization, but only one of them is acknowledged by industry.

The third element is *active involvement/commitment*; leaders should have a presence on the shop floor and show good behavior (i.e., during Gemba walks, coaching activities). We have not identified any asymmetry gap on this element; both health authorities and industry show the same understanding and expectations. For example, an industry representative stated: "*As a site quality director, I can talk about it all I want, but if people see me doing it. Then that is a very important reinforcement.*" The time invested by leaders in interacting with operators facilitates the transfer of cultural values and reinforces priorities for the employees. This forces leaders to expand their focus from their own tasks to active involvement in evaluating the outcome of manufacturing processes. They are informed on the current performance status on the site level, issues, and ongoing corrective initiatives. However, the involvement should not be understood as simple information gathering from the leadership as an established business routine, but as a crucial means to stay informed and demonstrate the willingness to guide and to be there where a problem has occurred.

The fourth element is *resource*, which has been argued by regulators. The pharmaceutical industry, like all other industries, faces trade-offs in resource allocation between multiple priorities. The investigators are aware that a quality department with extensive resources (e.g., number of employees) might be a signal that they must process numerous investigations. However, they recognize that quality units require an adequate number of resources to operate effectively. For example, company size has often been mentioned as an element influencing the decision on assigning resources: "*Big pharmaceutical companies [...] have all qualified personnel dedicated to different areas. So you do not have a person multiplying different functions [...] versus small companies who have to split their time into different areas of the quality.*" A company with many resources tends to have duplicates of things in the background. Nevertheless, officers pointed out that more resources do not necessarily result in better quality or culture, but rather in more flexibility compared to smaller firms. "*A small firm can have a really good quality culture and a really large firm can have one that is not like the result expected from the resources they have. For me quality culture would not be necessarily based on the firm size.*" The element of resources shows a gap between health authorities and industry. However, it seems that regulators are not setting expectations on the number of resources dedicated to quality; rather, they know that companies face resource constraints.

13.5.3 Business Continuity

The practice area of business continuity covers the elements related to a firm's ability to have in place a system to sustain and maintain operations during expected or unexpected disruptions (FDA, 2023a). We have identified three elements related to business continuity.

The first element is *manufacturing resiliency*; the focus for the health authorities in this context is the ability of a facility to set up a reliable manufacturing system: "*How ready is the firm able to respond in case something goes wrong, like a natural disaster or a cyberattack.*" The reaction plans must be based on a risk-based approach with a risk assessment and the prioritization of events based on their likelihood to happen and severity. On the one hand, the focus relies on the presence of backup plans for a broad range of events that might reduce the firm's capability of manufacturing their products, thus potentially leading to shortages. On the other hand, the ability of an organization to act against major events occurring unexpectedly by assessing: "*How they react to changes or issues that come up either by their own doing or when things just go wrong? How do they handle changes? How they handle interruptions?*" This assessment provides the regulators with a proxy to assess the ability to react to the establishment. The industry frames business continuity as the capability of acting proactively in case of changes and deviations. They did not mention backup plans as a key element of business continuity, but rather the fact that people are engaged and challenge the current status quo to identify weaknesses: "*It is a proactive approach to quality. In the broadest sense, not just product quality, but also the quality of service, and the quality of internal processes.*" We observe a slight gap between the expectations of the regulators and the industry's understanding of manufacturing resiliency. The industry understands resiliency as the capability of improving the system and keeping it running. Although regulators show the same expectation, they reveal an additional step towards the ability to react to unexpected elements and avoid disruptions in drug delivery. The industry should adapt its understanding of this element to be aligned with the health authorities.

The second element is *supply chain resiliency*. Both groups recognize that a close collaboration with the suppliers is necessary to support their internal manufacturing resiliency. Suppliers must be validated, and audits must be conducted to guarantee adherence to the regulations and quality standards. The supply chain must be screened to reduce the possible bottleneck generated by single key suppliers. We have not identified misalignment in supply chain resiliency between the two groups.

The third element is *change in key roles*. Although changes in personnel cannot really be predicted and avoided, they might have a severe impact on business continuity. The leadership is responsible for setting the tone and direction of one organization; the change in a central leader might result in a change in the reliability of the current system. A regulatory interviewee stated: "*Has there been any change in quality management since the last review? [...] you can have a very well-run company and then it changes management or ownership, and everything goes to pieces because they potentially might cut corners.*" The regulators suggested that this is an element to be considered when looking at the overall resilience of a manufacturing establishment. However, no formal protocol exists to measure this element. The industry agrees on the continuity that management provides to the company and that leadership changes might drastically modify the current situation. Again, we do not observe a gap between the two groups.

13.5.4 Advanced PQS

An advanced PQS combines quality principles and risk management approaches to reduce unintended events and disruptions in drug manufacturing (FDA, 2023a). We have identified six elements shaping the advanced PQS area.

The first element is *quality risk management*. Both health authorities and industry highlight QRM as an enabler of the PQS following the ICH (2008) guideline. The implementation of a quality risk management approach is required by regulations and agencies reviewing organizations' quality risk management results and processes. The review of quality risk management is used by investigators as an assessment of the PQS maturity: "*I expect to see quality risk management is something that's embedded within their quality system, a mature organization will use quality risk management across the organization.*" In addition, risk assessment documentation provides not only a good estimation of the general degree of maturity but also the level of proactivity of the establishments towards risk. The industry knows the importance of having risk assessments embedded in changes and processes: "*Most of our quality processes have risk-based components.*" However, having defined risk approaches is insufficient; they must be deployed in the operations and be part of the behaviors: "*A very strong and robust risk mindset and risk culture being able to identify and mitigate risk.*" We have not identified gaps between the two groups in quality risk management. Given the resonance of the topic over the last year, we observe that the efforts of ICH in creating an international understanding of quality risk management (ICH, 2023) and PQS (ICH, 2008) seem to have been successful.

The second element is *knowledge management*. Similar to quality risk management, health authorities and industry have the same understanding of this second pillar of an effective PQS (ICH, 2008). Thus, we have not observed any gaps between them. However, knowledge management plays an even more relevant role for the regulators than for the industry. In fact, investigators expect to review the documentation related to manufactured batches, performed investigations, and general documentation. This indirectly assesses the quality of knowledge management through documentation and the value that is given to documenting and reviewing information.

The third element is *investigations*. Regulators reveal that their focus in evaluating the maturity of the quality system is to assess the quality and accuracy of the investigations. During the interview, we identified three different points of interest for the investigation: corrective and preventive action (CAPA), biological product deviations, and quality defects in products. CAPA is a system for analyzing an issue and identifying the procedure to resolve the problem, mapping its origin, and deriving actions to prevent the problem from reoccurring (FDA, 2023). CAPAs are very resource-intensive activities that must be performed rigorously to identify the true root cause of a problem. Investigators focus on two aspects when looking at CAPA: the identified root cause and the quality of the investigation. First, the root cause analysis must identify the source of the problem and not only scratch the surface: "*I hate seeing retraining and human error. I mean, human error happens, but it should not be a recurring thing that the firm just leans on retraining*." Second, the time and resources invested in the investigation must be recognized. Mature firms tend to perform exhaustive investigations with rich documentation that shows their commitment to quality: "*The investigations; there are some sites they just come in and give you a couple sentences and then they call you off the day. So I think that [it] would be a good indication of how much quality culture is in their organization because if they care, they usually will put in that effort to actually give you a better report*." In addition, the effectiveness of CAPA is assessed by looking at the number of reoccurring deviations as possible indications for ineffective CAPA. Reoccurring deviations happen when the root cause has not correctly been identified or the preventive action cannot resolve the issue. The biological product deviations and quality defects in product reports are post-market actions to identify quality defects in distributed drug products (FDA, 2023a). In case of deficiencies in distributed products, the applicant of an approved product oversees timely reporting of the quality issue information to the agency. The reports might be reviewed as preparation for inspections and trigger an inspection depending on their content, gravity, and volume. Similarly to the CAPA, post-market officers reported that the quality of the reports (i.e., amount of information, undertaken actions) strongly varies between the establishments and provides a good estimation of their PQS maturity and the value they put into quality. The industry has not mentioned investigation as a core element of the PQS, thus suggesting a possible misalignment between the two groups. However, the investigations are required by the regulations and manufacturers cannot avoid them without facing sanctions. We assume that, in this case, the investigations are institutionalized expectations from the regulators that have been taken for granted by industry. Thus, the industry is not mentioning them because they are an obvious element of the PQS.

The fourth element is *standard operating procedures* (SOPs). Health authorities assess PQS maturity by reviewing the SOPs. The assessment is performed again on two levels; the first level focuses on the adherence of operators to the procedures described in the SOP. The second level analyzes how the standard is written as a signal of its effectiveness: "*You can draft a procedure that is effective that people can easily understand and train to and follow consistently or you can have an established written procedure that is not so well written so that you need more experienced people and they kind of learn on the job.*" This provides an indication to the investigator of what might be required in terms of skills and people to follow the SOP correctly. Similar to the investigations, the industry has not mentioned SOP as an element of the PQS. We assume that even in this case, SOPs are taken for granted as an element of the daily operations within the manufacturing facility. Manufacturers have institutionalized the existence of SOP in their PQS and probably did not mention them because they assume these standards to be part of the PQS.

The fifth element is *change control*. Regulators and industry mention change control as an element of a mature and robust PQS; thus, we have not observed any gap between them. Even the SOPs are not meant to last forever; they should be reviewed and updated on a regular basis. Change control guarantees that the changes in the quality processes are evaluated (risk assessment, notification, …) and correctly documented. The change control documentation is used by regulators as a proxy to assess the CI initiatives in establishments and its proactive mindset: "*They [mature establishments] actually have proactive changes rather than reactive changes and all that same thing with proactive preventive actions.*" The industry is aware of the importance of continuously improving but simultaneously staying compliant with the regulations and thus validating changes: "*You can change and continuously improve. But you have to do it within a well-defined framework and following defined processes of change management, for instance, where documentation, realization of decisions is key.*"

The sixth element is *monitoring*. The regulators recognize control as an element to evaluate whether processes are in a state of control and to estimate improvement potential. For example, one health authority's representative stated: "*The ability to actually measure [the quality system] and have a quality control procedure around that and a system in general that would be learning from errors and correcting based on them.*" Often, regulatory assessment relies on process performance and product quality monitoring systems. Similarly, drug manufacturers argue the importance of monitoring PQS. The industry operates with numerous performance measurement systems to monitor the status of the quality system. These measurement systems are hierarchically structured, and the metrics and frequency of review are defined based on the unit of analysis. For example, line-level supervisors work with real-time data and have daily review meetings. Instead, the site leadership team reviews aggregated data and schedules weekly meetings. The metrics must be representative of the situation and linked to products to identify changes in the process quickly. In addition, the industry uses the monitoring system as an indicator of potential gaps and areas to improve. Thus, a site quality manager recommends: "*Do not always measure the good ones [metrics/indicators] and then leave the not good ones hidden.*" We observe a clear alignment between regulators and industry on the monitoring element.

13.5.5 Technical Excellence

The fourth practical area is assessing the implementation of new technologies for manufacturing processes, data management, and information management (FDA, 2023a). We have identified three elements describing technical excellence.

The first element is *technological advancement*. The pharmaceutical industry, similar to all other industries, is constantly progressing in the adoption of new technologies, and this might result in changes to the current status quo. The health authorities recognize that technology evolves at high speed and manufacturers must stay updated: "*There is always new technology, new innovation, new things that can be brought into the process.*" The industry participants confirm what has been argued by the regulators. Manufacturing facilities must stay on the edge of technological innovation to guarantee product quality. Thus, we conclude that regulators and industry are aligned with this element.

The second element is *investment in equipment*. Both regulators and the industry highlight the importance of having constant investments in new equipment or in its maintenance to guarantee that the manufacturing processes are supported by stable machines. However, they recognize that investments in equipment require time to be approved and performed. This reduces the industry's capability to adapt to new technologies quickly. The facility flexibility is impacted not only by the volume of investment required for equipment, but also by the decoupling of corporate assignment of budgets to the sites. Again, we do not observe any misalignment between the two groups on this element.

The third element is *new skills/knowledge*. Regulators and industry are also aligned with this element of technical excellence. They mention the necessity for operators to become more familiar with new technologies (e.g., machine learning, natural language processing). The advancements in technology require operators to work with more complex systems; thus, they must be trained accordingly.

13.5.6 Employee Engagement and Empowerment

The last practice area focuses on the employees and their commitment to positively impact quality (FDA, 2023a). We have identified a total of four elements.

The first element is *accountability*. The health authorities highlight the importance of making shop floor operators accountable for what they are doing and expect them to be able to act on issues: "*They [shop floor workers] should be able to identify improvement ideas and to act on them without any fear of retribution or retaliation. The employees are the ones who see the process on a daily basis. So they are the most knowledgeable about like what may go wrong.*" Additionally, employees should not only be made accountable, but an error culture should be established to facilitate problem escalation to the leadership. Leaders are responsible for creating communication with employees; however, operators must have the opportunity to escalate problems to the leaders without fearing negative consequences for themselves: "*It really is important that people feel like they are heard like they can speak up and that those issues actually rise to certain levels where they can be addressed.*" Investigators want to observe a systematic process to collect, analyze, and answer concerns: "*Building an organization where people feel safe to express concerns where those concerns happen. You are never going to take everybody's complaint and turn it into a solution, but at least that you know that there is a process set up for managing all of that, and that it is taken seriously.*" The industry is completely aligned with what has been stated by the regulators and agrees on the fact that it is necessary to establish a culture of error. Thus, both groups are aligned with this first element.

The second element is *quality mentality*. Health authorities pointed out that employees must have and show a quality mentality on top of accountability. The quality-oriented mentality builds on the concept of total quality management and especially on the continuous improvement perspective. In fact, a regulatory interviewee stated: "*The quality culture piece and the improvement piece must go together because quality and quality culture almost becomes perishable if you are not continuously updating, learning and improving. So there is a requirement for continuously improving your processes.*" The employee must have a quality mentality leading to challenging the status quo, looking for improvements, and going beyond performing the assigned tasks. The industry is partially aligned with the regulators. Industry representatives recognize quality culture as part of the accountability element rather than creating quality behaviors (i.e., mentality). Since the industry did not explicitly make the link, we observe a small gap where the regulators should invest to communicate better their expectations on quality mentality to the industry.

The third element is *technical knowledge*. Most of the interviewees agree on the fact that operators must have a solid background on the current regulations to be enabled to perform continuous improvement. Compliance officers tend to challenge the employees in their area of responsibility by assessing the level of "*knowledge of good manufacturing practice and implementation.*" A good knowledge of them increases the employees' confidence and results in possibly more effective continuous improvement. The industry confirms the importance of training employees in technical knowledge as a basis for employee empowerment. Some corporate representatives assigned the technical knowledge to the elements characterizing the quality culture that they are deploying in the organization: "*Do you have a good culture in terms of regulation GMP practices […] How to undergo a process to make sure we are compliant?*" The employees must be aware of the steps that must be performed to make sure that changes are made according to the regulations and well documented. Thus, we conclude that there is no misalignment in this element.

The fourth element is *understanding and awareness*. Both health authorities and the industry stressed the importance that the employees understand what the impact of their jobs on the products and patients could be. Numerous investigators mentioned that by talking to the operators, they want to understand their awareness about the final product: "*When you are talking to a shop floor employee, can they explain like the true impact to the patient or are they just going through the motions that they were trained to do just to get through the day?*" Additionally, the regulators mentioned that this understanding might even be a good way to motivate people besides classical monetary recognition programs. By highlighting employees' impact on the patient, "*They feel that they are doing something that is really important for the help of the people*." The awareness of the employee might even be a further enabler towards continuous improvement: "*Like a reward and recognition program for staff across various levels of the establishment just to lead quality improvements*." The industry highlights that everyone is aware of the company's mission towards patients. Numerous representatives from plant and corporate mention this element as a baseline to build the company culture. Apparently, without having employees knowledgeable about the products and their contributions to helping patients, there are more difficulties in having employees engaged and motivated towards quality. Therefore, companies invest time and resources in informing employees on their impact on the patients: "*People understand what is a product we have been manufacturing, what could be the impact that we are delivering in terms of the patients' treatment*."

13.6 Discussion and Conclusion

The purpose of this study was to investigate whether quality management practices are part of the current regulatory overseeing activities (i.e., audits and inspections). The findings of this study contribute in numerous ways.

First, our research addresses the well-known challenge of overcoming the limitations in knowledge sharing between regulators and regulated (Gray & Silbey, 2014; Desai, 2016) by creating transparency on where the gap exists in detail. We show that the asymmetry is greater on the socially oriented components of the framework than on the technical ones. The social and cultural nature of these components makes them fuzzier and, thus, more likely to be open for subjective interpretations. For leadership commitment, we observe a misalignment in the understanding of the social values that should be acknowledged by leaders as part of the quality culture. Nevertheless, the industry prioritizes transparency and how to cascade down the sense of accountability to the operators. We assume that initiatives aimed at improving a culture of transparency will, directly and indirectly, establish better communication and trust. However, the industry should be aware that the regulatory expectation is to establish a culture based on all three aspects simultaneously. This asymmetry is further represented in the element of employee quality mentality.

Again, companies stress accountability as the baseline for the development of quality management practices and continuous improvement mindset, but regulators have a clear conception of how a quality-oriented mentality should look like. The analysis reveals little asymmetry in manufacturing resiliency. Regulators focus on unexpected major events that might impact drug availability to the patients. Instead, the industry is understanding resiliency rather as its ability to act on deviations. The gap between regulators and regulated on this element is on the scope of the resiliency. Health authorities expect a risk assessment that holistically considers all types of events, even low-probability, high-impact cases. The industry is rather focusing on how to react to the more likely deviations and how to guarantee that production is continuously running. Finally, we identify a gap in the investigation and SOP elements of the PQS. These elements are the baseline of regulatory inspections and are very objective to be assessed. Thus, regulatory focus is not surprising, since they are part of the requirements of the manufacturers. However, we imagine that the industry is taking investigations and SOPs for granted and therefore they must no longer be highlighted (DiMaggio & Powell, 1983, p. 9). In the case of less mature companies, the relevance of both elements should be greater since they first have to make sure to master all technical requirements. Despite having identified some discrepancies, we show that the elements acknowledged by the regulators and industry are mostly aligned, supporting the positive impact provided by regular exchange with the stakeholders during conferences as argued by Short and Toffel (2008). These elements can be leveraged by regulators to further foster the drug manufacturers (Edelman et al., 1999) to embrace internal learning on these quality management practices. In addition, we support the quantitative results from VanDuyse et al. (2021) and Fellows et al. (2022), who argued that advanced PQS has more mature quality management practices and results in higher operational stability. The interviewees described the quality management practices combined with an advanced quality system and provided examples of why these elements are crucial to managing, maintaining, and improving the system under the international regulatory framework.

Second, under the lenses of the compliance theory, we show that the exchange between regulators and regulated is not just a punitive relationship; rather, it consists of some coaching activities from health authorities to the industry. We provide a qualitative description of the quality management practices that are part of the regulatory collaboration between regulators and drug manufacturers. The regular exchange between investigators and industry covers multiple components affecting the manufacturing routines in the PQS and reducing the consequent risk of higher noncompliance generated by sporadic interaction with investigators (Anand et al., 2012). Manufacturers can develop complacency over time because of overconfidence in their current level of adherence to regulations, potentially leading to negligence. Investigations act as regulatory countermeasures to reduce the risk of insurgency of complacency in manufacturing establishments. The regulators should follow the so-called responsive regulation, meaning that the regulators' actions must be structured according to a hierarchical pyramid starting from very cooperative and ending in punitive approaches if the prior attempts fail (Nielsen & Parker,

2009). The regulatory inspection process covers a multitude of components, some of them rather cGMP related (CAPA, SOP, ...) and others more socially related (transparency culture, speak up, ...). The investigators visit sites not to serve as coaches actively, but to assess their compliance level. However, during the process, they encounter and observe situations that are compliant but not up to the officer's expectations. In this case, the observation has the scope of providing feedback to the sites and asking for improvement. If we consider the FDA's inspection outcome system, only one out of three classifications is enforcing changes in the system (i.e., OAI); the other classifications (i.e., NAI and VAI) are providing feedback without direct regulatory escalation. Establishments experiencing VAI and especially OAI results may be inspected with higher frequency due to the risk-based approach.[2] This is perfectly aligned with the responsive regulation approach that requires a more punitive action if the regulated fails to comply with previous feedback (i.e., experiencing two consecutive VAI results) or a less punitive one if the regulated follows the feedback (i.e., after a VAI addressing the point and having a following NAI) resulting from the interaction (Braithwaite, 2002). In addition, the regular interaction between regulators and regulated helps to reduce the information between the two actors. In the long run, the collaborative approach has been proven to provide a positive impact on the overall level of compliance and reduce the negative impact on organizational performance in case of punitive actions (Desai, 2016).

Third, we address the criticism from the industry on the establishment of quality management practice programs from regulators. According to Jaworski and Eaton (2023), the industry concerns are summarized in four major points. Our analysis provides some clarifications to two concerns expressed by the industry about the resources needed to administer the program and the need for a coordinated effort. The current inspection process covers directly or indirectly a broad range of elements connected to quality management practices, for example, quality culture and behaviors resulting from leaders' commitment and employee engagement. Manufacturing facilities often manufacture products for multiple countries and thus are under multiple regulatory regimes. As a result, they go through multiple inspections during the year. The quality team is already trained and used to engage with health authorities in terms of regulatory actions and collaborations. However, the regulatory pressure resulting from the exchange might set the focus of the quality team on additional elements. Instead, the health authorities might require additional resources to run such a program. Pioneers in the development of a quality management program already anticipated that the assessment will be performed independently from current investigation activities (FDA, 2023a). The concern of having coordinated global effort from health authorities seems to have already been addressed by the health authorities. The alignment of the quality management practices is supported by the standardization of cGMP provided by the PIC/S and the international alignment on the understanding of PQS (ICH, 2008). In addition, the close collaboration with the stakeholders is performed globally during many

[2] Compliance history (previous inspection results) is one of the many factors used by the CDER's Site Selection Model to define inspection priorities (FDA, 2023c).

international conferences and international inspections. Nevertheless, it is unclear whether a global attempt exists to establish comparable programs, or even one, between regulators to assess quality management practices.

Fourth, we add to the literature stream of quality management by providing more in-depth explanations of quality management practices from both industry and regulatory perspectives, thereby contextualizing existing quantitative studies. For instance, numerous elements in PQS, management commitment, and employee engagement can be found in the traditional construct of quality management practices and continuous improvement. The importance of quality management programs in management commitment and employee engagement has been widely empirically validated in multiple studies (Cua et al., 2001; Shah & Ward, 2003). Especially, our results show that continuous improvement is largely rooted in the technical and social components of the framework. The guideline ICH Q10 (ICH, 2008) stresses that an effective PQS facilitates continuous improvement during the entire drug life cycle (ICH, 2019). The continuous improvement mindset is established with specific enablers such as quality risk management and knowledge management that support the correct collection of various information, its documentation, and the proactive approach to evaluating and controlling risk to quality. The formal requirements must be enforced by the leaders and translated into a quality culture to be spread across the entire organization. The leaders' support for the social norms and their renewal through role modeling will result in the anchoring of these cultural values into quality behaviors from the employees (i.e., continuous improvement mindset). The positive relationship between quality culture and behaviors has already been discussed in two studies about pharmaceutical excellence. Patel et al. (2015) first demonstrated that companies with mature and well-defined quality practices show stronger quality behaviors and show that more advanced quality culture and behaviors will lead to better operational performance (Friedli et al., 2018). Our results reveal that the norms characterizing quality culture are deployed with a top-down approach. The continuous improvement mindset will arise as soon as multiple elements from the five components are ingrained in the employees' behaviors. Technical excellence provides support to the organization to manufacture products with reliable equipment. The stability resulting from effective and reliable equipment is necessary to challenge the system and continuously increase quality. Technical excellence, in combination with business continuity, defines the capability of a manufacturing facility to be resilient and compliant. In fact, resilient manufacturing activities are a fundamental component of the adherence to regulations. Poor planning and management across the entire plant might negatively impact resiliency (Kansteiner, 2022) and lead to rejection in the drug approval process due to non-resilient manufacturing (Liu, 2023). However, we show that resilience is generated not only in internal manufacturing but also in close collaboration with partners, as often argued by lean literature (Cua et al., 2001; Shah & Ward, 2003). Poor close collaboration with the key partners might lead to severe consequences, as in the case of Eli Lilly facing a drug approval rejection due to problems by the contract manufacturing organization of the drug (Kansteiner, 2023). This demonstrates that resiliency should be assessed from a broader perspective to stay compliant.

References

Anand, G., Gray, J., & Siemsen, E. (2012). Decay, shock, and renewal: Operational routines and process entropy in the pharmaceutical industry. *Organization Science, 23*(6), 1523–1783. https://doi.org/10.1287/orsc.1110.0709

Baker, D., Broadfoot, J., Mendivil, S., Caruso, A., & Lueken, S. (23 September 2021). *How to measure quality management maturity*. Retrieved August 31 2025 from PDA: https://www.pda.org/pda-letter-portal/home/full-article/how-to-measure-quality-management-maturity

Bernard, H. R. (2013). *Social research methods – Qualitative and quantitative approaches.* Sage.

Bernascon, M., Friedli, T., & von Dzengelevski, O. (2025). The impact of current good manufacturing practices inspections on continuous improvement mindset in the pharmaceutical industry. *Journal of Pharmaceutical Innovation, 20*, 1–19. https://doi.org/10.1007/s12247-025-10200-7

Bernasconi, M. (2024). *Advancing quality in pharmaceutical manufacturing: Analyzing regulatory-industry interactions*. Copy Blitz AG.

Braithwaite, J. (2002). *Restorative justice and responsive regulation*. Oxford University Press.

Chen, W. (2008). Determinants of firms' backward- and forward-looking R&D search behavior. *Organization Science, 19*(4), 497–668. https://doi.org/10.1287/orsc.1070.0320

Crilly, D., & Zollo, M. T. (2012). Faking it or muddling through? Understanding decoupling in response to stakeholder pressures. *Academy of Management Journal, 55*(6), 1429–1448. https://doi.org/10.5465/amj.2010.0697

Cua, O. K., McKone, K. E., & Schroeder, R. (2001). Relationships between implementation of TQM, JIT, and TPM and manufacturing performance. *Journal of Operations Management, 19*+(6), 675–694. https://doi.org/10.1016/S0272-6963(01)00066-3

Dean, J. W., & Bowen, D. E. (1994). Managmenet theory and quality: Improving research and practice through theory development. *The Academy of Management Review, 19*(3), 392–418. https://doi.org/10.2307/258933

Desai, V. M. (2016). Under the radar: Regulatory collaborations and their selective use to facilitate organizational compliance. *Academy of Management Journal, 59*(2), 636–657. https://doi.org/10.5465/amj.2014.0943

DiMaggio, P. J., & Powell, W. W. (1983). The Iron Cage Revisited: Institutional Isomorphism and Collective Rationality in Organizational Fields. *American Sociological Review, 48(2), 147–160.* https://doi.org/10.2307/2095101

Edelman, L. B., Uggen, C., & Erlanger, H. (1999). The endogeneity of legal regulation: Grievance procedures as rational myth. *American Journal of Sociology, 105*(2), 406–454. https://doi.org/10.1086/210316

FDA. (2019). *Drug shortages: Root causes and potential solutions*. U.S. Food and Drug Administration. Retrieved August 31 2025 from https://www.fda.gov/media/131130/download?attachment

FDA. (2022a). *Quality management maturity: Essential for stable U.S. supply chains of quality pharmaceuticals*. U.S. Food and Drug Administration. Retrieved August 31 2025 from https://www.fda.gov/media/157432/download?attachment

FDA. (2022b). *Chapter 56—drug quality assurance*. U. S. Food and Drug Administration. Retrieved August 31 2025 from https://www.fda.gov/media/75167/download

FDA. (28 March 2023). *Corrective and preventive actions (CAPA)*. Retrieved August 31 2025 from https://www.fda.gov/inspections-compliance-enforcement-and-criminal-investigations/inspection-guides/corrective-and-preventive-actions-capa

FDA. (2023a). *CDER's quality management maturity (QMM) program: Practice areas and prototype assessment protocoll development*. U.S. Food and Drug Administration. https://www.fda.gov/media/171705/download?attachment

FDA. (2023b). *Quality management maturity program for drug manufacturing establishments; Establishment of a Public Docket; Request for Comments*. U.S. Food and Drug Administration. Retrieved August 31 2025 from https://www.federalregister.gov/documents/2023/09/15/2023-20015/quality-management-maturity-program-for-drug-manufacturing-establishments-establishment-of-a-public

FDA (2023c) Understanding CDER's Risk-Based Site Selection Model Retrieved August 31st from https://www.fda.gov/media/116004/download
FDA. (2024). *Voluntary quality management maturity prototype assessment protocol evaluation program.* U.S. Food and Drug Administration. Retrieved August 31 2025 from https://www.federalregister.gov/documents/2024/01/25/2024-01423/voluntary-quality-management-maturity-prototype-assessment-protocol-evaluation-program
Fellows, M., Friedli, T., Li, Y., Maguire, J., Rakala, N., Ritz, M., … Viehmann, A. (2022). Benchmarking the quality practices of global pharmaceutical manufacturing to advance supply chain resilience. AAPS Journal, 24(111), 1–11. https://doi.org/10.1208/s12248-022-00761-7.
Frederick, T. (8 October 2020). *ISPE quality management maturity program: Advancing pharmaceutical quality.* Retrieved August 31 2025 from https://ispe.org/pharmaceutical-engineering/ispeak/ispe-quality-management-maturity-program-advancing-pharmaceutical
Friedli, T., Buess, P., Köhler, S., Chen, C., Mendivil, S., & Baker, D. (2018). The impact of quality culture on operational performance—An empirical study from the pharmaceutical industry. *PDA Journal of Pharmaceutical Science and Technology, 72*(5), 531–542. https://doi.org/10.5731/pdajpst.2018.008771
Gray, J. V., Anand, G., & Roth, A. V. (2015). The influence of ISO 9000 certification on process compliance. *Production and Operations Management, 24*(3), 369–382. https://doi.org/10.1111/poms.12252
Gray, G. C., & Silbey, S. S. (2014). Governing inside the organization: Interpreting regulation and compliance. *American Journal of Sociology, 120*(1), 96–145. https://doi.org/10.1086/677187
Haunschild, P., & Rhee, M. (2004). The role of volition in organizational learning: The case of automotive product recalls. *Management Science, 50*(11), 1545–1560. https://doi.org/10.1287/mnsc.1040.0219
ICH. (2008). *Pharmaceutical quality system Q10.* ICH. Retrieved August 31 2025 from https://database.ich.org/sites/default/files/Q10%20Guideline.pdf
ICH. (2019). *Technical and regulatory considerations for pharmaceutical product lifecycle management.* ICH. Retrieved August 31 2025 from https://database.ich.org/sites/default/files/Q12_Guideline_Step4_2019_1119.pdf
ICH. (2023). *Quality risk management (R1).* ICH. Retrieved August 31 2025 from https://database.ich.org/sites/default/files/ICH_Q9%28R1%29_Guideline_Step4_2025_0115.pdf
ICH. (30 August 2025). *Mission.* Retrieved August 31 2025 from https://www.ich.org/page/mission
Jaworski, D., & Eaton, J. (2023). *Industry remains divided on quality management maturity proposal.* Retrieved August 31 2025 from https://www.pda.org/pda-letter-portal/home/full-article/industry-remains-divided-on-quality-management-maturity-proposal
Kansteiner, F. (11 October 2022). Eli Lilly faces fresh FDA rebuke at a troubled plant in New Jersey: report. *Fiercepharma.* Retrieved August 31 2025 from https://www.fiercepharma.com/manufacturing/fda-uncovers-fresh-problems-eli-lilly-plant-subject-doj-probe
Kansteiner, F. (2. Oct 2023). Eli Lilly slapped with 3rd FDA rejection in 2023, this time for eczema treatment lebrikizumab. *Fiercepharma.* Retrieved August 31 2025 from https://www.fiercepharma.com/pharma/eli-lilly-slapped-third-fda-rejection-2023-time-eczema-treatment-lebrikizum
Knudsen, T., & Levinthal, D. A. (2007). Two faces of search: Alternative generation and alternative evaluation. *Organization Science, 18*(1), 1–164. https://doi.org/10.1287/orsc.1060.0216
Liu, A. (14 April 2023). FDA rejects Eli Lilly's ulcerative colitis prospect mirikizumab, citing manufacturing shortfalls. *Fiercepharma.* https://www.fiercepharma.com/manufacturing/manufacturing-shortfall-trips-eli-lillys-ibd-biologic-fda-rejection
Maguire, J., Fisher, A., Harouaka, D. et al. (2023) Lessons from CDER's Quality Management Maturity Pilot Programs. AAPS J 25, 14. https://doi.org/10.1208/s12248-022-00777-z
Nielsen, V. (2006). Are regulators responsive? *Law & Policy, 28*(3), 376–399. https://doi.org/10.1111/j.1467-9930.2006.00232.x
Nielsen, V. L., & Parker, C. (2009). Testing responsive regulation in regulatory enforcement. *Regulation & Governance, 3*(4), 376–399. https://doi.org/10.1111/j.1748-5991.2009.01064.x

Patel, P., Baker, D., Burdick, R., Chen, C., Hill, J., Holland, M., & Sawant, A. (2015). Quality culture survey report. *PDA Journal of Pharmaceutical Science and Technology, 69*(5), 631–642. https://doi.org/10.5731/pdajpst.2015.01078

PIC/S. (30 August 2025). *Introduction*. Retrieved August 31 2025 from https://picscheme.org/en/about

Saldaña, J. (2013). *The coding manufal for qualitative researchers*. Sage.

Shah, R., & Ward, T. P. (2003). Lean manufacturing: Context, practice bundles, and performance. *Journal of Operations Management, 21*(2), 129–149. https://doi.org/10.1016/S0272-6963(02)00108-0

Short, J. L., & Toffel, M. W. (2008). Coerced confessions: Self-policing in the shadow of the regulator. *The Journal of Law, Economics, and Organization, 24*(1), 45–71. https://doi.org/10.1093/jleo/ewm039

Van de Ven, A. H. (1986). Central problems in the management of innovation. *Management Science, 32*(5), 590–607. https://www.jstor.org/stable/2631848

VanDuyse, S. A., Fulford, M. J., & Barlett, M. G. (2021). ICH Q10 pharmaceutical quality system guidance: Understanding its impact on pharmaceutical quality. *AAPS Journal, 23*. https://doi.org/10.1208/s12248-021-00657-y

Weckenmann, A., Akkasoglu, G., & Werner, T. (2015). Quality management – history and trends. *TQM Journal, 27*(3), 281–293. https://doi.org/10.1108/TQM-11-2013-0125

Wu, S. J. (2015). The impact of quality culture on quality management practices and performance in Chinese manufacturing firms. *International Journal of Quality & Reliability Management, 32*(8), 799–814. https://doi.org/10.1108/IJQRM-12-2013-0199

Part III
How It Looks in the Field

Thomas Friedli, Matteo Bernasconi, and Fabian Klinkner

This section offers practical insights into how pharmaceutical manufacturers navigate constant change—from new technologies and evolving regulations to shifting market and patient needs. It explores what it takes to build, lead, and sustain a culture of operational excellence (OPEX) in a dynamic environment. Twelve seasoned industry leaders contribute chapters that chart the evolution of OPEX programs over the past decade across a range of biopharmaceutical companies, addressing key priorities such as improving product quality, boosting efficiency, adopting emerging technologies, standardizing processes, and enabling effective collaboration among increasingly global and cross-functional teams.

The opening chapters focus on shaping quality culture and behaviors that underpin true excellence.

They are followed by four chapters that trace recent transformations in OPEX programs, highlight the challenges presented by digital tools and new ways of working, and outline strategies for successful adoption and organizational design—including effective structures for managing dispersed teams.

Two case studies then provide concrete examples from a biologics manufacturer and an animal health company, illustrating how robust OPEX implementation delivers results.

Further chapters focus on the quality perspective: showing how data analytics drives both quality and continuous improvement, why quality management systems matter for more than just compliance, and what sets quality business leaders apart as partners in organizational success.

The section closes by expanding the OPEX lens beyond manufacturing to the entire supply chain, demonstrating that the principles and benefits of operational excellence can strengthen every part of the enterprise.

Culture as the Engine of Patient-Focused Excellence

14

Nuala Calnan

Abstract

In biopharmaceutical organizations, culture functions as a strategic control system influencing quality, compliance, innovation, and ultimately patient outcomes. This chapter explores organizational behavior, operational excellence, risk science, leadership theory, and high-reliability organization (HRO) practices to demonstrate how employee engagement, psychological safety, leadership behavior, and reward systems collectively shape patient-focused excellence. Drawing on academic research and direct industry experience, the chapter outlines how culture is the engine that drives employee engagement, risk transparency, and performance outcomes within regulated environments.

14.1 Introduction: Culture Is a Clinical Variable

Today, biopharmaceutical organizations operate within conditions of accelerated competition, increased scientific breakthrough (aka scientific and regulatory uncertainty), supply chain complexity, enhanced regulatory scrutiny, and high moral consequence. Failures in quality, issue escalation, collaboration, or data integrity may have direct patient impact and are as critical to performance outcomes as a clinical trial result.

As such, culture is not merely an optional extra—it is an operational determinant.

N. Calnan (✉)
TU Dublin, Dublin, Ireland
e-mail: nuala.calnan@tudublin.ie

T. Friedli et al. (eds.), *Striving for Operational Excellence in the Pharmaceutical Industry*, Future of Business and Finance,
https://doi.org/10.1007/978-3-032-26287-5_14

In BioPharma, culture influences whether:

- A deviation is escalated early.
- A scientist or subject matter expert (SME) challenges a flawed assumption.
- A site is inspection-ready.
- A lifesaving therapy reaches patients on time.

Organizations that have faced major compliance crises have learned this the hard way. The lesson is consistent: **Culture failures precede operational failures.**

Culture also influences whether leaders and managers reinforce integrity and right-first-time over speed, whether teams collaborate effectively across functional silos, and if continuous improvement is expected *and* habitual.

This chapter argues that patient-focused excellence can only emerge from a culture characterized by leader-led psychological safety, active engagement, accountability, and learning orientation.

14.2 Psychological Safety as a Foundation of Risk Transparency

The concept of **psychological safety** is foundational to patient-focused excellence. Amy Edmondson introduced psychological safety as:

> *"A shared belief that the team is safe for interpersonal risk-taking."*

Her seminal hospital research showed that teams reporting more errors were actually *safer*, not less competent—because they were more willing to report mistakes (Edmondson, 1999).

Subsequent research in this field confirms the following:

- Psychological safety increases learning behaviors.
- It predicts team performance in complex environments.
- It is critical in regulated, high-stakes industries.

Organizations that encourage curiosity, challenge, and dissent surface risks earlier, enabling proactive, preventative approaches rather than reactive, costly corrective actions.

Boeing's recent 737 MAX crisis provides a powerful culture and psychological safety lesson relevant to regulated industries, where investigations revealed:

- Management pressure for speed over safety.
- Escalation hesitancy.
- Cultural pressure.
- Communication silos.
- Safety concerns not elevated effectively.

Regulated industries share a common truth: when employees feel unsafe speaking up, risk compounds.

Implication for BioPharma

In regulated environments such as BioPharma and MedTech, higher deviation and issue reporting are often viewed as signs of failure. "Fear of the red" on the corporate dashboard can lead to underreporting, gaming, or ineffective issue resolution driven by calendar milestones rather than outcomes.

On the other hand, within a supportive learning environment, increased reporting of issues and risks can actually signal a stronger culture, not weaker performance.

This finding has direct implications, where:

- Proactive risk, near miss, and early deviation reporting rates are recognized as a sign of strength of the organizational culture.
- Climates of silence that increase latent risk accumulation are actively discouraged.
- Escalation hesitancy and decision paralysis are understood to undermine quality and compliance integrity and patient safety.

For industry-specific examples, we can first look at the Johnson & Johnson (McNeil Consumer Healthcare) high-profile product recalls that occurred between 2009 and 2010, which led to plant shutdowns and significant regulatory scrutiny. In this case, investigations highlighted systemic organizational culture weaknesses, including:

- Inconsistent deviation handling.
- Escalation delays.
- Documentation lapses.
- Weak site-level accountability.

Even a strong legacy mission/purpose for Johnson & Johnson (i.e., "Our Credo") was not enough without behavioral reinforcement and governance alignment.

More recently, during the rapid COVID-19 vaccine development effort, Pfizer demonstrated how culture acts as a competitive advantage, when they maintained regulatory rigor while accelerating timelines. Cultural success factors that were highlighted included:

- Clear mission alignment ("breakthroughs that change patients' lives").
- Cross-functional coordination at unprecedented speed.
- Strong governance oversight.
- High levels of discretionary effort.

Employees reported strong ***purpose*** activation—connecting their daily work to global impact. This is reinforced by research across multiple large-scale studies, including Gallup, Google's Project Aristotle, Harvard Business Review, and Zenger

Folkman, that consistently links the leadership behavior of *providing clear direction and purpose* to higher employee engagement by:

- Communicating a compelling vision.
- Connecting daily work to broader organizational goals.
- Setting clear expectations and priorities.

The lesson for BioPharma is when purpose and discipline align and where employees feel that their work matters and understand how to succeed, employee engagement accelerates execution without compromising compliance or reputation. High engagement correlates directly with transparent, proactive risk management. Culture, therefore, becomes a preventive risk control.

Patient-focused excellence is not achieved solely through strategy, process, or compliance systems, and purpose statements alone do not build culture. It is achieved when **culture aligns behavior to purpose,** consistently, visibly, and predictably. It requires employees at every level to internalize one core belief:

> "My daily decisions and behaviors affect patient lives."

14.3 High-Reliability Organization (HRO) Principles

Early research on high-reliability organization (HRO) theory provides a complementary lens, yet provides powerful parallels for BioPharma. Weick and Sutcliffe, in their 2001 book *Managing the Unexpected*, identified five principles associated with reliable performance under hazardous conditions:

1. Preoccupation with failure.
2. Reluctance to simplify interpretations.
3. Sensitivity to operations.
4. Commitment to resilience.
5. Deference to expertise.

These principles closely align with patient-focused excellence in BioPharma. For example:

- Preoccupation with failure parallels vigilance in quality systems that seek to learn from mistakes coupled with a strong escalation culture.
- Deference to expertise does not mean overreliance on SME opinion; rather, it supports the need for scientific debate and cross-functional respect.
- Commitment to resilience reinforces the need for robust continuous improvement practices.

An often-quoted case study for a *high-reliability culture* is the Mayo Clinic's **patient-first** model, which is deeply embedded in its culture through:

- Team-based decision-making.
- Integrated collaboration.
- Clear hierarchy during emergencies, but collaborative norm otherwise.
- Peer accountability.

Other healthcare-based research studies consistently link higher staff engagement to lower patient safety incidents and higher quality outcomes. In their 2025 *Transforming Health Care Through Quality and Safety* paper, Ramar et al. (cf. Ramar et al. 2025) provide the National Academy of Medicine definition for a *learning health system* as:

> One in which **science, informatics, incentives, and culture** are aligned for continuous improvement, innovation, and equity with **best practices and discovery** seamlessly embedded in the delivery process, individuals' and families' active participants in all elements, and new knowledge generated as an integral by-product of the delivery experience (The National Academies (2013), Ramar et al., 2025).

The paper describes how the Mayo Clinic has invested heavily in developing HRO capabilities by creating the *Kern Center for the Science of Health Care Delivery*, with over 150 scientists, analysts, healthcare engineers, project managers, and support staff. It purposefully brings together cross-functional experts to create and evaluate data-driven solutions that transform healthcare for patients, clinicians, and communities. The center has a diverse range of programs including applied operation research, diversity science, pragmatic trials and evaluation, human factor engineering, science of quality measurement, clinical and economic evaluation, knowledge synthesis, surgical outcome research, and applied machine intelligence. These experts work directly with the frontline practitioners to co-create solutions focused on improving quality and safety.

A recent case published by the Kern Center (cf. Ramar et al., 2025) illustrates the strong collaboration between patient safety, practice, and research centers to address recurring issues related to intravenous infusion pumps. Certain intravenous pumps were found to malfunction, with occlusion alarms resulting in medication infusion errors, leading to significant near-miss events, particularly in intensive care units and emergency departments. In response, the organization adopted a high-reliability framework to achieve consistent, high-quality performance in complex and high-risk environments, emphasizing deference to expertise, sensitivity to operations, and reluctance to simplify. Specific steps included conducting root cause analyses to identify device issues, developing targeted staff training on identifying and responding to alarm malfunctions, and implementing enhanced data tracking to monitor device performance. Multidisciplinary collaboration among nursing, quality, patient safety, healthcare technology management, and pharmacy teams ensured the ability to adapt and address vulnerabilities effectively, preventing recurrence and fostering a safer care environment for their patients.

Implication for BioPharma

A patient-focused culture in BioPharma mirrors these HRO principles, especially in terms of vigilance, escalation, and a commitment to continuously improve product and process knowledge. The HRO literature suggests that reliability and continuous improvement must be culturally embedded within the organizational mindset, top-down and bottom-up, rather than procedurally prompted through standard operating procedures (SOPs) or light-touch training programs.

The lesson here is that employee engagement and patient safety performance correlate strongly in high-stakes environments, but it must operate cross-functionally and across all levels of the hierarchy, from the shop floor or lab bench to the boardroom and all points in between.

14.4 Employee Engagement as a Driver of Enhanced Performance

So, what are we talking about when we refer to *employee engagement* and why should we care?

First and foremost, **employee engagement is a risk and innovation indicator**, not a morale score. High-engagement organizations typically show:

- Higher early risk reporting.
- Higher idea generation.
- Higher recognition frequency.
- Lower regretted employee attrition.
- Lower poor-quality recurrence.
- Higher internal mobility and career progression.

In a regulated environment, high employee engagement becomes a *risk control mechanism*:

- Engaged employees escalate deviations earlier.
- Engaged scientists challenge flawed assumptions.
- Engaged teams collaborate across silos.
- Engaged managers coach rather than command.

Low engagement, by contrast, increases silent risk accumulation. If organizations want to actively reduce risk to the business and the patient, then investment of time and resources in enhancing employee engagement is the route to sustainable success.

Employee engagement is defined as a measure of *discretionary* individual effort, commitment, and emotional investment in an organization's vision, values, and goals. So how can leaders foster this discretionary commitment?

Findings from my own research and industry experience confirm that *purpose activation drives discretionary effort*. This means ensuring that every employee, right across the product life cycle, has a direct line of sight to the patient. Connection to the patient creates purpose and strengthens emotional connection. Practical daily strategies, such as patient storytelling, product news shared at team or tier meetings, getting team members to pick a product they have worked on to share their connection to that product, lunch and learn sessions led by product owners, and after-action reviews following a customer complaint, are just a few of the ways leaders can actively sponsor and foster emotional connections.

Annual town halls with patients or patient advocacy groups are impactful, but we cannot expect to forge a deep-seated emotional connection with once-a-year events. Leaders need to emphasize the expectation, through their own daily actions and behaviors, that employees act as advocates for the patient, every batch, every day. This is good not only for the patient, but also for the business.

Studies have consistently demonstrated the relationship between employee engagement and performance outcomes. Harter et al. (2002) found that effective engagement predicts productivity, profitability, and customer loyalty across industries. However, Christian et al. (2011) further demonstrated that engagement predicts both task performance and organizational citizenship behavior. Adam Grant's research highlights that task significance (purpose), i.e., the degree to which work positively impacts others, is the primary driver of employee motivation and productivity. In *The No. 1 Feature of a Meaningless Job* (2014), Grant shows that when employees, from fundraisers to lifeguards, see the tangible, positive impact of their work on beneficiaries, they work harder, longer, and with more dedication.

Ultimately, in quality-critical environments, effective employee engagement correlates with:

- Reduced quality and safety incidents with higher reporting rates.
- Lower voluntary employee turnover, i.e., better employee retention rates.
- Higher productivity and profitability.

Visible connection between daily work and patient outcomes reinforces meaning and resilience. These findings suggest that effective employee engagement functions as a latent energy source enabling proactive behavior, innovation, risk reduction, and enhanced outcomes for the business. So, who has a vested interest in delivering these enhanced outcomes?

14.5 Leadership Behavior and Transformational Influence

Leadership behavior significantly shapes engagement and culture, more than any other factor.

Early *Transformational Leadership* theory by Bernard M. Bass identified four "I" dimensions of a transformational leader (Bass, 1985): *Idealized* influence, *Inspirational* motivation, *Intellectual* stimulation, and *Individualized* consideration. Bass proposes that leaders who use these four dimensions reduce resistance to change by fostering trust and communicating a positive, future-oriented vision.

By providing a clear vision (*inspirational motivation*), acting as role models (*idealized influence*), and challenging the status quo (*intellectual stimulation*), leaders motivate employees, creating engaging work environments that directly influence innovation, improve job performance, and increase commitment. Leaders who practice active listening (e.g., during routine GEMBAs) and demonstrate empathy (*individualized consideration*) foster trust and respect, leading to higher levels of employee attachment and commitment to the organization.

Subsequent evidence confirmed that transformational leadership is a predictor of employee satisfaction and performance (Judge & Piccolo, 2004). More recent studies on the mediating effect of organizational culture and knowledge sharing have reinforced that employees are more committed to supporting initiatives from leaders who lead by example and personally demonstrate what needs to happen (Aminah et al., 2022).

What is clear, whether from academic studies or from direct observation in the field, is that **culture is shaped most powerfully by leadership behavior**.

ISPE's *Advancing Pharmaceutical Quality (APQ) Cultural Excellence Guide* (ISPE, 2022) provides the 5V Model of Leader Influence on Culture, encouraging high-engagement leaders to consistently:

- Communicate a compelling patient-centric vision and provide clear expectations by committing to consistently act in alignment with that vision.
- Put the patient first in the hierarchy of organizational values. Prioritize decisions that best benefit the patient and emphasize the values of transparency and openness to protect the patient and the business.
- Use their voice to offer frequent, specific recognition to reinforce desired behaviors. Encourage curiosity and constructive challenges to the status quo by coaching for success rather than micromanaging milestones.
- Create "listening posts" between employees and management, through the use of routine GEMBA (a key aspect for employee engagement) to enhance visibility and enable real-time, frontline evaluation of a site or function's performance.
- Use vigilance to ensure that actions continue to be aligned with the vision and values, which is essential to stay the course and endure the ups and downs of a culture and performance improvement journey. Leaders must ensure that they are holding themselves and the rest of the organization accountable for performance outcomes and continuous improvement targets.

Within BioPharma and high-safety contexts, leadership inconsistency produces cultural fragmentation and heightened tensions between functions, which are destructive and exhausting. On the other hand, leadership behaviors that reinforce purpose, reward risk transparency, and model accountability are central to patient-focused excellence *and* business growth (See example in Table 14.1).

David Ricks stands as a rare exception because he has successfully bridged the gap between high-stakes scientific moonshots and disciplined operational execution. He has mastered the middle ground where many leaders fail during periods of rapid disruption. His success proves that transformational leadership in BioPharma requires more than just a bold vision; it demands intellectual honesty and high trust to commit existing revenue streams to future innovation, all without losing the confidence of the short-term market, where shareholders rather than patients take precedence.

Table 14.1 Transformational leadership exceptional growth

Transformational Leadership: Exceptional Growth
No transformation has been as impressive as that of Eli Lilly in recent years. David Ricks, CEO, utilized transformational leadership to transition the 150-year-old company into the world's most valuable pharmaceutical firm, primarily by dominating the GLP-1 market (Bigman, D. and Bililies, T., 2025). Named as Time CEO of the Year in 2025, a review of Ricks' success factors shows how he consistently applied the four dimensions of Bass's theory to drive this change:
1. Inspirational Motivation: The "One Billion People" Vision
Ricks redefined Lilly's mission from simply managing chronic diseases like diabetes to a generational opportunity to impact one billion people globally with obesity and related metabolic conditions. He set a bold vision to address the root of the tree of chronic disease (obesity) rather than just treating its downstream effects (heart disease, sleep apnea). This high-standard vision shifted the internal culture toward aggressive growth and massive scale.
2. Intellectual Stimulation: Disrupting the Pharma Model
Ricks challenged established industry norms, specifically regarding how patients access care and how drugs are developed. He launched LillyDirect, upending the traditional model by allowing patients to order GLP-1 medications directly from the company. He also pushed for *Day 1 Digital* integration, partnering with NVIDIA on AI to halve traditional drug development timelines. He also allegedly uses AI in every meeting to ask scientific questions, modeling the importance of continuous learning.
3. Idealized Influence: Courageous Risk-Taking
Ricks built immense credibility by taking high-stakes, ethically grounded risks that set new industry standards. He made the controversial decision to slash insulin prices to $35, a move that pressured the entire industry to follow suit, and then committed nearly 25% of sales to R&D ($14 billion), funding moonshots in Alzheimer's while running clinical trial studies in parallel to stay ahead of competitors.
4. Individualized Consideration: Empowerment and Respect
Ricks is reported to foster a culture that balances rigorous data-driven systems with deep respect for qualitative judgment and people. He maintains a hybrid decision-making model where the leadership team debates issues with an equal voice. By being a Lilly veteran of over 25 years, Ricks uses his deep institutional knowledge to mentor leaders across marketing, sales, and R&D, ensuring a culture of "vision, collaboration, and inclusivity."

14.6 Learning, Continuous Improvement and Knowledge Sharing

The concept of the learning organization, first introduced by Senge (1990), involves *entities that continuously expand their capacity to create desired outcomes and* share several cultural characteristics. They embody a shared vision, deploy systems thinking, encourage experimentation, and practice reflective dialogue afterwards to gain deeper insights and share knowledge.

More recently, Richter explored how biopharmaceutical companies can learn from their mistakes and embrace a culture of learning in order to reduce quality defects and drug shortages by deliberately and continuously innovating to improve operational performance and reduce patient risk. This research (Richter, 2024) identified that learning organizations are skilled at five main activities:

1. Systematic problem-solving.
2. Experimentation with new approaches.
3. Learning from their own experience and history.
4. Learning from the experiences and best practice of others.
5. Transferring knowledge quickly and efficiently throughout the organization.

An important lesson for BioPharma is that a patient-focused learning culture does not aim for *perfection*; it aims for *progress*.

Organizations that normalize small, daily improvements create learning opportunities which build resilience and adaptability, all essential capabilities in an industry shaped by scientific uncertainty and regulatory evolution.

For continuous improvement (CI) to become routine, leaders must set the expectation that it is a daily practice through their own active sponsorship, behaviors, and actions. In addition, employees must first feel competent, supported, and empowered to:

- Identify current inefficiencies.
- Challenge legacy processes.
- Suggest simplifications.
- Run improvement experiments, e.g., "just-do-it's" and local CI projects.
- Share lessons learned.
- Reflect, reiterate, and repeat.

And they must have the time and space to practice these skills and competencies, not just as a training exercise for a belt certificate program, but included in the flow of their daily work.

Routine CI programs should provide a practical means for teams to undertake proactive risk reduction, optimize CAPA effectiveness, engage in cross-functional problem-solving in order to create learnings from mistakes (or near misses), and facilitate knowledge sharing, with the expectation that they are seeking small improvement opportunities, *every batch, every day*.

More recent research in a UK-based research laboratory within a global high-tech corporation identified **leadership's actions and behaviors that** encourage open knowledge sharing and learning **as critical** by shaping policies, structures, and social norms that make it common practice. Practices like open communication, collaborative workspaces, and mentoring all help to build this culture. This growth mindset must be reinforced by supportive leaders and a shared belief that collaboration and learning benefit everyone (Jasimuddin & Saci, 2022).

On the other hand, in Richter's case study (cf. Richter, 2024), within an early-stage biopharmaceutical company, **an inability by leaders to provide strong sponsorship** was identified as a key barrier to learning or "creating the desired [performance] outcomes." The team observed many instances of a sponsor myopically focused on project execution, timing, and accountability to the schedule (*getting it done*) rather than on delivering the business and patient outcomes targeted by the project (*getting it right*). Richter reports that:

- The sponsor issued constant reminders during report-out presentations that the timing of the project was getting close to being overdue.
- Sponsor behaviors that were not observed included acting as an advocate or coach or using their influence to direct a positive outcome.
- The team involved in this case study provided feedback indicating that this type of sponsor behavior was a barrier to truly achieving the goal of integrating organizational learning and continuous improvement to improve organizational performance in their supply network (Richter, 2024).

From direct experience, unfortunately many sponsors and leaders can miss the opportunity to reinforce right-first-time quality and transformation by prioritizing speed and short-term schedule adherence over long-term outcomes for the business and the patient. How do we overcome this tendency to prioritize progress against short-term goals (what leaders are measured against) over longer-term outcomes?

The ISPE's Advancing Pharmaceutical Quality Assessment Guide on Cultural Excellence may offer the answer when it advises that:

> *You don't attempt to change culture—you change behaviors! It is the daily practice of the new 'desired' behaviors that changes the culture* (ISPE, 2022).

The key to transformational change that sustainably delivers performance improvement is to systematically identify, measure, and reinforce the behaviors (both leader and employee) that matter through coaching, feedback, *and* timely recognition.

14.7 Recognition: The Engagement Engine

> *Definition of Recognition: Psychological boost that lets employees know how they are doing and whether their behaviors and performance meet and exceed expectations and are valued.*

Recognition plays a pivotal role in shaping how work gets done and how people feel at work. When thoughtfully designed, it highlights the behaviors, results, and values that matter most to the organization, clearly aligning day-to-day actions with strategic goals. At the same time, genuine recognition, expressed through timely praise, visible appreciation, and meaningful acknowledgement, can meet core human needs for competence, progress, and belonging.

In such environments, employees feel recognized, celebrated, and valued, which in turn encourages them to go beyond minimum requirements and offer the elusive discretionary effort needed for active engagement. This creates a positive performance spiral, called *the recognition effect*: good performance is noticed and rewarded; employees feel valued, and they respond with even higher levels of quality, initiative, and ownership.

Bradley notes that in 2026, the companies that get recognition right will have a distinct competitive edge: higher retention, stronger culture, and employees who genuinely want to contribute. However, he also warns that recognition does not exist in a vacuum. Its impact grows when it is shared. Public recognition (whether in team meetings, Slack channels, or company feeds) activates the *social proof effect*: when others see positive behavior being celebrated, they are more likely to emulate it (Bradely, 2025).

Remember, recognition is not about perks or praise; it is about psychology. It is about understanding (by asking them and actively listening) what truly makes people feel valued, connected, and motivated. Employees no longer want generic rewards; they want meaning, connection, and authenticity. The formula is simple:

- Make it meaningful.
- Make it frequent.
- Make it visible.
- Make it easy for everyone to participate.
- Make sure that it is directly linked to your organizational values.

Recognition is best when given to employees of varying levels of performance when they **demonstrate desired behaviors**. It is ideally given in **real time to reinforce** the qualities and behaviors demonstrated to achieve a given result. It is a good lever for **retaining talent** but can be misused when only given to top performers or only when the result is good for the company and not for the individual (Saunderson, 2021).

Saunderson also cautions that although the two terms are often used interchangeably, **recognition** is not the same as **reward**, as they have entirely different purposes and different effects.

> *Definition of Rewards: Associated with objectives or target outcomes and are usually tangible and transactional. Used for attracting talent, encouraging productivity, or acknowledging tenure or service.*

Rewards refer to programs that measure performance and aim to motivate employees on either an individual or a group level. They are normally considered separate from salary but might be monetary in nature or otherwise have a financial aspect tied to them.

Recognition programs are intended to provide a psychological boost by focusing praise on the employee and their work or accomplishments, rather than on a physical prize for those accomplishments. In designing your program, you may wish to consider Table 14.2 to maximize the effects.

Ultimately, recognition is the engine of employee engagement, reinforcing the sense that “I matter here” and “my work has impact,” strengthening emotional connection to the organization. When done right, it builds loyalty and reduces turnover, as employees who feel valued are less likely to leave, especially high performers. A culture of recognition supports reputation, improving the employer’s brand, which helps to attract talent in tight workforce markets like BioPharma. If all that is not enough to convince you to actively recognize effort on a daily basis, remember … it is free!

Table 14.2 10 Distinctions between reward and recognition (Saunderson, 2021)

10 Distinctions between reward and recognition (Saunderson, 2021)	
Rewards	Recognition
Rewards are tangible.	Recognition is intangible
Rewards will always be transactional	Recognition should always be relational
Rewards are simply consumed	Recognition is mostly experienced
Rewards are transferable	Recognition is nontransferable
Rewards are certainly conditional	Recognition happens to be unconditional
Rewards are expected	Recognition is a surprise
Rewards are economical	Recognition is emotional
Rewards are outcome driven	Recognition is focused on behaviors
Rewards are fixed	Recognition is flowing
Rewards can be impersonal	Recognition is personal

14.8 Integrating Behaviors, Engagement, and Patient Outcomes by Measuring What Matters

When governance systems integrate engagement metrics with quality and retention indicators, culture (through behaviors) becomes measurable and manageable.

When measuring culture through behavior, be careful not to rely on sentiment alone. Annual employee engagement or periodic pulse surveys are important, but culture becomes visible through behaviors (what we say and what we do), which are tangible, observable, and therefore quantifiable.

Most corporate dashboard and metric reports are built on Key Performance Indicators (KPIs), traditionally set by global or site leadership to chart progress on strategic goals and objectives. By their nature, KPIs are typically lagging in nature and are historical measures of performance, reported *after* the facts or data has been evaluated over a given period (monthly, quarterly, annually). They reflect a result (outcome) and offer little opportunity to *recalibrate* or *course correct* in real time.

On the other hand, a leading indicator can be defined as a performance measurement that occurs *before* a process begins, often measures behaviors, and is predictive, as behaviors are typically precursors of results. Key Behavioral Indicators (KBIs) are far more responsive and can be quantified in much shorter timeframes, i.e., hours, days, and weeks, allowing for actions to be taken to adjust the course of travel in real time, at a local level. As KBIs count actions, they are typically observed, measured, and reported at a local level, e.g., at local tier boards, where progress is visible and reflection on what is working well or not can be shared and discussed with the team directly.

The premise is that when KBIs are intentionally designed by the team, for the team, and tracked on a daily/weekly basis, they will positively impact the long-term delivery of the targets set by the associated KPI. For those wishing to learn more about a practical tool (called the IMPACT tool) on how to derive KBIs from leader-led KPIs, please see the ISPE's APQ Cultural Excellence Assessment Guide (ISPE, 2022). Below are some leading Key Behavioral Indicators to consider:

Engagement and Quality/Safety

- Near-miss reporting rates.
- Proactive risk identification.
- Early deviation identification.
- Participation in reflective practices, after-action reviews, lessons learned, mistake-proofing activities.
- Issue escalation velocity.
- Peer-to-peer recognitions given in a period.

Continuous Improvement

- Employee improvement ideas per FTE.
- Improvement suggestion-to-implementation ratio.
- Ratio of corrective actions (CAs)/preventive actions (PAs).
- CAPA initiation timeliness (e.g., number commenced within the first 48 hours).
- Improvement in RFT outcomes for critical processes/products.

Leadership Activation

- GEMBA activities completed in the period.
- Number of improvement opportunities identified on GEMBA.
- Coaching conversations completed in the period.
- Leader recognition number and distribution (formal and informal).
- Team-level engagement variance (to identify best practice sharing opportunities between teams).

Commitment and Retention

- Voluntary attrition of critical team members.
- Internal mobility rate.
- High-performer retention rate.
- Number of succession planning and personal development conversations completed with direct reports.

In high-performing organizations, engagement data is not just tracked by HR. Its impact is integrated into the quality metrics, safety metrics, innovation or new product introduction (NPI) launch metrics, and employee attrition metrics, and in this way, culture becomes measurable, visible, and actionable.

14.9 Conclusion

Culture must be treated as an operational determinant, not a campaign.

Culture functions as *the* critical risk control strategy for BioPharma organizations and should be considered as a key business capability, not an optional extra. It shapes whether employees escalate concerns, collaborate across silos, and continuously improve processes. Academic research has consistently demonstrated that psychological safety, engagement, transformational leadership, and learning orientation predict performance and reliability in complex systems.

Patient-focused excellence, therefore, is not solely a strategic aspiration. It is the outcome of disciplined value-led design, evidence-based leadership, and systematic reinforcement of behaviors that protect and serve patients and deliver enhanced business performance.

Organizations that operationalize culture through behavioral metrics and accountability mechanisms are more likely to achieve sustained patient and regulator trust, which in turn will drive shareholder value and business growth. The reverse is not always true; organizations that put shareholder and market satisfaction before patient trust will struggle to foster a culture of excellence, retain talent, and learn from its mistakes. The transformation starts with you.

References

Aminah, H., Lin, P.-K., Susita, D., Helexandra, L., & Moslehpour, M. (2022). How does servant leadership affect public employees organizational citizenship behavior? The mediation effect of organizational culture and knowledge sharing. *International Journal of eBusiness and eGovernment Studies, 14*, 361–387.

Bass, B. M. (1985). *Leadership and performance beyond expectations*. Free Press.

Bigman, D., & Bililies, T., CEO of the Year: A Conversation with Eli Lilly's Dave Ricks *Chief Executive*, 21 Oct. 2025.

Bradely, J. (2025). *The Psychology of Recognition: What Actually Works in 2025/2026*, Juno, 25 October 2025. Accessed at https://www.withjuno.com/blog/the-psychology-of-recognition-what-actually-works-in-2025-2026, October 8, 2025.

Christian, M. S., Garza, A. S., & Slaughter, J. E. (2011). Work engagement: A quantitative review and test of its relations with task and contextual performance. *Personnel Psychology, 64*(1), 89–136.

Edmondson, A. (1999). Psychological safety and learning behavior in work teams. *Administrative Science Quarterly, 44*(2), 350–383.

Grant, A. M. (2014). The No. 1 Feature of a Meaningless Job. *Psychology Today*, Posted 30 January 2014.

Harter, J. K., Schmidt, F. L., & Hayes, T. L. (2002). Business-unit-level relationship between employee satisfaction, engagement, and business outcomes: A meta-analysis. *Journal of Applied Psychology, 87*(2), 268–279.

ISPE. (2022). *Advancing Pharmaceutical Quality (APQ) Cultural Excellence Guide*. ISBN: 978-1-946964-59-5.

Jasimuddin, S. M., & Saci, F. (2022). Creating a culture to avoid knowledge hiding within an organization: The role of management support. *Frontiers in Psychology, 13*, 850989. https://doi.org/10.3389/fpsyg.2022.850989

Judge, T. A., & Piccolo, R. F. (2004). Transformational and transactional leadership: A meta-analytic test. *Journal of Applied Psychology, 89*(5), 755–768.

Ramar, K., Oxentenko, A., & Dowdy, S. (2025). Transforming health care through quality and safety. *Mayo Clinic Proceedings., 100*(8), 1385–1401. https://doi.org/10.1016/j.mayocp.2025.04.012

Richter, L. (2024). *Enhancing organisational performance through patient-focused learning excellence in the biopharmaceutical industry*. Taken from https://arrow.tudublin.ie/context/schschpsdis/article/1004/viewcontent/Enhancing_organisational_performance_through_patient_focused.pdf

Saunderson, R. (2021). North Star Meetings Group, Incentive post, *Top 10 Differences Between Employee Rewards and Recognition*, October 2021, access at: https://www.north-

starmeetingsgroup.com/Incentive/Strategy/Top-10-Differences-Between-Rewards-and-Recognition?inc=1
Senge, P. M. (1990). *The fifth discipline: The art and practice of the learning organization*. Doubleday.
The National Academies Press Institute of Medicine. *Best Care at Lower Cost: The Path to Continuously Learning Health Care in America*. Washington, DC; 2013. https://doi.org/10.17226/13444
Weick, K. E., & Sutcliffe, K. M. (2001). *Managing the unexpected: Assuring high performance in an age of complexity*. Jossey-Bass.

Linking Culture and Behaviours to the Essence of Work

15

John Quirke

Abstract

Embedding a lasting culture of excellence in any organisation depends on deliberately shaping everyday behaviours and designing supportive business systems. Many organisations struggle with cultural transformation because what they say does not match what employees do, their systems make the right behaviours hard or leadership is not effective. Excellence happens when everyone knows what ideal behaviours look Ike, and these are supported, encouraged and practiced consistently by all—from the shop floor to senior leadership. Leaders should build environments that encourage curiosity and focus on processes and ongoing improvement while addressing any issues openly and respectfully. It is vital to involve all levels of the organisation, especially middle managers, and make sure that business systems help people succeed. Lasting excellence requires what is called 'transformative participation'—when everyone understands the organisation's goals and works together to embody shared values through strong systems and purposeful leadership. Rather than presenting a theoretical model, the content of this chapter relies on observed behaviours, lessons learned and concrete recommendations, making the path to a culture of excellence relatable and actionable.

The team at S A Partners have supported businesses across sectors and across continents helping these organisations unlock the excellence that is already there. During our consulting work, we see varying approaches to defining and developing

J. Quirke (✉)
S A Partners LLP, Newport, Wales
e-mail: john.quirke@sapartners.com

T. Friedli et al. (eds.), *Striving for Operational Excellence in the Pharmaceutical Industry*, Future of Business and Finance,
https://doi.org/10.1007/978-3-032-26287-5_15

culture and values within organisations. Some organisations, dissatisfied with current levels of performance, embark on programs of *'cultural transformation'* which focus on the development of a sustainable *'culture of excellence'*, *'operational excellence'* or *'quality culture'*. While some organisations achieve measurable success and some external recognition, sadly many of these efforts lose momentum allowing the status quo to re-establish itself.

There is, we believe, a need to refocus on what excellence really is and how that excellence becomes embedded into an organisation's culture in a way that is genuinely 'lived' and positively impactful. To begin, it is worth reflecting on what organisational culture is and how it links to the tangible behaviours that make the difference between successful and suboptimal outcomes.

Culture, good or bad, in an organisation is the accumulation of observable behaviours within that organisation. They are the actions we see happening as employees and leaders go about their work. The output or result of any work activity, good or bad, is therefore dependent or the expressed behaviours of those individuals are directly involved with that work activity. We define a good result as an outcome delivered with the minimal necessary effort and resources needed to deliver that result. You may eventually get a 'good result' following rework or retesting, but this is not a sustainable excellent result. It is a suboptimal result that weighs heavy on our pressing environmental and sustainability obligations.

An ideal culture, therefore, is one that is defined by intentional behaviours that deliver optimal and ever-improving results for the organisation.

In our practice, we see three main reasons for failure in efforts to develop and sustain the necessary culture to support business strategy and ongoing improvement in performance.

- Failure to align cultural aspirations to the actual behaviours that gets work done in the best way.
- Business systems, which instead of making it easier to do the right thing, often through complexity, poor design or lack of system improvement, create conditions and work practices that drive the wrong behaviours and the wrong outcome.
- Leaders lack the sensitivity and/or skill in fostering behaviours that make a real difference to performance and engagement, with the focus being on 'frontline' behaviours rather than the ability of leadership to support and encourage these behaviours.

15.1 Linking Culture to the Work That Must Be Done

As mentioned, most organisations spend time defining culture and values for their teams. However, in far too many cases, these values tend to be for external audience rather than internal consumption and focus. The corporate values may appear on websites or corporate reports, but they have little real day-to-day impact on how the work of the organisation at every level is done.

By looking out rather than looking in, organisations fail to link the desired culture to the work that is done. This not only is the direct operational work, but also includes support services, maintenance, sales, engineering, research, product development and customer interactions.

So, what are the behaviours needed to deliver a good outcome from work? Some might argue, particularly in pharmaceutical and life science sectors, that getting a 'good result' is dependent on following the agreed '*Standard Operating Procedure (SOP)*'. But what if the SOP is overcomplex and does not reference the actual detail and nuances of work activity effectively? We may also look for and measure *'good catches'* where employees observe safety or quality concerns or *'suggestions'* where employees put forward ideas for improvement. Or maybe we measure the behaviours of leaders. Has the leader completed their 'go see' or 'gemba' activity this week/month?

While these behaviours are certainly part of what could be considered a 'good culture', they fall far short of a pervasive culture of excellence which enables optimal results. To explain the necessary depth and focus of behaviours needed, let me share a story that my colleague and I experienced while working in a European capital not so long ago.

A colleague and I were walking from a restaurant one evening following a good day with a client. The food and service at the restaurant were excellent. The pricing was not too high, but we both agreed that the experience had been exceptional. Being consultants on the road, we are lucky in that we visit many restaurants and many hotels in various parts of the world, yet we both agreed that this was a special experience.

Being interested in behaviours and culture, we reviewed what we had observed and what had made it such a memorable experience. The restaurant was recommended to us by the client. We rang earlier in the day to book a table. On that call, we were asked in perfect English for our preference for location in the restaurant, outside non-smoking, outside smoking, inside by the window or just inside. As this was our first time in this city and it was a summer's evening, we opted for outside non-smoking. We were also asked if we had any specific dietary requirements. After a brief and pleasant conversation, no background noise and no rushing off the phone, all was good, and the table was booked.

On our arrival, we were greeted by a friendly hostess. She quickly looked at the booking. *'Ah yes outside seating non-smoking, and I see this is your first time in our beautiful city!'* Her tone was engaging and genuine. The conversation continued as we walked to the table, *'Where did we plan to visit and what we had seen since our arrival?'* Our hostess was shocked at our usual excuse of '*Sorry we have no time. We are on a business trip*'. *'But you must see something!'* She persisted, and after asking where we were staying, she suggested some interesting sites in walking distance from our hotel.

On presenting our menus, our hostess proceeded to describe the menu and her personal favourites and what was new to the menu and the specials for the day. But she also explained the local origin of the vegetables and meats and that local organic sourcing was a basis for the restaurant.

As we reviewed the menu, we overheard similar conversations with other guests by other members of the waiting staff. The restaurant was filling up, yet each new group was met by the same knowledge of their stay or the fact that they had visited the restaurant previously. Each guest group was given the same detailed review of the menu and the sources of the food. We also overheard a detailed conversation around allergies and what was suitable and what was not.

As we ate, we noted the continuation of the behaviours we had experienced from our server with other servers and their guests. We also noted and remarked on how well the staff got on with each other.

Throughout the evening, our food was excellent with service that flowed seamlessly. As each course was cleared, there was a pleasant exchange each time which focused on how a particular item on the menu tasted. And here again, phrases like *'Yes, it is just the right season for that vegetable'* or *'Yes on that farm they really care for the ducks, and it shows in the quality of the meat'* were heard.

One could be cynical and feel that all this '*niceness*' and '*foodiness*' were in place by staff to ensure a nice tip! But when we went to pay, while we offered a tip, we were met by a 'Thank you, but we have a no tip policy here'. *'Our food and our staff speak for themselves. We are happy you enjoyed the service. Please recommend us to your friends'.*

As we walked back to our hotel, my colleague and I reflected on our experience.

We observed fantastic behaviours from a group of employees who delivered excellent service. But we recognised that we had only seen a small snippet of the ideal behaviours that must be in place throughout the organisation to enable such an excellent and holistic customer experience.

Who selected and hired the staff? What criteria were set for joining what was clearly a high-functioning team? Who trained the staff, and how were they trained? And how was that training updated based on new menus and sources of ingredients?

The quality of the food was exceptional. The individual sourcing the food must have had an in-depth knowledge of the local organic food supply network. They must have had a deep interest in food and pride in the fact that it is locally sourced. These buyers most likely passed this enthusiasm and knowledge to the waiting staff with the expectation that this is an important conversation to have with customers.

The food presentation was exceptional, and the consistency we observed would require a high level of control and clear standards. These standards must undergo constant revision and update as recipes and serving standards changed. The systems worked so well for the staff, and they were so engaged in the activity that our belief was that they must have some input into developing standards within the restaurant.

From our experience and our observations, it was clear that ideal behaviours were intentionally integrated into how the work works at the restaurant. From our limited observations, it was also clear that business systems within the restaurant were intentionally designed to deliver the essence of this organisation: providing exceptional food in an exceptional way at a good price point while supporting local organic food growers.

System, we could point to, could include the training system, the food procurement system, the new meal development system, and the documentation and standards system. Even the booking system allowed for entry of information that allows staff to have an initial engagement with their guests like *'So, I see this is your first time in our beautiful city'*, allowing staff to quickly connect at a personal level with the restaurant clients.

The engagement communication and leadership systems were focused on ensuring that staff were aware that they were part of something bigger, not just a restaurant, but a food movement, a city or a destination, and that the staff made it happen. As an organisation, this allowed the team to focus on a purpose and align behaviours, improvement activity and business systems to that purpose.

The result is a reasonably priced exceptional meal in a wonderful atmosphere of efficiency and friendliness.

15.2 The Essence of What You Do

In business and particularly in the pharmaceutical and life science areas, successful outcomes depend on a lot more than the technical aspect of the drug manufacturing or device assembly processes. How the product is stored, packaged and tested and the quality and accuracy of paperwork and records associated with the maintenance of equipment and environment in which the product was produced, all have a potential to impact product supply. Records of training and the skills of the individuals involved in the process if incorrect or neglected can also cause the delay or loss of product batches. The ultimate success or failure of the outcome from each of these activities is dependent on the expression of a human behaviour at a point in time.

Both the FDA and European Medicines Agency regularly report on common failures within pharmaceutical manufacturing and packaging operations. The annual report of the European Medicines Agency (2024) captures the top reasons for the recall of centrally authorised products as follows:

Manufacturing laboratory control issues: These include out-of-specification results obtained during quality control testing.

Product contamination and sterility issues: These include chemical, microbiological or physical contamination of the medicinal product.

Product label issues: These include issues related to labelling of the medicinal products (e.g. a missing or incorrect batch number).

Product packaging issues: These relate to physical issues (e.g. a mix-up or a damaged container).

Product physical issues: These relate to incorrect product physical properties (e.g. friability, size/shape, leakage).

Pharmaceutical and medical device manufacturers, while operating in apparently highly controlled environments, are regularly *'caught on the hook'* over what could be considered basic oversights. Putting in another way, these issues are often caused by the absence of the appropriate behaviour at a point in time.

If we really want to focus on the true culture of excellence, then we must have a hard conversation around the actual behaviours that enable the results we require. It is not enough to simply refer to the instances listed in the European Medicines

Agency report as 'human error'. To do so only perpetuates the current cycle of firefighting and compliance complexity experienced by many life science organisations today.

This time, let us take an imaginary example that is close to a personal experience observed within a pharmaceutical manufacturing operation.

Many modern biopharmaceutical processes rely on a separation step involving large-scale column chromatography, the basis of which is a chromatography substrate consisting of activated resin bead separation process. The quality and consistency of column packing with this substrate can have a major impact on product quality and process efficiency.

In our example, despite the presence of highly automated process control, column packing remains a manual activity where operators decant active resin into a prepared column and pack it in preparation for initial activation. Resin is supplied in fifty litre sealed drums. Operators open the drums under cleanroom conditions and decant the contents into a stainless steel column. The material arrives with testing and functionality certificates from the supplier. These are reviewed by incoming inspection and quality control prior to release into production. The final product arising from the process is currently on allocation, and every batch is critical. These are lifesaving drugs!

On this occasion, the operator after opening the container noted that the top surface of the resin in the container was at an angle. It was as if the container had been stored on its side for some time rather than standing on its base. Our operator has not seen this before. The operator now has an option as regards the expression of a behaviour. Some questions would go through his mind.

> *'The paperwork and testing results are OK so that looks fine. The batch number must be OK, or I would not have been able to scan it into the batch record.*
>
> *Is this unusual enough to warrant stopping what I am doing and escalating this to my supervisor? I will need to stop the process, and I am almost finished my shift. That will put us under pressure to meet the numbers this week. I will have to raise a deviation and get involved in that process. It will be a lot of paperwork and hassle!'*

Of course, we are assuming that our operator notices the subtle difference in the consistency of this raw material in the first place. We must acknowledge that in many cases, this behaviour may not even be present!

If we continue our example based on the operator noticing the difference in the material, there is a choice. Stop activity, and inform the supervisor or continue based on evidence from testing and what the procedure says is acceptable.

Let us assume that the operator is experienced enough to know that there is something unusual here and the decision is taken to inform the supervisor.

Now the choices are with the supervisor who will no doubt feel the pressure of results and 'meeting the numbers' greater than the operator. What are the expressed behaviours available to the supervisor?

Option one:

The supervisor tells the operator that the issue sounds fine. QCs have passed the raw material as good, and we cannot second-guess the incoming testing team.

Without observing the issue directly, the supervisor tells the operator to proceed with production.

Option two:
The supervisor listens to the operator carefully knowing how experienced the operator is. The supervisor goes to the shop floor and observes what the operator observes and reviews the data on testing. In conjunction with the operator the supervisor begins to assess the situation. How many barrels look like this? Is it only this batch number? Can we check other material in the warehouse? Based on these observations the supervisor with the input of the operator makes a call.

Like any good TV series, we will leave it hanging there! The outcome and ultimate decision we would hope would be better based on the expressed behaviours of option two.

Option one may have worked out OK in the end. But what are the messages the operator is getting in our two scenarios above? Which expressed behaviour from the supervisor is better for the long-term performance of the business?

In option one, the operator's curiosity and process experience are ignored. If the behaviours in option one continue over time, the operator will 'shut down'. Over time, the operator's discretionary effort and curiosity will be lost.

Also, we must wonder if the outcome of option one was bad, what would be the consequences within the team? Who could or would end up with the blame? Would there be a detailed enough root cause investigation? How would that investigation proceed, and who would be involved?

In option two, the curiosity and process focus of the operator are rewarded by the interest and activity of the supervisor. Both support further investigation, and both are involved in the final decision on whether the issue should stop production and the suspect material be put aside for further examination. If a problem was avoided due to the observation of the operator, an ideal observable behaviour would see the supervisor having no hesitation in celebrating the operator's proactive behaviour and process focus and recommending some form of recognition for the operator. However, even if ultimately there was no issue, the supervisor would still thank the operator for his diligence and process focus.

We have positioned two scenarios above which could be described a dominant organisational culture. Scenario one describes culture that we can describe here as type one. In scenario two, we will call culture type two. It could be argued that culture one is a result of poor leadership from an individual supervisor. Whereas in the case of culture two, we have a 'good' supervisor who is a 'good' leader.

Regardless of the outcome, the behaviours in culture type two are far closer to ideal than culture type one.

Maybe we were unlucky to have a bad leader in culture type one and lucky to have a good leader in scenario two.

The quality and impact of leadership within the critical environment of a pharmaceutical business cannot be left to luck. Leadership, good or bad, should not depend on a lucky or unlucky hiring process. The chance that we have the right leader in the right place at the right time will not sustain any attempt at creating a sustainable culture of excellence. Good leadership is an intentional focus of a business as is the detection and management of poor leadership.

So, let us assume that the culture and behaviour of the leader in culture type two are intentional. How would that happen? As we described in our restaurant example, the hiring process would be designed to find and select for the qualities of leadership in culture type two. During early induction process, clear expectations would be set as regards the type of leadership and the need to constantly respect and develop the individuals in their team. Phrases such as 'leaders don't own the problem, but they do own the quality of conversation, skills and behaviours around the problem' would resonate and be constantly reinforced and supported by the leader of the leader.

The development of coaching and listening skills would be a critical part in the leader's development. Internal peer-to-peer coaching would be active and valued. It is acknowledged that coaching and listening are not an activity preserved for the annual review process. Instead, it is viewed as an integral part of daily interactions, tier meetings, problem-solving and improvement activity.

The leader of the leader would regularly check in as a supportive coach to the leader. Humble intentional servant leadership is experienced by the leader from their leader. There is an expectation that this is how we lead. Not only would the operator be recognised and celebrated for the behaviours they expressed in scenario two. The leader's leader would recognise the fact that the operator's leader seized the coaching moment to reinforce respect for the operator while fostering curiosity, process focus and application of scientific thinking as the operator and his peers investigated the extent and possible cause of the issue.

Unfortunately, based on the evidence from reports such as the European Medicines Agency and similar summary reports from the FDA and our own experience working across a wide range of pharmaceutical and life science organisations, the ideal behaviours highlighted in culture type two are not expressed as frequently as they need to be.

Due to the highly automated nature of modern manufacturing processes, there is a real need to foster behaviours based on curiosity, process focus and constant learning. Every manufacturing process and the enabling activities prior to and following the process contain opportunities for variation that impacts the overall result of the activity. We often speak to the need to listen to the voice of the process. But it is often the whispers and murmurs of the process that give us the early indications that something is not as it should be. These whispers and murmurs tend to be outside anything an SOP could capture. Curiosity and process focus do not necessarily mean constant change and revalidation. Behaviours associated with curiosity and process focus must be balanced with caution, critical thinking and scientific approach.

15.3 Ideal Versus Not Ideal Behaviours

When it comes to identifying what are the behaviours needed for success, it is necessary to have conversations around behaviours that are not ideal. In our scenario above, the operator saying nothing about his concerns is not ideal. The behaviour of the supervisor in culture type one is not ideal. Both will result in possible consequences for the business. However, the behaviours of the supervisor will determine the real and 'felt' culture within the organisation.

Each department in an organisation must define the actual behaviours that are needed and those that must be avoided if the group is going to produce optimal work. If an issue arises, it may indicate the absence of an ideal behaviour or the need to reaffirm what is needed for success. But the behaviours must be associated with the work of that specific team. Telling a group how they should behave is a parent-child conversation. It will shut people down and generate resistance.

Asking what people need and respecting what is said is an adult-to-adult conversation, which engages teams in the detail of the work they do. They are the experts, they know. Sure, thinking and positioning can be challenged and clarified, but it must form part of a two-way respectful conversation held between adults. Maintenance tools clean and in the right place. Report information that is accurate and on time so that necessary decision can be made. Inspections should be carried out with a focus on the process rather than a focus on the check sheet. There should be sensitivity to what is not right or out of place.

It is up to the leaders of the group to listen for signs and signals in relation to behaviours. Is workload being shared fairly, and is the full team present when they need to be? Are there hints of inappropriate behaviour, bullying, isolation or favouritism? While often these issues are not mentioned outright, careful questioning and direct observation may draw them out. If such poor behaviours are present, it is down to the leadership community to address them. If they do not, leaders will lose the respect of the team.

Conversations within organisations around necessary or ideal behaviours can sometimes feel overwhelming. Leaders experience despair when they realise the thousands of potential ideal behaviours that are needed to deliver a consistent optimal result. But the presence of multiple ideal behaviours is exactly what is necessary to deliver consistent excellence. A culture of excellence is built upon thousands of observable 'micro' or 'atomic' behaviours that are expressed at a point in time and ensure a consistent optimal result. All as basic as *'will I stop the process and inform my supervisor?'*, which accumulates into a palpable 'felt' culture of sustainable enterprise excellence.

In many cases, these atomic behaviours can be aligned under umbrella value statements, but the detail of the micro behaviours necessary to enable 'good work' must not be lost. Discussions around behaviours that matter must be part of regular team review meetings to highlight what is working and what may need to be improved.

The good news is that in most cases, the frontline teams are fully capable of defining and managing the necessary behaviours to ensure success.

Leaders need to acknowledge two important elements of their role in supporting these behaviours:

1. A consistent focus on the quality of the way work is done in the business and the ability of business systems to support that work.

2. The quality of their leadership and their ability to ensure an environment where these ideal behaviours can be expressed and encouraged.

15.4 Are Business Systems Enabling or Disabling the Behaviours You Need?

When there is a mention of a business system, the default in many groups is to think of various IT systems. However, every business is made up of systems that must work together to enable the overall purpose of the organisation.

Organisations often define and manage certain systems, but others perhaps do not fit comfortably into the normal consideration of a system. Training, maintenance, quality management, manufacturing and planning may be familiar as a system. However, communication, recognition, engagement, leadership development, improvement and strategy deployment may not naturally fit into the category of a business system. Yet, these systems are as fundamental to aligning and managing the culture and behaviour to deliver a desired result, as are the manufacturing and quality systems.

Our business systems either enable or disable the desired behaviours we seek. Take for example training. What is the desired outcome of any training activity? The development of a skill? Awareness of a necessary procedure or way of doing things, e.g. raising a corrective action or filling in a request for holidays? Or retraining in the event of a change or update to a work activity or an admin process?

The hope is that any training activity has a real value and purpose for the trainee and allows them to participate fully and safely within the activities of the organisation. Training activity should therefore be presented in a way and at a time that works and does not become burdensome and valueless.

Unfortunately, in many life science and pharmaceutical companies, the weight of compliance results in an avalanche of training and retraining for the average employee. In a sample survey carried out during our consulting activity, when employees were asked how much of the actual training hours they logged related directly to the work they did, the figure was less than 10%. Employees were left to wade through and sign off on procedures that were unrelated or only remotely related to their work.

So, what is the behavioural impact of the system here? Basic good manufacturing practice teaches us that our signature is our personal bond. It confirms that a task has been carried out or that we have read and understood a document or procedure. Yet, due to the amount and make-up of much of the training burden and the fact that it is flagged to the 'boss' that training is overdue, the tendency is to basically glance over the procedure, and if there are knowledge questions, we guess them or keep trying until we get the right answer!

The system through poor design and management is forcing individuals to fake their knowledge and understanding. The system is teaching, enabling and rewarding the wrong behaviours.

15.5 Developing the Quality of Leadership for a Culture of Excellence

The ideal scenario to allow organisations cope with the ever-changing business landscape (along with the drive to constantly challenge and improve on the status quo) is the need to develop a genuine deep-rooted culture of excellence that is supported by business systems and good leadership.

The term transformative participation has been used in recent years in reference to the need to actively involve and engage minority groups in the design and implementation of legislative reform to protect vulnerable habitats around the world. To engage these indigenous groups, their real needs and day-to-day interactions with their environment must be clearly understood. Their voices and contributions must be included in any proposed changes to how a habitat is managed and protected. Transformative participation has also become a key element in linking disinterested public in the political mechanisms that influence their lives or engaging younger generations in national policy and environmental justice.

Adopting a similar approach within business, transformative participation creates a cultural expectation on those who join the business and those who work, manage and lead the business. It is a culture that is expressed in aligned and focused continual improvement across the organisation, a relentless call to action to make tomorrow better than today in all aspects of the way work is executed, resources are consumed and goals are achieved.

But true transformative participation is fundamentally rooted in the quality and consistency of leadership across an organisation. Developing leadership skills and measuring the quality and impact of those skills are the basis of a culture of excellence.

15.6 Senior Leaders Must Practice What They Intend to Preach!

Given that the concept behind transformative participation applies to the whole business, senior leadership must consider the behaviours **they** need as **a leadership team** to ensure that **they** work effectively together to support the desired culture (see Fig. 15.1).

Starting with a list of current not ideal behaviours is a sobering moment and allows the leadership team to speak openly on the personal and organisational impact of these suboptimal behaviours. Flipping a list of not ideal behaviours into a list of 'draft ideal behaviours or commitments to each other is a powerful bonding activity. The team must be prepared to apply and review their adherence to and hold each other accountable to them. Together, they practice as a team in the work that they do together.

One leadership team we worked with admitted that '*too much stuff was not talked about*'. As a team, they were not having the '*right conversation*' at the right time. One of their adopted behaviours was simply …

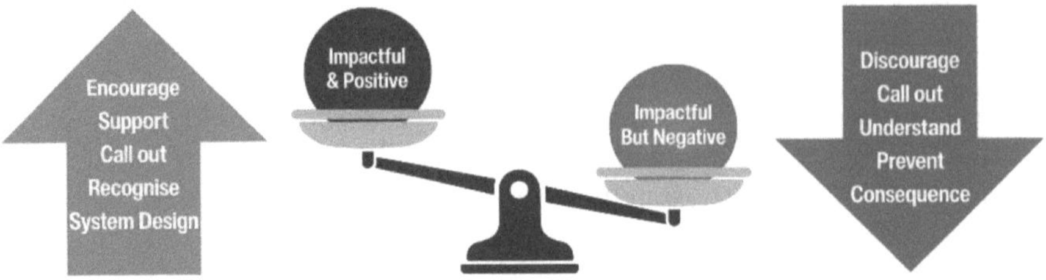

Fig. 15.1 Key role of leaders is to build the skill and confidence to support positive behaviours and manage inappropriate ones

'Let's ensure we have the right conversations at the right time'.

So, what did this mean in practice?

Members of the leadership team needed to ensure that they had data and facts to support a position. If data or information were not available, the conversation would be suspended until more in-depth information was available.

Uneasy silences were noted and 'poked' to get to the bottom of any concerns.

Unanimous approval and head nodding were followed by the question: '*Are we sure we all understand the full picture here*?'

If an issue was raised late or in the wrong forum, the question was asked: '*What stopped us talking about or being aware of this when we were in the tier two meeting?*'

At the end of each meeting, the question was asked: '*Is there anything left unsaid?*'

While visibly awkward at first, the leaders soon realised how important for themselves personally, and as a team, it was to ensure that the right conversation was happening at the right time. It began to significantly improve communication amongst themselves. It also significantly improved the alignment and prioritisation of activity between departments. There was no space for a misunderstood priority. If priorities were unclear, why did we not have the right conversation?

While these draft behaviours were not deployed to the wider organisation, the leadership team soon began to see the impact of their behavioural standards on their own teams. The leadership team met regularly to review and share their learnings and support and coach each other. On occasion, a leader might admit that they had not handled a situation with their team the way he or she would have liked to, given the behaviours the team were aspiring to achieve. Through peer-to-peer conversations and coaching, the leader could replay conversations and 'practice' a new way asking questions and interacting more effectively with their teams.

Being part of a leadership team can be a lonely and stressful place. It is a fantastic thing to observe a leadership team begin to help and guide each other into a new way of working together.

As the leadership teams develop their 'behavioural muscles', there can be a rush to engage the wider team in conversations around current state behaviours and culture. Unfortunately, in many situations, enthusiastic leadership teams begin to dive straight down to the shop floor completely bypassing the middle management group. Over-enthusiasm can lead teams straight into the minefield that is *'leader standard work'*. This is a mistake. First, it must be recognised who the real influences of organisational behaviour are.

15.7 The Meat in the Sandwich: Middle Management

We often hear discussions about the 'squeezed middle layers'. In some bizarre situations, all the woes of the business seem to be due to the lack of management and leadership skill or alignment at the middle management levels. Leadership teams can often refer to a consistent lack of ownership and accountability in this group: *'They just don't seem to be engaged or take ownership for the current state'*.

The middle management teams did not suddenly appear in the organisation. The issues or problems and their links to behaviours and perceived poor attitudes again did not suddenly appear. They have developed over time or may have always been there. Now unfortunately, they are part of how the 'work works around here'. If we want to see greater ownership and accountability, we must first create the desire and interest to do so.

Middle managers, supervisors and line leaders, through their direct and consistent contact with frontline teams, have the greatest impact on the felt culture in which daily work takes place. They are the engine of any cultural transformation.

To begin to have a real impact on improving the behaviours around the work of an organisation will need to engage the engine that is the middle management community. We do this in a similar way to the conversation we had earlier with the senior teams. But as we are often dealing with a larger cohort of individuals, we may need to break the activity out after an initial positioning session led by site senior leadership into smaller working groups.

The organisation must be prepared for some difficult conversations.

> We may for example make statements as to the importance of safety. But the reality on the shop floor is that there are lots of safety issues that have been highlighted but not addressed effectively.
>
> We may make statements about respect and the importance of inclusion within our workplace, yet there are obvious behaviours among leaders and managers and some teams that do not reflect the value of respect and inclusion.
>
> We may speak to the importance of the customer or patient focus; in reality, the teams know that the processes they work with are broken and are not providing the best outcomes for the customer or patient.

Leaders need to be mindful of the true realities and challenges their teams deal with each day as they try to do their job as best they can, given the current reality of doing their work.

One of the ground rules we put in place for these conversations is that we look at process not person. Conversations that have a risk of getting personal are raised back up to a process viewpoint. *'What may be wrong with our process to cause individuals to behave in that way?'* Through a series of workshops, some of which are facilitated by members of the leadership team, another list of not ideal observable behaviours is formed.

By some amazing force of nature, this list is often 80–90% the same as the list generated by the leadership team. However, it is important for the leadership team to listen to and explore the differences. These are often the not ideal behaviours that the middle management group observe **within the leadership team**. These are often the critical conversations that allow the teams to gain trust in the entire process.

Do not be afraid to explore the difficult issues. Remember process focus.

> Why is it that people feel ill-informed? We have not clearly laid out how we communicate important information to all levels.
>
> Why is there a lack of trust with leadership? Priorities keep changing. The process of aligning and reviewing priorities is either non-existent or not adequate. We see actions that are at odds with the messaging from the leadership team.
>
> People do not feel engaged; their opinions are not heard! We had a process for regular one-to-one sessions and to meet with area leads, but this habit has fallen away when we got busier!

Once we are honest and genuine in our interest and efforts to improve the way we all need to work together, teams will listen. But the conversations need to be real. If the current environment is tense and difficult, we strongly recommend professional facilitation for these working sessions. A facilitator can set ground rules and keep conversations away from personal attacks and finger-pointing. A good facilitator keeps bringing the conversation back to the bigger picture and ensures that the team remains grounded and focused on this.

15.8 Measuring the Quality of Leadership

There are many metrics in our business that at first glance are considered Key Performance Indicators (KPIs) but which have a strong underlying behavioural element. Teasing these out allows us to drive conversations at leadership levels in every area of our business. For example, as an organisation, we want to encourage change and improvement, yet it takes us four months to get a revised document signed off, then how will a culture of improvement be sustained?

So, what are the behaviours we could observe around document change? Do we require multiple signatures from people who do not understand the process? Will managers and supervisors leave their documentation change review to a Friday evening? When documents enter the document update department, are they treated based on priority or first in first out? Or are they batched until someone is available to review them and get them on the system? Or is there a constant process of expediting 'urgent' document update due to a pending audit or corrective action close date?

All these situations give us indications of what positive and negative behaviours could be observed and possibly measured within different local elements of the document change process.

The beauty is that we should not dictate any of these measures to the individuals and teams involved in process improvement and document update. We could simply state that flow is one of the key values in our business and ask two questions.

What are the key things that need to happen to ensure that we get flow in our process improvement and document change activity?

What are the things we should keep an eye on (measure), and where should we look for these?

Very quickly, the team will define what needs to be measured and where. These measures can then form part of any tier meeting or department huddle. If they are not having the desired impact, then maybe we need to review as a team and find different measures to ensure flow in our process improvement and document change activity.

In the examples of the restaurant and the behaviours expressed by the supervisor in option two, we can see the impact of good leadership on the expression of positive behaviours at the frontline.

By focusing on the behaviours needed to ensure the optimal outcome of work, we must focus and measure the ability of our leadership community to identify and support the behaviours that enable positive outcomes.

In our earlier examples, an operator stopping a process due to a concern about a raw material does so because they feel safe in that they are trusted in their ability to make that decision regardless of the outcome. That feeling of 'safeness' and the interest in process detail come about from the quality and purposeful intent of that individual's direct leader. The restaurant staff who automatically support colleagues to share workload do this because their mangers see this behaviour and intentionally celebrate and reinforce it.

15.9 Pulling It Together

Delivering for the business involves the constant interplay between the identification, standardisation and continual improvement of critical work methods and the systems that support them. These systems must be designed and managed purposefully to enable the behaviours and culture necessary to ensure ongoing and sustainable improvement of the businesses' performance.

When an organisation embarks on a culture change initiative, it must be based on three core realities:

What are the core elements of our work that provide value to our customers?

Are our leaders aware and supportive of the behaviours we need to see expressed in our teams, which allows us to provide this value in the most effective and efficient way possible?

Are our business systems enabling or disabling the behaviours we need for an optimal outcome of the work we do?

Unless leadership teams can have real open discussions around these three questions, there is no point in moving forward.

Sustainable cultural transformation towards ever-increasing levels of excellence requires a different way of looking at engagement. We use the term transformative participation to describe an environment where every person understands at a very personal level what the organisation is trying to achieve and what are the base values that the organisation stands for.

Power in effect is ceded to majority of the workforce. They manage and ensure that the culture and behaviours they have agreed as necessary to support values and objectives are in place and are upheld. The team **want** to be part of the story of the organisation.

While power to manage culture is ceded to frontline teams, senior leaders and middle managers must constantly verify and actively support the desired culture through the purposeful design and management of key systems within the business. Business systems provide architecture by which leaders can either talk in and recognise desired behaviours or they can talk them out.

Systems of work, supporting systems, and leadership systems are core to supporting and enabling the desired culture and behaviours within a true culture of excellence.

Reference

European Medicines Agency. (2024). *Annual report 2024*. European Union. https://www.ema.europa.eu/en/annual-report/2024/index.html

Operational Excellence in Pfizer Global Supply (PGS)

16

Binyah Kesselly, Joseph Watson, and Woon Lit Ong

Abstract

Pfizer is a global leader in biopharmaceutical innovation, committed to delivering breakthroughs that change patients' lives. With a legacy of scientific excellence and a robust pipeline of transformative therapies, Pfizer's mission extends beyond drug development. It encompasses the optimization of how medicines are manufactured and delivered. At the heart of this operational transformation lies the Integrated Manufacturing Excellence (IMEx) system, a cornerstone of Pfizer Global Supply (PGS) strategy to streamline production, enhance quality, and embed a culture of continuous improvement across its global supply network.

IMEx is not merely a framework; it is a philosophy that aligns people, processes, and performance. It empowers teams to work smarter, faster, and more collaboratively, ensuring that Pfizer stays agile and resilient in a rapidly evolving healthcare landscape. From the deployment of digital tools to the standardization of visual management and leader standard work, IMEx represents PGS' commitment to operational excellence at every level of the organization.

B. Kesselly (✉)
Pfizer, Montvale, USA
e-mail: binyah.kesselly@pfizer.com

J. Watson
Pfizer, Kalamazoo, USA
e-mail: joseph.w.watson@pfizer.com

W. L. Ong
Pfizer, Singapore, Singapore
e-mail: woonlit.ong@pfizer.com

T. Friedli et al. (eds.), *Striving for Operational Excellence in the Pharmaceutical Industry*, Future of Business and Finance,
https://doi.org/10.1007/978-3-032-26287-5_16

16.1 What Does Operational Excellence Look Like in the Field?

Pfizer is a global leader in biopharmaceutical innovation, committed to delivering breakthroughs that change patients' lives. With a legacy of scientific excellence and a robust pipeline of transformative therapies, Pfizer's mission extends beyond drug development. It encompasses the optimization of how medicines are manufactured and delivered. At the heart of this operational transformation lies the Integrated Manufacturing Excellence (IMEx) system, a cornerstone of Pfizer Global Supply (PGS) strategy to streamline production, enhance quality, and embed a culture of continuous improvement across its global supply network.

IMEx is not merely a framework; it is a philosophy that aligns people, processes, and performance. It empowers teams to work smarter, faster, and more collaboratively, ensuring that Pfizer stays agile and resilient in a rapidly evolving healthcare landscape. From the deployment of digital tools to the standardization of visual management and leader standard work, IMEx represents PGS' commitment to Operational Excellence at every level of the organization.

16.2 Our Production System: High-Level Design and Scope

Pfizer Global Supply (PGS) has undertaken a transformative journey through the deployment of Integrated Manufacturing Excellence (IMEx). This initiative has not only standardized operations across sites and functions but also embedded a culture of continuous improvement, leadership accountability, and data-driven decision-making. The deployment of IMEx followed a structured, phased approach. It begins with design, followed by deployment, embedding, and, ultimately, sustainability. At the site level, IMEx is deployed through a series of ecosystems:

1. Active Pharmaceutical Ingredient (API) Ecosystem

   Focus on the physical assets and resources used for manufacturing of APIs for small molecules. It emphasizes standard work and effective scheduling, with improvements driven by process robustness, total productive maintenance (TPM), and continuous improvement (CI). Metrics tracked include safety, quality, supply, financials, and people. These metric pillars apply across all ecosystems.
2. Drug Substance (DS) Ecosystem

   Like API, this ecosystem centers on manufacturing drug substances for large molecules. It uses standard work and scheduling to manage resources and assets, aiming to reduce downtime and improve yield through TPM and CI.
3. Drug Product (DP) Ecosystem

   This covers the manufacturing of drug products such as solids, liquids, and powders. It aims to reduce variability and increase manufacturing flow using standard work, manual process scheduling, and runtime efficiency improvements.

4. Packaging (PACK) Ecosystem
   It focuses on packaging drug products. It targets micro stop reduction and improved runtime efficiency of packaging lines through TPM programs and CI activities.
5. Quality Control (QC) Ecosystem
   This organizes laboratory analysts and equipment using standard work and scheduling. CI loops are used to drive improvements.
6. Quality Assurance (QA) Ecosystem
   It manages the review and approval of product-related documents. It uses standard work and scheduling, supported by CI loops to enhance performance.
7. Material Supply (MS) Ecosystem
   This includes warehousing and planning processes. It identifies and monitors material flow disruptions, focusing on lead time attainment, plan stability, and schedule adherence.
8. Leadership Ecosystem (for both At-Site and Above-Site)
   It defines standards, governance processes, behaviors, talents, and capabilities needed for business success. This ecosystem is foundational for enabling and sustaining IMEx across the organization.
9. Engineering Ecosystem
   It supports physical asset management and maintenance and uses TPM and CI to improve reliability, reduce downtime, and optimize performance.
10. Generic Ecosystem
   This is used in functions like Operational Excellence (OPEX), business planning, and digital operations. It emphasizes goal alignment, reflection, and horizontal integration.

Each ecosystem is built on a modular system of elements that function across it. These elements are not standalone; they are interdependent and designed to reinforce each other in pursuit of Operational Excellence.

1. Process-Centric Team (PCT) Element: The Anchor of Execution
   Every ecosystem begins with a PCT, a cross-functional team responsible for executing IMEx elements. The PCT is the operational nucleus, ensuring that standard work, visual management, CI loops, and other elements are implemented consistently. PCTs receive structured Gemba coaching. They are trained to execute TPM, CI, and standard work routines. Their performance is tracked through leader standard work and visual boards.
2. Standard Work (SW) Element: Foundation of Consistency
   Standard work codifies the "one best way" to perform tasks. It ensures repeatability and reduces variability. SW is executed by PCTs and reviewed during Gemba walks and visual management (VM) meetings. SW enables CI loops to identify deviations. It feeds into capacity planning and process robustness. Horizontal integration is achieved by aligning SW across ecosystems.
3. Visual Management and Meeting Cascade (VM and MC) Element: Driving Transparency

VM boards display metrics, goals, and actions. They are updated regularly and reviewed in tiered meetings. VM ensures that performance is visible and escalations are tracked. VM integrates with leader standard work to guide daily reviews. It supports goal setting and CI prioritization. VM boards are customized by ecosystem but follow common standards.

4. Continuous Improvement Loop (CI) Element: Engine of Innovation

   CI loops are structured processes for identifying, prioritizing, and executing improvement projects. They are aligned with goals and tracked through PDCA cycles. CI projects are selected based on sources of loss identified via VM. CI feeds into TPM, process robustness, and digital operations center (DOC).

5. Leader Standard Work (LSW) Element: Enabling Accountability

   LSW defines tactical, strategic, and people-oriented activities for leaders. It ensures that leaders reflect, coach, and align their teams with IMEx goals. LSW is reviewed in VM meetings and Gemba walks. It includes horizontal integration activities with other teams. LSW documents are adapted by role and ecosystem.

6. Structured Gemba (SG) Element: Coaching in Action

   Gemba walks are structured coaching sessions where leaders observe work, ask questions, and provide feedback. SG reinforces SW, VM, and CI. SG cadence is built into LSW. Observations feed into CI and VM updates. SG is tailored to ecosystem maturity and team needs.

7. Total Productive Maintenance (TPM) Element: Asset Reliability

   TPM ensures that equipment is maintained to prevent breakdowns and improve efficiency. It is executed by PCTs and tracked via VM boards. TPM findings often initiate CI projects. TPM routines are standardized across ecosystems. TPM supports process robustness and system care.

8. Capacity Planning (CP) Element: Matching Demand and Resources

   CP aligns production capacity with demand. It uses data from SW and VM to forecast needs and identify constraints. CP worksheets are reviewed in VM meetings. CP informs CI prioritization and TPM scheduling. It ensures that ecosystems are not over- or underutilized.

9. Process Robustness (PR) Element: Ensuring Stability

   PR focuses on reducing variability and improving process control. It uses data from CI and TPM to identify weak points. PR projects are tracked in CI loops. PR metrics are displayed on VM boards. PR supports quality assurance and engineering ecosystems.

10. System Care (SC) Element: Sustaining the Gains

    SC ensures that improvements are maintained over time. It includes audits, checklists, and governance routines. SC is embedded in LSW and VM. It supports certification readiness. SC is critical for sustaining CI gains.

Each IMEx ecosystem, whether drug product, engineering, or generic, deploys these elements in a tailored but standardized way. For example:

- In generic ecosystems, elements like VM, CI, and SW are adapted for non-asset-based work such as business planning or digital operations.
- In API and Drug Substance ecosystems, TPM and PR are emphasized to ensure equipment reliability and process stability.
- In leadership ecosystems, LSW and SG are central to coaching and governance.

16.3 Introduction of Digital

The digital operations center (DOC) is a transformative platform that has become central to IMEx deployment, enabling a harmonized, data-driven approach to manufacturing excellence across the PGS network. DOC is not just a suite of applications; it is the digital backbone of IMEx, designed to operationalize its principles, streamline workflows, and empower teams with real-time insights.

1. DOC as the Digital Thread of IMEx

   DOC sits at the heart of the IMEx digital thread, serving as the delivery platform for all digital solutions aligned with IMEx standards. It integrates core IMEx components, visual management (VM), standard work (SW), continuous improvement (CI) loop, and incident boards, into a single, accessible interface across devices including desktops, tablets, smartphones, and Microsoft Surface Hubs. This integration allows teams to move away from manual processes and whiteboards, replacing them with digital dashboards that provide real-time visibility into operations.
2. Visual Management and Standard Work in DOC

   DOC's visual management application delivers a collaborative data visualization platform that supports tiered meetings and ecosystem-specific views. It enables teams to monitor metrics, trends, actions, and initiatives in real time, fostering transparency and accountability.

   Standard work is digitized through role cards and batch execution trackers, allowing teams to track weekly production plans and execution status without relying on paper-based systems. This digital standard work ensures consistency, reduces variability, and supports agile decision-making.
3. Continuous Improvement Loop and Analytics

   The CI loop within DOC collects and presents process data to identify improvement opportunities. It supports plan-do-check-act (PDCA) thinking, a core principle of IMEx, and enables teams to experiment, learn, and refine their processes continuously.

   DOC insights, a specialized dashboard, empowers CI specialists to optimize performance by providing analytics on role card execution, issue tracking, and CI loop effectiveness. These dashboards are designed to be user-friendly and centralize performance data, making it easier for teams to act on insights and drive improvements.

4. Leadership Enablement and Accountability
   Leaders can use DOC to monitor their teams' progress, conduct structured Gemba walks, and ensure that goals are connected to CI initiatives. This visibility fosters a culture of ownership and continuous improvement, reinforcing IMEx "One Best Way of Working" philosophy.
5. DOC's Impact on Site Operations
   DOC has transformed site operations by enabling:

   - Real-time collaboration: Teams can interact with dashboards during meetings, reducing reliance on static reports.
   - Data-driven decisions: Integrated metrics and analytics support faster, more informed decision-making.
   - Operational efficiency: Automation of routine tasks and elimination of manual data entry reduce waste and free up time for value-added activities.
   - Scalability: DOC's modular design allows it to be tailored to different ecosystems, drug product, packaging, quality control, etc., while keeping consistency across the network.

6. DOC and the Future of IMEx
   As Pfizer continues its PGS transformation, DOC is positioned as the enabler of future digital capabilities. It supports the evolution of IMEx by:
   - Embedding digital maturity into site readiness assessments.
   - Supporting infrastructure upgrades for wireless connectivity and device integration.
   - Facilitating the transition from paper-based to digital-first operations.
   - DOC's role in shaping the "PGS user experience" is clear in its ability to deliver the right information at the right time, empowering teams to act with speed and precision.

16.4 Health Checks/Certifications: Setting Up the System

To ensure that IMEx is not only deployed but also sustained effectively, PGS has developed a rigorous IMEx Health Check and Certification process. This process evaluates the progress of IMEx implementation at sites and functions, identifies gaps, and guides teams toward higher levels of operational capability.

IMEx Health Checks and Certifications are not audits; they are developmental assessments. Their purpose is to:

- Evaluate the maturity of IMEx deployment across ecosystems.
- Identify strengths and opportunities for improvement.
- Align teams on goals, priorities, and next steps.

IMEx Health Checks and Certifications are essential mechanisms for embedding Operational Excellence at Pfizer. They provide a structured, transparent, and developmental approach to assessing maturity, aligning teams, and recognizing sustained performance. Through rigorous assessments, leadership engagement, and digital enablement, Pfizer ensures that IMEx is not just deployed, but lived, across its global network.

16.5 Pfizer's Journey with IMEx

Pfizer's journey with the Integrated Manufacturing Excellence (IMEx) system began in 2016. Since then, the company has been continuously deploying and modifying the system to meet its evolving needs. To date, Pfizer has deployed, qualified, and certified over 1600 process-centric teams. These teams are spread across various ecosystem types, reaching sites around the globe and eventually supporting over 180 markets worldwide.

IMEx is not limited to manufacturing sites; it extends to functional groups such as global supply chain, quality, and external supply, to name a few. Leadership teams at all levels of the organization, from the shop floor to the PGS leadership team, benefit from having IMEx in place.

The current challenge is how to sustain IMEx. Sustaining a Lean production system is extremely complex, as any Lean practitioner will attest. Let us look at the typical headwinds faced when attempting to sustain IMEx.

16.6 Headwind: "Why?"

Over time, it was noticed that IMEx deployments that were the easiest to deploy and sustain were established by teams with a burning platform. Their goal was clear, and the way to get there was even clearer: leverage the production system to drive continuous improvement (CI). Once they did, they saw the value of the system and continued to leverage it for the long run.

A leader once mentioned that while this success was great, the idea of a burning platform sounded terrifying, and they would not ever want to be in such a situation. They challenged the team to find a way to establish the system and leverage it without a burning platform. This challenge was pondered for years, as it was an enigma. Lean production systems were always found to be logical, and it was baffling why establishing and sustaining these systems were so difficult.

An epiphany finally occurred, but it took a visit to a Toyota manufacturing site before the light bulb went on. Before explaining that, let us first share some additional thoughts. Without a burning platform, how can the success rate of an IMEx deployment be improved? First, what is the importance of a burning platform? It brings a clear "why." Why is this system being put in place? It is important to be careful of deploying just to comply with a directive that everyone has to follow IMEx. Instead, the most meaningful "why" for a team should be found before even starting the deployment. Once the team has a clear purpose, the path to a successful deployment begins. However, that only gets the team so far. Yes, they can rally around a common goal, but they need to be taught how they might leverage the system to achieve that goal.

16.7 Headwind: Blind to Waste

Teams can be taught about how the system works, but that is not enough. They must see the waste in their processes that gets in the way of achieving their goal. Teams must apply this new lens to look through. To do this, they are often sent on waste walks of their processes, and they usually come back with numerous opportunities for CI. Now they are getting it. They begin to apply this learning within the system, seeing the waste in their processes and how these losses relate to their overall goal. Success! They begin to progress toward achieving their goal through root cause analysis and CI. They are beginning to see the value of leveraging the system, but this still does not guarantee that they will sustain the system. Why? This still perplexed the team.

16.8 Headwind: Hidden Flow

This is where the trip to Toyota helped us "see" why this is so difficult. Over the years, countless books have been written clearly explaining the simplicity of production systems, the principles of Lean, and even the common mistakes of unsuccessful deployments. So, what was missing? What was to be seen at Toyota that was not already described in all this literature? The answer was tangible "flow." The instant the doors leading to Toyota's manufacturing floor were walked through, it hit like a ton of bricks; this is the tangible flow that is "hidden" at Pfizer and probably throughout the pharma industry.

There were cars flowing along the assembly line and eventually rolling off the line as close to the desired takt time as possible. The true importance of standard work was quickly understood, one of the most difficult parts of a production system to establish. If standard work could not be done within the allotted time, the flow would stop. If 5S was not in place, there would not be time to find tools and still meet the standard work in time, and the flow would stop. If materials were not in the right quantity at the right place at the right time, the flow would stop, and the list goes on. In a very hands-on environment like Toyota, each worker can understand the importance of their role and their standard work to the end-to-end flow of the product.

Of course, all the tools and flow disruptions mentioned above apply to Pfizer as well, but the problem is that the end-to-end flow is hidden. This is not referring to the vials going down the conveyor belt, but the flow of product end-to-end, often taking 300–400 days from API production to finished goods leaving the warehouse, passing through bottlenecks like quality control, inspections, and quality assurance release, along the way. The challenge is how to make this flow more visible. Once visible, teams need to understand their role in the flow. Then and only then can a team understand what waste is preventing the flow of product, which then leads to what they need to focus their CI efforts on to unlock end-to-end flow.

16.8.1 Headwind: Time Commitment

One of the most vital enablers of Lean and IMEx success is the presence and engagement of leaders. At Pfizer, we have seen that when leaders are actively involved from the start—helping define the clear "why," supporting teams in building their systems, and guiding understanding of each IMEx elements—momentum builds quickly. Sustaining that momentum, however, requires ongoing leadership presence and support.

From shop floor leaders to senior leadership, being present on the Gemba is more than symbolic—it is transformational. It reinforces not just the importance of the IMEx way of working but also nurtures the IMEx way of thinking. This mindset, rooted in plan-do-check-act (PDCA), empowers teams to align, execute, learn, and continuously improve—striving to make each day better than the last.

Leaders play a pivotal role in removing barriers and enabling progress. By helping teams understand the "why," connecting their work to the broader product flow, aligning goals, identifying and eliminating waste, and guiding performance, leaders create an environment where continuous improvement becomes second nature and is driven by leveraging IMEx.

Moreover, great leaders inspire a mindset of curiosity and ambition—never settling for the status quo but always seeking the next opportunity to enhance flow and efficiency. This mindset not only sustains IMEx but also accelerates the maturity of Operational Excellence across the PGS network.

In conclusion, while systems and tools are essential, it is the consistent, visible, and empowering presence of leaders that truly unlocks the potential of IMEx. When teams see their leaders coaching, supporting, and celebrating progress, they are inspired to take ownership, drive improvement, and embrace the IMEx way of working and thinking.

16.9 Forward Looking: Importance of OPEX Maturity

The importance of OPEX maturity cannot be understated. It allows organizations to build resilience, rapidly adapt to the changing competitive environment, and seamlessly achieve long-term sustainability to better:

1. Optimize processes.
2. Develop a culture of continuous improvement.
3. Identify and eliminate waste.
4. Drive innovation and growth.
5. Improve customer satisfaction.

A mature OPEX ecosystem is adept at uncovering the largest sources of loss (LSL) and identifying continuous improvement (CI) opportunities through structured evaluations and transparent data analysis. This will help organizations pinpoint inefficiencies that may otherwise go unnoticed and implement targeted interventions for sustainable impact.

16.9.1 Not Just a Measure of Standard Adherence

True OPEX maturity extends beyond mere compliance, checklists, and measuring strict adherence to production system standards. It delves into the foundational behaviors, mindsets, and operational principles of an organization and enables them to consistently deliver value. And it reflects the organization's integration of best practices into standard processes that are consistently implemented across the network. This mindset shift ensures that OPEX becomes deeply embedded in the organizational culture, driving continuous improvement from the ground up and creating an "institutional muscle memory" for OPEX across the enterprise.

16.9.2 Benchmarking

One great way to gauge progress in an organization's OPEX journey is to utilize internal and external benchmarking. Internal benchmarking allows organizations to identify high-performing teams or units and replicate their best practices across the enterprise. Similarly, external benchmarking offers a reality check by comparing performance with industry leaders, highlighting gaps and opportunities for advancement.

16.9.3 Improved OPEX Maturity = Improved Performance

Organizations that invest in advancing their OPEX maturity often find that their performance metrics improve in direct proportion. As OPEX capabilities strengthen, processes become more stable, risks are managed more effectively, and agility increases throughout the enterprise. These higher levels of maturity translate into measurable gains—greater productivity, enhanced quality, and stronger financial results—creating a foundation for sustained excellence that is easily replicated at scale.

16.10 Uncovering the LSL and CI Opportunities

16.10.1 Productivity Hunting

Proactive productivity hunting is a hallmark of advanced operational maturity. This involves putting a methodology in place to relentlessly pursue ways to streamline processes, eliminate waste, and maximize resource utilization. One way to accelerate this process is to leverage digital technologies and innovative thinking. This can empower organizations to make smarter, faster, and more impactful improvements across the enterprise.

16.10.2 Leveraging AI

Artificial intelligence (AI) offers unprecedented opportunities to boost productivity by automating routine tasks, enhancing decision-making with predictive analytics, and uncovering complex patterns in huge amounts of operational data in a matter of seconds. At present, this process has historically involved manually scrubbing chunks of carved-out data and then utilizing spreadsheets or minitab to analyze the data for patterns. Then, based on the findings, the analyst can infer probable root causes and make some determinations on which CI projects to undertake in order to drive productivity. With AI, all of this can be done in a matter of seconds, with the bulk on the time spent on actual value delivery, as opposed to that time being spent on problem identification.

And, if the AI solution is developed properly, one of the most valuable aspects of this mature OPEX ecosystem is the institutionalization of the learning process—replication. By documenting and cataloging solutions, and then sharing this knowledge across teams, organizations ensure that once a problem is solved, it remains solved, thus reducing recurring issues, accelerating innovation, and enhancing overall operational reliability.

16.11 Self-Sustaining CI Culture

16.11.1 The Leader's Role

Leaders contribute to operational excellence by enabling teams, recognizing accomplishments, and encouraging practices that support ongoing improvements. Leadership plays a role in long-term organizational outcomes and establishment of a continuous improvement (CI) culture. This includes promoting appropriate mindsets, integrating improvement behaviors into daily activities, and developing leadership skills throughout the organization. Teams often reflect the actions and approaches of their leaders; as a result, leaders' interest or lack thereof in fostering such a culture will influence the success of CI initiatives.

16.11.2 Mindset

A growth-oriented mindset is at the core of the CI journey. Teams are encouraged to view challenges as opportunities, embrace experimentation and failures, and learn from both successes and setbacks. This psychological safety fuels innovation, empowers creativity, and drives sustained excellence.

But the significance of mindset in the pursuit of OPEX maturity cannot be overstated. While tools, systems, and methodologies create the infrastructure for improvement, it is the mindset that determines whether these structures thrive or wither. A shared commitment to curiosity and adaptability ensures that teams do not simply implement prescribed solutions but actively seek out new ways of creating value, continually questioning the status quo and exploring untapped potential.

Organizations that invest in cultivating a CI mindset foster resilience in the face of change, challenge, or ambiguity. Individuals feel empowered to voice ideas and concerns, confident that their contributions matter. This ownership transforms improvement from an episodic event into a sustained cultural norm, where every person—regardless of role or seniority—acts as an agent of positive change.

Crucially, mindset acts as the invisible thread that weaves together all OPEX capabilities. It connects strategy with day-to-day behaviors, ensuring alignment and energy across the enterprise. When the collective mindset values learning, adaptability, and collaboration, organizations are not only equipped to reach new heights of performance but also able to sustain these gains in an ever-evolving environment. In this way, mindset is not a soft factor or an afterthought—it is the very foundation of lasting operational maturity.

16.11.3 Behaviors

OPEX maturity is reflected in everyday behaviors—attention to detail, willingness to challenge the status quo, and commitment to data-driven decision-making. These behaviors reinforce a collective discipline that underpins ongoing improvement efforts. When individuals consistently demonstrate openness to feedback, respect for diverse perspectives, and a proactive approach to solving problems, they create a ripple effect that amplifies across teams and departments.

Consequently, small actions—such as following up on process changes, celebrating incremental wins, and embracing transparency—become the building blocks of organizational culture. Over time, these behaviors establish trust and psychological safety, empowering all members to contribute their insights and drive innovation. Furthermore, embedding such practices transforms improvement from isolated initiatives to a shared, ongoing responsibility.

Ultimately, behaviors serve as the visible expression of an organization's commitment to OPEX maturity. They move the concept from theory to practice, ensuring that CI principles are not just espoused but lived daily. Without the right behaviors, even the most sophisticated systems and strategies will falter; with them, organizations unlock the full potential of their people and processes, sustaining excellence regardless of changes in technology or market conditions. This is why behaviors are an indispensable facet of OPEX maturity, linking intention and execution in the pursuit of lasting value.

16.11.4 Importance of Gemba

Gemba, the practice of "going to the place where work happens," is far more than an observational tool—it is a cornerstone in the advancement and sustainability of

CI across the entire enterprise. By purposefully stepping into the heart of operations, leaders and teams gain direct, unfiltered access to the realities of processes, uncovering both latent potential and hidden obstacles that are often missed from a distance.

Regular Gemba walks create a dynamic feedback loop, encouraging open dialogue between frontline workers and leadership. This fosters a culture of mutual respect, shared ownership, and psychological safety, where every individual feels seen and heard. Employees are empowered to surface issues, suggest improvements, and contribute insights born of hands-on experience. In this way, Gemba acts as a catalyst for engagement, transforming passive participants into active architects of progress.

Moreover, Gemba is essential for embedding CI as an enterprise-wide norm rather than a fleeting initiative. When leaders model curiosity and humility by engaging directly at the source, they signal a commitment to transparency and collaborative problem-solving. This visible leadership presence supports the cross-pollination of ideas, accelerates the identification of best practices, and enables rapid course correction where needed.

Organizations that leverage Gemba as a strategic discipline are uniquely positioned to sustain momentum in their CI journey. Challenges are addressed authentically and in real time, enabling faster innovation cycles and more resilient processes. Gemba ensures that the voice of the shop floor is integrated with strategic decision-making, aligning operational realities with organizational aspirations.

Ultimately, Gemba is not simply a ritual; it is an essential ingredient in cultivating a culture where CI thrives. By weaving Gemba into the fabric of daily operations, organizations unlock deep insights, nurture engagement, and build the collective capability needed to achieve—and sustain—enterprise-wide excellence.

16.12 OPEX Capabilities: Toward Sustainable Value and Impact

The strategic development of OPEX capabilities stands at the heart of a transformative CI journey. While behaviors and rituals like Gemba walks lay the cultural foundation, it is the deliberate cultivation of organizational skills, systems, and mindsets that truly propels CI from a series of disconnected projects to a sustained enterprise-wide movement. Embedding these capabilities ensures that teams are not only equipped to solve immediate challenges but also positioned to anticipate and adapt to future demands.

A well-orchestrated approach to capability building acknowledges that excellence is not achieved by chance—it is engineered through intentional investment in people, processes, and enabling technologies. This means designing learning pathways, fostering cross-functional collaboration, and deploying tools that make data-driven problem-solving an everyday habit. As organizations develop these strategic muscles, they become more resilient and agile, gaining the capacity to innovate with speed and confidence in an ever-evolving environment.

Ultimately, capability building is the linchpin that aligns vision with execution, ensuring that the pursuit of CI is sustained and scalable. By systematically nurturing these competencies, organizations create a self-reinforcing cycle of improvement, where success is not a single milestone, but a continuous ascent toward greater value and impact. In this way, capability building transforms operational excellence from a theoretical aspiration into a living, dynamic force that drives long-term organizational success.

16.12.1 Looking Ahead

By embracing the OPEX journey, organizations do more than refine existing processes. They ignite a culture of perpetual growth, innovation, and adaptability. The deliberate actions of fostering Gemba-driven insights and cultivating robust operational capabilities lay the groundwork for transformation that extends far beyond incremental improvement. Equipped with these foundational practices, leaders and teams become architects of their own destiny, resilient in the face of uncertainty and ready to seize emerging opportunities.

Ultimately, those who commit to operational excellence are not just preparing for the challenges of tomorrow—they are shaping a future where enduring success, sustainable value, and industry leadership are the natural outcomes of a relentless pursuit of excellence.

Operational Excellence (OPEX) Needs an Upgrade: Reinventing OPEX in the Age of Digital/AI

17

Daniele Iacovelli

Abstract

In the pharmaceutical industry's new AI era, Operational Excellence (OPEX) is evolving beyond lean and Six Sigma to embrace the power of Industry 4.0 technologies. Continuous Improvement 4.0 integrates classic OPEX principles with digital and AI capabilities, shifting organizations from incremental gains to transformative value and agility. Yet, many companies struggle with siloed functions, fragmented budgets, and isolated projects; this not only limits innovation but also prevents lasting change and scalable impact. Integrated strategies are essential for success. Combining digital, AI, IT, and OPEX within unified teams and operating models allows organizations to radically simplify processes, optimize resources, and accelerate time to market. Key elements include adopting modular technology, integrating end-to-end data, automating and agentifying processes, enhancing human skills with advanced tools, and keeping patient value at the center. Joint accountability, cross-functional teams, and scale-up governance create an environment, where both operational and digital excellence drive sustainable improvement and resilience. Throughout the chapter, the author presents practical frameworks, real-world examples, and lessons learned from large-scale transformations. The chapter combines strategic analysis with actionable recommendations, illustrating how integrated approaches unlock exponential impact.

D. Iacovelli (✉)
F. Hoffmann-La Roche AG, Basel, Switzerland
e-mail: daniele.iacovelli@roche.com

T. Friedli et al. (eds.), *Striving for Operational Excellence in the Pharmaceutical Industry*, Future of Business and Finance,
https://doi.org/10.1007/978-3-032-26287-5_17

How the integration of OPEX and digital/AI creates a competitive advantage in pharma operations and paves a new paradigm for continuous improvement.

17.1 Introduction: The Evolving Role of OPEX in the Tech and AI Era

In today's rapidly evolving tech and AI landscape, **Operational Excellence (OPEX)** in the pharmaceutical industry **can no longer rely solely on lean tools, Six Sigma, or traditional continuous improvement methods**. The emergence of Industry 4.0 technologies, particularly AI, machine learning, IoT, and advanced analytics, has paved the way for a new paradigm: **Continuous Improvement 4.0**. This new approach integrates classical OPEX disciplines with digital and AI capabilities, enabling organizations not only to sustain improvements but also to achieve exponential gains in performance, agility, resilience, as well as customer/patient and employee experience.

Yet, despite this potential, most companies **still treat OPEX and digital as separate functions**—managed by different organizations, with distinct teams, reporting lines, operating models, and budgets. This fragmentation and organizational separation create silos and inefficiencies, preventing organizations from fully capitalizing on the synergies between process excellence and digital innovation.

Moreover, digital and AI initiatives without embedding OPEX often result in **flashy tools (applied on non-harmonized processes) and isolated pilots** that fail to drive lasting value or change in ways of working. Conversely, OPEX initiatives without the power of digital and AI tend to deliver only incremental improvements, often limited to "low-hanging fruit" that, in most cases in previous lean programs, has already been harvested—the result: **lean programs tend to lose "oxygen" to survive**. To unlock the full value of Continuous Improvement 4.0, organizations must break down these barriers and pursue a truly integrated approach—aligning business, technology, and people in a unified approach to transformational and sustainable performance.

In this chapter, I introduce the concept of CI 4.0 and its relevance in today's pharmaceutical environment. Drawing from two decades of experience, including my current role as SVP and Chief Digital Officer for the Pharma Medicines business at Roche, and my prior work as a Partner at McKinsey & Company, advising multiple companies across industries and leading large-scale transformations combining OPEX and digital/AI, I aim to present the case for change, main lessons learned, and practical frameworks that can help organizations make the leap from incremental improvements to exponential impact.

17.2 The Case for Change: Why the Integration of OPEX and Digital Is Essential

Digital and AI initiatives often focus on developing impressive, but isolated, tools or pilot projects. These may deliver short-term wins but often fail to integrate with frontline processes and rarely scale sustainably. Meanwhile, OPEX programs, although grounded in process rigor, tend to yield only incremental improvements and struggle to address systemic challenges due to limited access to data and digital tools.

Despite the transformative potential of technology, digital and AI, most pharmaceutical organizations still treat OPEX and digital as separate domains. Typically, OPEX is anchored in manufacturing or quality units, while digital and AI are embedded within central functions or IT departments or even dispersed in every business function. These silos result in misaligned priorities, inefficiencies and not fully converted investments, duplicated efforts ("mushrooming"), unhealthy competitions, and missed opportunities to scale improvements.

I fundamentally believe that OPEX needs an upgrade. Based on my experience, I see the two following realities in companies while applying digital/AI and lean/OPEX programs:

(I) State of industrial AI: A persistent "pilot purgatory" and the allure of impressive technology vs. tangible business value

Despite considerable investment, widespread enthusiasm, and mobilization, the adoption of digital/AI and value generation remain challenging. Multiple white papers of the World Economic Forum (WEF) Global Lighthouse Network indicated that over **60% of all digital transformations fail to deliver tangible value**, frequently becoming mired in a frustrating "pilot purgatory."

Besides, "The GenAI Divide: State of AI in Business 2025" report from MIT's NANDA initiative shows that while generative AI holds vast potential, most corporate efforts to capture rapid revenue gains are falling short. Despite the rush to adopt cutting-edge models, only about 5% of AI pilots deliver significant financial value and around 95% see no P&L impact. Drawing on 150 executive interviews, a survey of 350 employees, and an analysis of 300 public AI deployments, the research highlights a stark contrast between a handful of success stories and a wide field of stalled initiatives.

A primary driver behind these widespread failures is often an **isolated (over-) emphasis on technology itself—I would call it the "shiny object syndrome."** This pervasive focus on the novelty and potential of new digital and AI solutions frequently overshadows the fundamental need for integration with OPEX. Without a robust connection to OPEX, organizations often struggle to:

(a) Clearly define the business need: Digital and AI initiatives frequently prioritize the development of impressive, technologically advanced solutions. While technological sophistication is certainly valuable, it often becomes an end in itself rather than a means to achieve concrete business outcomes. This can manifest for instance in **focus on features over functionality**;

that is, the development team may prioritize adding a wide array of features without adequately considering whether these features directly address a business problem or enhance operational efficiency.

(b) Redesign business processes to create a new way of working: The successful adoption of new digital and AI solutions necessitates a re-evaluation and often a significant redesign of existing operational processes (process simplification, harmonization, and redesign). Without this crucial step, the new technology is merely overlaid onto inefficient processes, negating its potential benefits and often creating new bottlenecks—almost no value/no scale.

(c) Maximize adoption and avoid the "Build it and they will come" mentality: There is an assumption that if a powerful AI solution is developed, it will automatically be adopted and utilized effectively by the organization, overlooking the critical change management and integration efforts required. **Ensuring the effective use and scale of the new digital/AI solution is key.** A lack of integration with OPEX means that user adoption, training, and ongoing operational support/coaching are often neglected, leading to underutilization or abandonment of the new tools.

A disregard for user experience and operational workflows may lead to solutions that are technologically sound but difficult to use or they may disrupt existing workflows without providing a superior alternative, leading to resistance from end users.

(II) State of OPEX/Lean Programs: The Depletion of Vitality of OPEX/Lean Initiatives

OPEX/lean programs, although grounded in process rigor, tend to yield only incremental improvements and struggle to address systemic challenges and disruptive improvements due to limited access to data and digital tools.

Organizations that fail to integrate OPEX and digital risk underperforming in a world where real-time decision-making, predictive analytics, and autonomous systems are fast becoming the norm. The WEF Digital Lighthouse Network initiative shows that lighthouses which are adopting and integrating Fourth Industrial Revolution technologies are outperforming competitors, e.g., in speed to market, customer experience, sustainability, or productivity. Lighthouses are demonstrating a clear competitive advantage, setting a new benchmark for OPEX and digital transformation. This outperformance is clearly evidenced by their achieved improvements in efficiency, productivity, sustainability, and agility, allowing them to navigate complex market dynamics and respond more effectively to customer demands than their less digitally mature counterparts.

In several organizations I have observed, a concerning trend emerges: **a gradual depletion of vitality within lean/OPEX initiatives**. This often leads to their dedicated OPEX teams feeling increasingly sidelined and disheartened, eventually becoming irrelevant to the company's core strategic direction and operational improvements.

The fragmentation of OPEX and digital is a critical barrier. Organizations that fail to integrate OPEX and digital risk underperforming in a world where real-time decision-making, predictive analytics, and autonomous systems are fast becoming the norm.

Ultimately, OPEX is crucial for scaling digital programs and fostering new ways of working. Conversely, digital and data analytics are essential for revitalizing OPEX/lean initiatives, providing the impetus not just for incremental gains but for achieving the next S-curve of continuous improvement.

17.3 One Strategy | One Team | One Operating Model: Building Blocks for an Integrated OPEX and Digital Organization

When I joined Roche Pharma, we made a bold decision: to create a fully integrated organization that brings together digital, AI, IT, and OPEX under a single leadership structure, directly embedded within the business. Its purpose was to align these functions with a common vision and shared business objectives, operating as a peer to the various global heads of business functions.

This organizational redesign was not cosmetic; it was foundational. Our goal was to shift from siloed innovation to **enterprise-wide and sustainable engine for data-driven and systematic improvement to deliver better outcomes and experience for patients, the business, and our employees.**

In pharma technical operations, we established a shared mission: pioneering for excellence and radical simplification to deliver transformative treatments and value for our patients worldwide. This involves eliminating complexity and inefficiencies within its processes, technology, and infrastructure to deliver transformative treatments and value for patients. Key aspects of this vision include leaving lean/OPEX every day, integrating AI and digital technologies, modernizing infrastructure to be more modular and composable, and empowering employees to achieve more with less—ultimately: **fast to patients, robust processes (right every time), and efficient operations (no waste, no wait, no touch)**. We aim to accelerate time to market for new medicines and optimize costs by transforming the way drugs are developed, manufactured, and supplied, ultimately enabling Roche to do "now what patients need next."

Key Elements of Radical Simplification (See Fig. 17.1)

- Digitalization and AI Integration
  Implementing AI and digital tools to transform and automate processes, create innovative insights, and significantly speed up the time it takes to bring new drugs to market.
- Modular and Composable Technology
  Rethinking and redesigning the tech infrastructure to be more flexible and adaptable, moving away from rigid, monolithic solutions toward more modular and composable architectures.

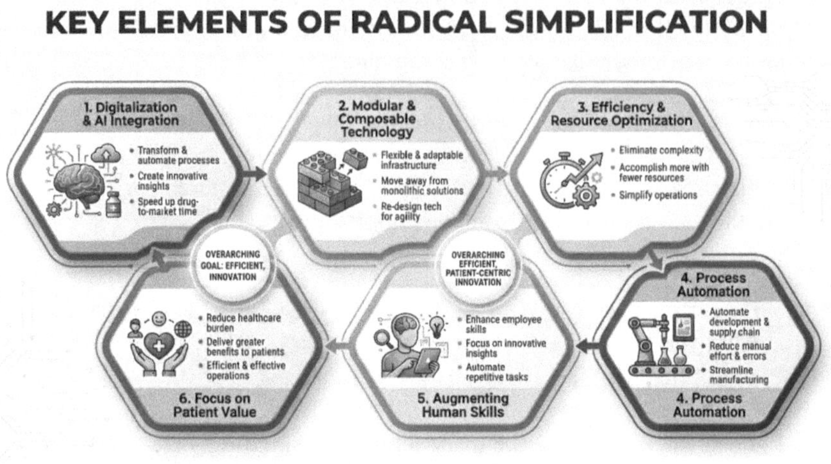

Fig. 17.1 Key elements of radical simplification

- Efficiency and Resource Optimization
  Simplifying operations and processes to eliminate complexity, which allows teams to accomplish more with fewer resources.
- Process Automation
  Automating processes within development, manufacturing, and supply chains to reduce manual effort and errors.
- Augmenting Human Skills
  Using technology to enhance the skills of employees, allowing them to focus on creating innovative insights rather than on complex, repetitive tasks.
- Focus on Patient Value
  The overarching goal is to reduce the healthcare burden and deliver greater benefits to patients and society by making operations more efficient and effective.

The revitalization and integration of the OPEX group require more than only capability building; it requires a comprehensive shift in governance, roles and responsibility, operating model, culture, and capabilities. I see key design principles in the way the integrated organizational setup should be crafted along the lines of "one strategy I one team I one operating model" (see Fig. 17.2). More specifically:

One Strategy:

- An integrated digital and OPEX strategy linked to the business strategy and priorities, a clear road map, and the articulation of expected value:
  This means moving away from siloed digital experiments and isolated process improvement projects toward a unified approach directly supporting the company's overarching goals. The road map must clearly delineate milestones for both technological implementation and process redesign, ensuring that they happen in concert rather than sequentially. Furthermore, defining and tracking expected

Fig. 17.2 Revitalizing OPEX through integrated design principles

business value—whether in cost reduction, efficiency gains, or revenue growth—are crucial for maintaining executive support and demonstrating return on investment.

- Integrated budgets and strategic planning cycles together with digital/AI and OPEX:
  Disconnected financial planning is a primary barrier to integration; digital tools are frequently funded separately from the necessary process change management efforts required to make them work. By merging these budgets and aligning planning calendars, organizations ensure that resources are allocated holistically to tackle business problems with the right mix of technology and process expertise. This synchronization prevents the common scenario where technology is purchased but lies dormant because the corresponding operational changes were never budgeted for or scheduled.

One Team/Culture:

- An OPEX and digital/AI leadership team with joint accountability:
  Success cannot be achieved if the Head of Digital and the Head of OPEX are measured on conflicting KPIs or compete for resources. Instead, leadership must share a single set of objectives and outcomes, forcing collaboration at the highest level and sending a unified message to the organization. This joint accountability ensures that decisions are made for the benefit of the total integrated system, rather than optimizing individual functions at the expense of the whole.
- A culture of co-ownership with all the business units and manufacturing sites:
  The integrated digital/OPEX group cannot function merely as an external consultancy that "delivers" solutions to "passive" business units. True transformation requires local teams to actively participate in designing and testing the solutions, ensuring that the end result addresses their on-the-ground realities and

pain points. This sense of shared ownership fosters genuine adoption and ensures that continuous improvement becomes embedded at the site level long after the initial project team has moved on.

One Operating Model:

- Cross-functional squads with an aligned operating model, combining operational/process excellence, data, AI, and tech (IT/OT, automation, etc.) expertise: The traditional baton-pass approach between process designers and technology implementers is too slow and ineffective for today's pace of change. Instead, agile squads must bring together process experts, data scientists, and engineers into a single unit focused on solving specific business challenges from day one. This aligned operating model ensures that technology solutions are rooted in process reality and that operational workflows are designed from the start to maximize new digital capabilities.
- A scale-up-focused governance enabling the move "from toys to tools":
  Too many organizations suffer from "pilot purgatory," where successful proofs of concept fail to achieve scaled impact due to a lack of standardized processes and infrastructure. Governance mechanisms must shift focus from purely encouraging initial innovation to rigorously managing industrialization and replicability across the enterprise. This means establishing clear criteria and dedicated resources for moving a solution from an experimental "toy" phase to a mandatory, industrialized "tool" embedded in daily operations.
- Radically rethinking the role, capabilities, and mandate of the OPEX team:
  Traditional lean and Six Sigma toolkits, while still valuable, are no longer sufficient on their own; the modern OPEX practitioner must also possess high data literacy and a solid understanding of current digital capabilities. The team's mandate must expand from a narrow focus on waste reduction and standardization to becoming the primary "translators" between business needs and technological potential. Ultimately, they must evolve into change agents capable of guiding the organization through the complex human and cultural shifts required by digital transformation.

17.4 Conclusion: Upgrading OPEX to Create Value at Scale with Digital and AI

The convergence of AI, digital technologies, and OPEX presents a generational opportunity for pharma operations. CI 4.0 is not about choosing between digital and OPEX—it is about harnessing the power of both in a cohesive, strategic, and human-centric way. The uncomfortable truth: Most of the reports on AI (MIT, World Economic Forum, etc.) reveal that the majority of companies do not generate significant (and tangible) value with their (Gen) AI programs. Why? Because technology is accelerating exponentially (e.g. full-stack applications can be developed in days, without significant coding experience), but our organizational structures and operating models are stuck in the past. Consider this: – When did you last introduce a new AI tool? Likely weeks ago. – When did you last rethink your organizational structure or operating model? Likely years ago. – How does "span-of-control" evolve when a middle manager oversees a handful of humans alongside dozens of autonomous agents and robots? It is time to bend the curve between technical and organizational evolution and reimagine our organisation and operating models to be able to deliver value at scale from AI and Technology.

As demonstrated by our transformation at Roche and supported by experiences across industries, integrating these disciplines drives not just better outcomes but a fundamentally new way of working. The shift from isolated tools and incremental gains to system-wide, exponential impact is within reach.

For leaders willing to break down silos, invest in capabilities, and lead with purpose, the rewards will be profound. The future of pharma operations will be shaped by those who operate at the intersection of business, technology, and people—and who are bold enough to reinvent the system for the age of AI.

Operational Excellence Leadership Lessons

18

Thomas J. Burns

Abstract

This chapter presents a collection of practical, experience-based lessons, each anchored in real situations and personal stories from years of Operational Excellence leadership. Lasting improvement comes from involving everyone in the organization, especially those closest to the work, in both running and improving the business. By structuring the content as actionable insights supported by concrete anecdotes, the chapter shows that Operational Excellence is not a destination or a grand program—rather a discipline built through thousands of honest, everyday actions.

Early in my career, I learned a truth I have come back to again and again: *there is extraordinary, often invisible potential sitting inside every organization*—potential that only becomes real when you engage every colleague, every shift, and every function. When continuous improvement becomes part of the identity of the organization, not a project or an initiative, everything changes. People stop thinking of "doing the work" and "improving the work" as separate activities. Instead, improvement becomes an instinct.

But getting there requires a shift many leadership teams underestimate. In this chapter, I have highlighted some key lessons I have learned over the years that have stuck with me and helped define my strategy and approach as an operational excellence leader.

T. J. Burns (✉)
CSL, Takeda Pharmaceuticals, Shire Pharmaceuticals, and AstraZeneca, Needham, MA, USA

T. Friedli et al. (eds.), *Striving for Operational Excellence in the Pharmaceutical Industry*, Future of Business and Finance,
https://doi.org/10.1007/978-3-032-26287-5_18

18.1 Lesson 1: The Untapped Engine Is the Front Line

I have seen too many companies pour time and money into massive, top-down transformations, only to leave 80–90% of their workforce standing on the sidelines. It is like trying to row a boat with only two people paddling while ten others watch.

When you build a system where everyone contributes ideas, solves problems, and takes ownership of improving their work, you unlock performance gains that dwarf what any single initiative can deliver. It becomes, in a sense, pure physics: turning *potential* energy into *kinetic* energy.

The mechanism does not have to be complex. Kaizen events are one avenue. Make daily tier meetings or weekly cross-functional huddles—spaces where ideas are captured, prioritized, and acted upon—this creates a drumbeat. Some ideas become projects. Others are "just-do-its." And the most powerful ideas often come from the person who sees the problem every day. Whenever possible, empower them to take the lead in finding the solution. This builds capability faster than any training program. Recognize and reward both successful attempts and failed experiments, and free up time for your team to work on improvement ideas.

If done well, you create a flywheel you will never want to stop. Do not underestimate the power of a true grassroots CI culture: everybody, every dose, every day.

18.2 Lesson 2: Operating Systems Only Work as Systems

I have had the privilege of building and deploying Lean operating systems across four global companies. Each time, the system improved—not because I am smart, but because the team carried forward what worked, challenged what did not, and adapted to the company culture.

Across biopharma, many operating systems now look similar—and that is good. It means the industry has cross-pollinated and matured. But one leadership trap persists: viewing the operating system as a *toolkit*, rather than a *system*.

Consider a few examples of how these elements build upon each other:

- *Strategy deployment* aligns the entire organization to the critical few priorities.
- *Value stream analysis* reveals the highest-impact improvement opportunities across the end-to-end supply chain.
- *Tiered accountability and visual management* elevate issues quickly and transparently.
- *Structured problem-solving* gives teams the ability to resolve those issues at the right level.
- *Leader standard work and process confirmation* ensure that the designed ways of working are actually being followed and provide leaders firsthand insights into what is working—and what is not.

These elements are intentionally designed to function as a *system*, not a disconnected set of tools.

In the past, I have used the concept of a *lighthouse*—a model area where several operating system elements are successfully deployed together. The lighthouse serves as the visible proof point of what good looks like, allowing the organization to attribute real performance improvements to the operating system in action.

18.3 Lesson 3: Tools Do Not Transform; Leaders Do

An OE colleague once handed me a roll of duct tape—his good-natured jab at how often we are asked, "Can you just give us a tool?"

But tools are not magic.

A carpenter can give you her hammer, but that will not make you a builder. Skill is developed through coaching, repetition, and the occasional mistakes you never want to repeat.

The same is true for continuous improvement. Tools are necessary, but leadership behaviors—humility, curiosity, and discipline—are what transform organizations.

18.4 Lesson 4: Structure Matters More Than We Admit

Many Lean deployments focus on new processes but leave the organizational structure untouched. When companies do rethink structure in support of new ways of working, it can be transformational.

Early in my OE career, I led a Lean transformation that included a full process-centered reorganization into what were called *process execution teams (PETs)*—autonomous, self-sufficient teams organized around a core value stream. Each PET included all the roles needed to run their part of the business: operators across shifts, dedicated mechanics, planning and materials staff, quality specialists, a learning lead, and an OE coach. PETs typically ranged from 30 people in focused value streams to 300 in complex, multiline operations.

The benefits were considerable. Decision-making was accelerated. Priorities were aligned. People finally saw the customer at the center of the work. And perhaps most importantly, teams felt ownership. They no longer had to navigate a maze of matrixed stakeholders to solve a simple problem.

The lesson: *if you want new behaviors, sometimes you need a new structure.*

18.5 Lesson 5: CI Champions Are the Cultural Accelerators

Embedding one or two senior CI champions into site leadership teams has consistently been one of the best investments I have seen organizations make.

These are not OE staff. They are leaders who carry the torch—connecting the why, reinforcing the how, and ensuring that the operating system is not viewed as "someone else's work."

To be effective, they should sit on the leadership team; positioned too low, they lose visibility and influence.

The role should be rotational. When a leader completes a 2-year tour as a CI champion and then moves into a line job, the ripple effects are tremendous.

They evolve into custodians of the culture you are building.

18.6 Lesson 6: Sponsorship Makes or Breaks Projects

Active, engaged sponsorship is one of the most important differentiators in project success. Too often, a "sponsor" exists in name only, responsible merely for periodic status updates to senior leadership.

To set the right expectations, clearly define sponsor responsibilities in a role description. Choose someone with true ownership of the outcome—someone with real skin in the game.

A sponsor is accountable for the ultimate project deliverable. He/she should:

- Ensure that the team has sufficient resources—capital, expenses, and people.
- Remove obstacles quickly.
- Maintain regular check-ins with the project lead.
- Drive timely escalations when needed.
- Include the project in his/her annual goals to reinforce accountability.

When leaders sponsor well, teams rise to the challenge. When they do not, improvement stalls.

18.7 Lesson 7: Go to Gemba, But Go with Purpose

The Gemba is the place where value is created—whether that is a manufacturing floor, a QC lab, or an office cubicle where someone is analyzing data.

For leaders, visiting the Gemba is essential. It allows them to see firsthand what employees experience. This is a cornerstone of operating system sustainability. *Leader process confirmation* is the "secret sauce" of sustaining Lean.

Importantly, leaders must go to Gemba with a *clear purpose*. Early in a deployment, wait until the operating system has sufficient elements in place so leaders have something meaningful to confirm. Casual walk-arounds focused on pleasantries can become distractions if not grounded in purpose.

Effective Gemba behaviors include humble inquiry, walking the process, asking the right questions, and avoiding the temptation to take ownership of problems away from the team.

18.8 Lesson 8: Strategy Deployment Is about Alignment, Not Wordsmithing

Strategy deployment is where everything begins. Its purpose is simple: ensure that the entire organization is clear about strategic priorities and is working on the right things.

I have worked with varying approaches—from traditional Hoshin Kanri with x-matrices to hybrids using objectives, goals, strategies, and measures (OGSM), objectives and key results (OKRs), or other frameworks. The structure matters less than the *clarity* of priorities and the *alignment* from the executive team to the front line.

Across companies, I have seen leaders agonize over the phrasing of breakthrough objectives while avoiding the hard part: *making goals SMART and truly committing to them.*

True stretch targets drive higher performance. Conservative ones maintain the status quo.

18.9 Lesson 9: Performance Dialogue Builds Organizational Maturity

Publishing a scorecard is a common practice, but it is just data; the real value comes from a disciplined monthly performance dialogue with the leadership team.

A robust dialogue explores the following:

- Do we understand the root causes for underperformance?
- What countermeasures are in place?
- What support does the team need from leadership?
- What insights can be drawn from the data beyond the obvious?

The goal is not to turn the scorecard green. The goal is to ensure that you are measuring the right indicators and surfacing real issues so they can be addressed. If a metric is always green, replace it with one that drives improvement. There is *always* room to get better.

18.10 Lesson 10: End-to-End Value Stream Analysis Provides Insights No Other Tool Can

Second only to strategy deployment, end-to-end value stream analysis (VSA) is one of the most critical—and most underutilized—methodologies in pharma operations.

Few methodologies have as much leverage—and as little consistent application—as full end-to-end VSA. I have witnessed VSAs reshape supply-chain design, avoid costly mistakes, and uncover bottlenecks invisible in functional silos.

The ideal time to conduct a full E2E VSA is as a biopharmaceutical product enters pivotal Phase III clinical trials. At this point, establishing baseline data across supply nodes helps inform critical launch and future commercial decisions. Without this, companies frequently make well-intentioned but suboptimal choices that are later difficult to change—sometimes sending product back and forth across continents, adding unnecessary cost and lead time.

Despite the perception that "we already have the data," the true power of VSA comes from consolidating and visualizing it end to end: lead times, uptime/downtime, inventory at each step, and information and system inputs and flows.

When done rigorously, the insights can be transformative. At Shire, for example, a three-day E2E mapping exercise for their first ophthalmic product uncovered bottlenecks we had not recognized, leading to strategic changes that significantly improved commercial performance.

Update VSAs at least every 12 months; the initial mapping is the heavy lift, and subsequent updates are far easier. Insights compound.

18.11 Lesson 11: Tiered Accountability Is the Spine of the Organization

Tiered accountability is the backbone of any operating system—the equivalent of the body's spine and nervous system. It provides a structured communication channel from the shop floor to the C-suite, cascades strategic priorities, surfaces issues quickly, and accelerates resolution.

Daily management, typically conducted through tier boards, works best when supported by strong visual management. A well-designed board answers key questions at a glance:

- Are we winning or losing today?
- Where are the issues, applying the 30-second rule—meaning you should be able to spot the important issues in 30 s or less?
- What are the trends?
- What actions are underway?

The board should follow a logical flow: operational status → KPIs (safety/quality/delivery/cost/people) → trend charts → Pareto of causes for misses → actions. This creates a clear performance story and ensures that the team focuses on meaningful issues.

In the era of digitalization, it is easy to lose the power of physical boards. There is something profoundly reinforcing about picking up a marker and writing a red number for a daily performance metric. To counter this, digital boards must remain highly visible throughout the day—ideally on a large screen in the operations area—not hidden behind a laptop tab.

18.12 Lesson 12: 5S Is Simpler—And More Powerful—Than It Looks

5S is deceptively simple and yet frequently executed poorly. Many organizations stop at "2S"—sorting and setting in order—without thoughtful consideration of having the right quantities of consumables or the sustainability of the system deployed.

A strong 5S program requires discipline and ongoing ownership, but the payoff is enormous. We consistently underestimate the waste created by searching for tools, materials, or information.

A good 5S deployment should perpetually be focused on what goal you are trying to achieve, e.g., improving turnaround time on a test or improving safety, and should incorporate a kanban system for consumables.

18.13 Lesson 13: Problem-Solving Is a Capability, Not a Niche Skill

A clear, simple, structured problem-solving methodology—DMAIC or some other variant—is an essential element of a robust operating system. Your methodology should be simple and easy to use. Limiting each phase (define, measure, analyze, improve, control) to two or three core tools and providing straightforward templates will help you achieve that desired simplicity.

Organizations that excel in problem-solving train a critical mass of their workforce—60% as a minimum, ideally 100%. Most issues encountered by an organization are small to mid-sized; giving everyone the skills to solve these efficiently pays major dividends.

The foundational level of training is often dubbed yellow belt. I have seen the company brand color used to denote this foundational level as well. A well-structured Six Sigma, problem-solving ecosystem will build upon this foundation with more advanced training at green belt, black belt, and master black belt levels, where more advanced tools, training, and rigor are applied at each level. Yellow belt should focus primarily on special cause variation and eliminating defects in the organization. Green and black belt should be positioned to address common cause variation and more complex special cause problems.

My learnings from owning and managing Six Sigma programs are that they can suffer from a bad reputation if they are not focused on the company's most significant and pressing problems and its strategic imperatives. Candidates should have an active sponsor for their project and that project should be expected to deliver a minimum of cost savings or avoidance and should be governed by a robust stage-gate process.

Further, coaching capacity can be a significant constraint if not planned appropriately as all successful Six Sigma programs have a good coaching model.

18.14 Lesson 14: Human Error Is Not the Root Cause

One of the most profound courses I ever took made one idea incredibly clear: *human error is a symptom, not a cause.*

The course entitled "Causes and Prevention of Human Error" started with a biological and anatomical primer. As a biologist from my undergraduate education, I found this context fascinating, one of the key tenets being that human brains are wired for speed, not accuracy. While modern humans have a highly developed prefrontal cortex for complex thought, the primitive brain (comprising the brain stem and amygdala) continues to influence behavior, most notably the "fight or flight" response, and this sometimes leads to unhelpful automatic reactions in modern situations.

The bottom line: we are wired for speed, pattern recognition, and instinct. If someone makes a mistake, leaders must ask:

- Was the environment designed well?
- Were instructions intuitive?
- Was the person fatigued or distracted?
- Were controls adequate?

Human-centered design is not a luxury—it is a necessity. Your design should include not only processes and systems but also maintenance.

18.15 Lesson 15: Inventory as a Strategy

Early in my Lean journey, I learned the importance of understanding supply and demand variability when setting inventory targets—from raw materials to WIP to finished goods. Because carrying inventory is costly, OE teams often face pressure from leadership to reduce it, but the key is how to optimize.

It is important to understand that the right inventory level balances:

- Supply variability (reliability and lead times).
- Demand variability (including seasonality affects).
- Organizational risk tolerance.

Inventory optimization is a discipline—one that rewards rigor.

18.16 Lesson 16: There Is Only One Culture: A Culture of Excellence

We often talk about "safety culture," "quality culture," or "CI culture," as if these are discrete things, but in reality, there is only one: a culture of excellence.

The same principles underpin all of them—discipline, consistency, rigor, and pride in doing things right the first time. A process executed correctly is by definition safer, more compliant, more efficient, and of higher quality.

For someone who is in the latter years of my career, it has become very apparent that Operational Excellence is not a destination but an ongoing discipline—one built daily through the choices leaders make, the systems they reinforce, and the behaviors they model. The lessons in this chapter reflect years of successes, missteps, recalibrations, and, above all, privilege of learning from teams who were willing to challenge the status quo. When leaders commit to engaging every person, shaping structures that enable ownership, and cultivating a single culture of excellence, the results extend far beyond business metrics—they reshape how work feels and what people believe is possible. My hope is that these reflections not only affirm what you already know to be true but also encourage you to push further, lead with greater intention, and continue building organizations where continuous improvement is lived, not simply taught.

Beyond Distance, Towards Excellence: Leading Operational Excellence Programs in Dispersed Teams

19

Beatrice Pimlac Loesch and Alexis Baptiste Martin

Abstract

This chapter examines how to maintain strong engagement and cohesion in Operational Excellence (OPEX) programs within globally dispersed teams. It relies on Takeda's transformation and the eAGILE program to show how focusing on both technical and social skills helps organizations thrive despite the distance and constant change. The chapter showcases how to build team commitment and drive business results during the organizational restructuring, especially, the importance of clear communication, shared goals, and introduction of new team behaviors and norms to foster trust and team performance.

"To improve is to change; to be perfect is to change often," by *Winston Churchill.*

As customers, patients, and market demands are volatile, organizations are continuously changing. Innovation in science and technology adds to this complexity, requiring the workforce to continuously change. Furthermore, organizations in the pharmaceutical and biotechnological industry are regularly undergoing organizational restructurings and transformations. According to a recent study, the demand for not only technological skills but also social and emotional skills is increasing (McKinsey, 2024). This poses a challenge for the current workforce in organizations.

To support Takeda's transformation journey in Global Manufacturing Supply (GMS), global teams are dedicated to enabling sustainable transformation by

B. P. Loesch (✉) · A. B. Martin
Takeda Pharmaceuticals International AG, Glattpark, Switzerland
e-mail: beatrice.loesch@takeda.com; beatrice.loesch@tu-dortmund.de; alexis-baptiste.martin@takeda.com

T. Friedli et al. (eds.), *Striving for Operational Excellence in the Pharmaceutical Industry*, Future of Business and Finance,
https://doi.org/10.1007/978-3-032-26287-5_19

identifying and supporting value creation, capability building, and increasing overall Operational Excellence (OPEX) maturity in the respective teams and sites. The overarching transformation program is called eAGILE. Within eAGILE, there are teams composed of "Factory of the Future," "Business Process of the Future," and "Global Operational Excellence." This setup reflects Takeda's ambition to transform GMS as well as upskill the GMS workforce to be fit for the future. The eAGILE organizational structure tackles, among others, the challenge of increasing skills. However, the focus of capability building has been more on technological transformation as well as the OPEX and problem-solving capability. The AGILE Operating System (AOS) is Takeda's overarching system for continuous improvement, putting the business performance at the center. AOS helps to drive the transformation in eAGILE. This AOS is designed to provide a common language and common standards that support all colleagues in building their capabilities as Lean/Six Sigma problem solvers throughout the Takeda network and delivering value for Takeda. AOS is not a toolbox, but an interconnected, dynamic system that focuses on the most relevant and impactful elements to improve and deliver on business needs. The AOS training curriculum is categorized into so-called AOS elements. Although there are some elements such as "colleague engagement" and "leader standard work" in Takeda's eAGILE program, challenges to increase social and soft skills in GMS still exist.

The emphasis placed on change management can differ significantly across organizations. In some cases, it is a central focus with dedicated resources and programs, while in others, it may be integrated under different initiatives or receive less direct attention. At Takeda, for example, there is an AOS element specifically dedicated to change management, also referred to as colleague engagement. This AOS element can bring significant synergies when combined with elements that have measurable business performance and financial impact such as single-minute exchange of die (SMED), overall equipment effectiveness (OEE), business process management (BPM), or value stream analysis (VSA) deployments. As organizations are continuously changing, it is necessary to sustain colleague engagement in times of change and transformation.

An approach from the Global Operational Excellence (GOE) team to ensure colleague engagement was to have regular visits on-site. However, travel restrictions and travel budget constraints have significantly changed our ways of working. Instead of hands-on, face-to-face visits (Gemba walks) through the shop floor, most OPEX deployments are being led and deployed from our laptops in virtual meetings. However, virtual settings face numerous challenges; for example, it is challenging to keep participants engaged on the topic and guarantee it for the entire duration of the meeting. It requires mutual trust. Other examples are virtual OPEX trainings and site maturity assessments of AOS elements. The lack of control over how well an OPEX standard is deployed within global teams and among local site teams is challenging. It leaves us with only one option: build relationships and sustain team engagement and thus team cohesion. This leads us to the following questions below:

1. How to sustain team cohesion in dispersed global OPEX teams?
2. How to enhance team engagement by fostering a continuous improvement culture in global and local OPEX teams despite geographical distance?

This chapter aims to answer these questions on how to sustain team engagement in OPEX programs in dispersed teams by outlining best practices. The first part is dedicated to best practice within a global OPEX team spread across three continents, focusing on how to enhance team cohesion after an organizational transformation. The second part describes team engagement between the latter global OPEX team and local OPEX teams on sites, emphasizing the positive impact of aligned global teams which cascade on team engagement with local sites.

19.1 Part 1: How to Sustain Team Cohesion in Dispersed Global OPEX Teams

Takeda's structure revolves around its therapeutic areas (TAs) and geographic regions, with its global operations organized to serve these areas and focus on key strategic functions. Takeda's GMS organization currently operates more than 25 manufacturing sites worldwide, covering small molecules, biologics, plasma-derived products, and vaccines. Each site is committed to identifying efficiencies that improve our processes, so that we contribute to improving our patients' quality of life with the purpose for them to live better lives.

The GOE team is spread out across three continents in Takeda's global hubs, which are Tokyo (Japan), Greater Boston (USA), and Zurich (Switzerland). Figure 19.1 depicts the global hubs on the world map, illustrating the distances and different time zones. There are GOE teams in each global hub. The GOE teams operate out of global hubs to serve sites in their respective regions.

Prior to an organizational restructuring in 2024, the GOE team was organized in a matrix structure according to the four operating units comprising of small molecules, biologics, plasma-derived products, and vaccines along with differentiating OPEX between Lean and Six Sigma. Figure 19.2 illustrates a simplified organization chart of the GOE team. Please be informed that this is not the exact organization chart due to confidentiality agreement. The organization chart on the left side describes the structure prior to the reorganization in 2024. The GOE teams are set up by geographic location and distinguished between Lean and Six Sigma capability or projects. From a local perspective, a manufacturing site had two business partners (BPs) from the GOE team. This meant one person of contact for Six Sigma and another person of contact for Lean principles. Furthermore, the person responsible for Six Sigma and Lean was not based in the same global hub location by default.

Figure 19.2 also depicts the complexity of responsibilities within the GOE team itself and the sites.

This section is divided into two main parts to provide a comprehensive understanding of team dynamics within the GOE. The first part focuses on the GOE team

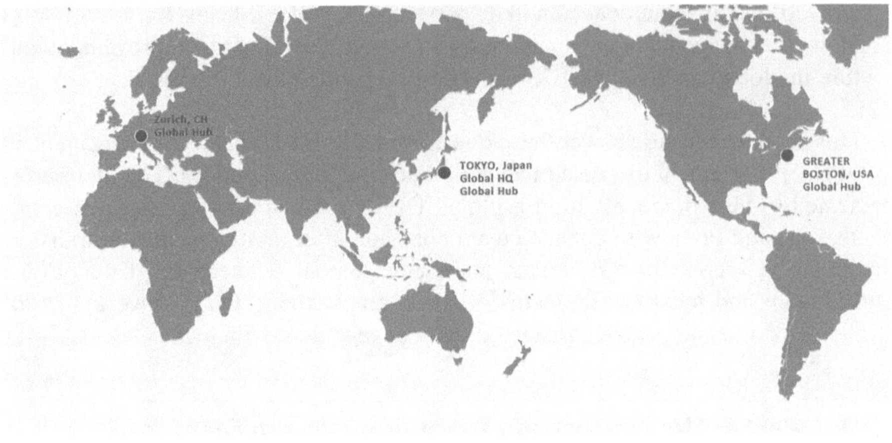

Fig. 19.1 The Global Operational Excellence team is spread across three global hubs

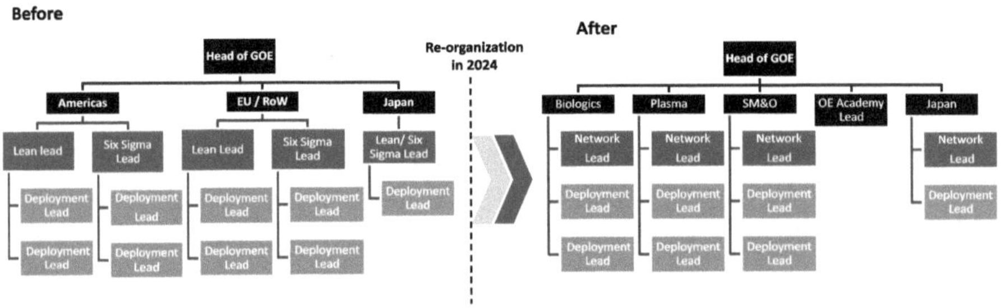

Fig. 19.2 GOE organization chart before and after reorganization in 2024

itself, examining their challenges and the approaches they used to strengthen team cohesion and engagement, serving as role models for the broader organization. The second part explores the specific challenges and interactions between the GOE team and local teams, highlighting how these relationships influence overall team effectiveness.

On comparing the left to the right organization chart, it is obvious that there was not a dedicated OPEX Academy Lead. Instead, every team member had assigned Lean Six Sigma (LSS) expertise and would train and deploy the respective LSS module on demand. Each member of the GOE team was required to train, coach, and deploy by geographical region and not by operating unit. As a result, there was a lack of knowledge in specific operating units such as small molecule and oncology (SM&O), plasma, or biologics. Another significant challenge the GOE team encountered was effective knowledge sharing and management. The existing organizational structure, with its regional and functional divisions, often led to the formation of silos. This made it difficult to ensure that valuable information and best practices were consistently shared across all parts of the team or between different regions.

19.1.1 Organizational Structure and Knowledge Management

Under the 2024 organizational transformation of GMS, the GOE organization chart also changed to focus more on operating units. As illustrated in Fig. 19.2, the reorganization shifted the GOE structure from a geographical distinction to three main operating units within GMS. Each operating unit now has a dedicated GOE team, and within these, there is one deployment lead per global hub (except in Japan). This setup enables each sub-team to focus on operating unit-specific training, coaching, and deployments. As a result, knowledge management and sharing within the GOE team are encouraged and can be improved.

19.1.2 Emotional Impact of Organizational Change

Like many organizations, Takeda's transformation brought emotional challenges. Ambiguity during transitions can create uncertainty, fear of unemployment, and confusion—affecting both individuals and teams. These factors can reduce commitment to the team and organization, and high turnover is often linked to a lack of team cohesion. Research also shows that uncertainty and diminished team spirit during organizational transformation can lead to decreased performance (Heim & Sardar-Drenda, 2020).

19.1.3 Employee Engagement and Areas for Improvement

Beyond business performance KPIs, Takeda conducts an annual global employee engagement survey with nearly 50,000 participants. The GOE team's results were suboptimal, revealing strong dissatisfaction in several categories and highlighting significant room for improvement.

19.1.4 Data-Driven Problem-Solving and Team Cohesion

Leveraging their problem-solving expertise, the GOE team analyzed the survey results together. They distinguished between factors within their control (such as how the team presents itself to stakeholders) and those outside their control (such as shifting stakeholder priorities or long software rollout times). By focusing on what they could influence, the team identified improvement opportunities and recognized that the success of these efforts depends on strong team cohesion.

Team cohesion is defined as a shared attraction, belonging, and connection or sense of pride among team members that is driven by social and behavioral elements associated with team membership (CaseyCampbell & Martens, 2009). Research shows a positive relationship between team cohesion and team performance (Beal et al., 2003; Grossmann et al., 2021). The goal for the GOE team was to strengthen team cohesion and thus improve team performance. To achieve this

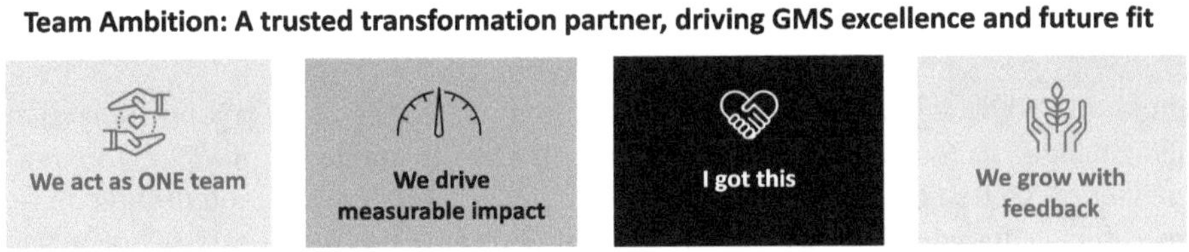

Fig. 19.3 GOE team ambition statement and four new behaviors

goal, the leadership team brought the entire GOE team together for a face-to-face strategy workshop in one location in the beginning of 2025. The 5-day strategy workshop with the entire team included stakeholders, business partners, as well as representatives from operating units and sites. With the help of external facilitators specializing in high-performing teams, we identified our common team's ambition. Building on the GOE team's ambition statement, we defined four new team behaviors that we hold each other accountable for. These are depicted in Fig. 19.3.

While holding each other accountable for these new team behaviors, also called norms, we set up new ways of working resulting in a governance model on meeting structure, knowledge management, and regular check-in for well-being. One member out of the team is assigned to organize and coordinate topics on team cohesion. This includes personal growth development events such as a psychological profile assessment for each team member. After each team member had accessed his/her profile, a framework for the entire team profiles and personalities was outlined together. Further learning and development opportunities for the team are ongoing and planned.

In sum, the efforts to increase high performance in a dispersed GOE team start with strengthening social and soft skills. This is reflected in newly identified behaviors and regular interactions. Moreover, team cohesion – comprising open communication, psychological safety, trust, and collaboration, as well as bonding (team building) – is a prerequisite to achieving measurable impact and thus team engagement in service of higher performance.

19.2 Part 2: How to Enhance Team Engagement by Fostering a Continuous Improvement Culture in Global and Local OPEX Teams despite Geographical Distance

In addition to their value-adding contributions to local OPEX teams and sites, the GOE team serves as a "bridge" connecting local OPEX teams with global functions and programs. The primary purpose of the GOE team is to empower local OPEX teams by building capabilities and increasing OPEX maturity, ultimately driving business performance. Capability building encompasses training in Lean Six Sigma (LSS), coaching LSS belts, and supporting deployments, all tailored to the OPEX proficiency and available resources of each site.

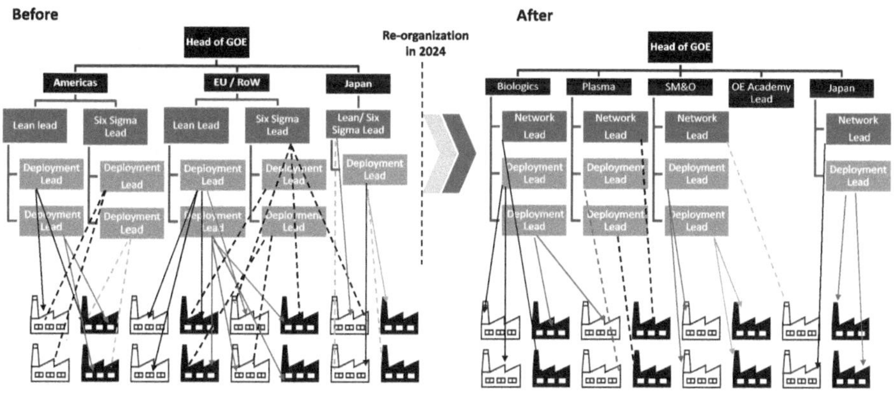

Fig. 19.4 GOE business partners as the person of contact for local OPEX sites

Importantly, decision-making within this framework is decentralized and operating unit focused. Requests for capability building from other site functions—such as quality, manufacturing, supply chain, or engineering—are evaluated based on strategic goals, annual initiatives, and value creation at the operating unit level. Decisions to support or not support are made locally, ensuring alignment with the specific needs and priorities of each site.

Regular connections between the GOE team and sites are essential for understanding business needs and aligning them with broader global initiatives. In this decentralized model, the GOE team often acts as the "voice of the customer" for Takeda's more than 25 manufacturing sites, representing local OPEX teams and sites to global functions and playing a pivotal role within GMS.

In parallel with the global organizational transformation of GMS in 2024, significant changes also took place at the site level. As previously discussed, such transformations have a direct impact on team relationships and engagement. The GOE team leveraged this period of change to adopt a more customer-oriented approach, aiming to enhance the quality of service provided to each site. One of the key improvements was the assignment of a single GOE business partner as the primary contact for each site. This means that, regardless of whether the need relates to Lean or Six Sigma, sites now have a clear and direct point of contact within the GOE team. Figure 19.4 illustrates the evolution of the organizational structure, highlighting the streamlined contact points between sites and their respective GOE business partners.

Comparing the organizational chart on the left with the right one in Fig. 19.4, the responsibilities towards the sites are now clearer and simplified. For example, there is no longer a distinction of either Lean or Six Sigma business partner for the sites. As a result, the local OPEX teams and sites only have one GOE business partner who is the person of contact for other global functions. Building on this simplification, communication and decision-making are more straightforward.

As in other pharmaceutical and biotechnological organizations, we also have challenges between global and local functions. We are continuously learning and improving on all levels, such as leadership and talent development. Despite challenges, we can conclude that the quality of collaboration between global and local levels is built on team engagement and trust. Since the new organization chart, many sites perceive the GOE business partner as part of the local OPEX team. This is the result of building meaningful collaboration with proven outcomes that add value to the business operations. We see it as an improvement when the global OPEX teams and local OPEX teams have a shared ambition, which is to sustain a culture of continuous improvement.

With the new organizational structure and streamlined connections between global and local teams, we focused on standardizing our ways of working by revamping the operating model. Teams were organized into agile squads, each composed of GOE members and their respective local OPEX partners. One of the core tools supporting this approach is the value creation and capability-building roadmap, collaboratively developed by each squad. This roadmap establishes mutual agreement on annual objectives, aligning OPEX expertise and capability-building efforts.

Similar to how a project charter is essential for project success, these roadmaps are fundamental for effective collaboration and team engagement between GOE and local OPEX squads. Once developed, each site's roadmap is presented to stakeholders for approval and alignment, providing a holistic view of initiatives across GMS and helping to break down silos. By mutually agreeing on these roadmaps, we also address the perception that global functions only create administrative work, demonstrating clear business value through targeted, squad-driven efforts.

In line with our approach to fostering alignment, the GOE team and each local OPEX team jointly developed a mutual ambition statement to clarify shared goals. Together, they identified the critical business objectives for each site—for example, ensuring uninterrupted production with objectives such as "don't miss a batch!" To support these goals, the teams conducted a gap analysis, comparing current OPEX skills with the capabilities required to achieve the business objectives. This OPEX proficiency assessment allowed the teams to align skills development with strategic priorities and incorporate identified gaps into the capability-building roadmap. This collaborative process not only ensures targeted capability building but also further strengthen team engagement.

Regular meetings between the GOE and local OPEX teams play a vital role in maintaining momentum and sustaining engagement. Consistency, mutual trust, and reliability have proven to be key drivers of measurable impact. These qualities are reinforced through ongoing meetings and active knowledge sharing. Furthermore, all stakeholders have demonstrated a commitment to breaking down silos between global and local teams, further enhancing overall team engagement.

While focusing on what is in our span of control, the GOE team put efforts in re-creating their image towards the sites by delivering tangible added value to the business onsite. This also strengthens relationships within the teams and enhances team engagement. Our best practice shows that when the GOE team shows up in unity, which is one of our new behaviors, it creates transparency and consistency, contributing to building trust.

Finally, the GOE contributes to building a culture of continuous improvement by training the local OPEX teams to take over more responsibility in OPEX capability building. It is the concept of train-the-train and thus empowering the local OPEX teams to be self-sustaining in OPEX. Due to resource availability, local OPEX teams still rely on GOE to support them.

However, it is the mutual understanding and ambition for the local OPEX capability to increase in the mid- and long term.

19.3 Conclusion

Our experience demonstrates that strong team cohesion is a critical driver of sustainable results and high performance. By fostering a culture where every team member takes responsibility and embraces continuous learning, we have been able to navigate organizational transformation effectively. The GOE team's success is rooted in behavioral change, adoption of new ways of working, and shared ambition.

These efforts have not only strengthened engagement within dispersed teams but also translated into measurable improvements in collaboration and business outcomes. Ultimately, our journey underscores that investing in team cohesion delivers lasting value, especially in a continuously changing environment.

References

Beal, D. J., Cohen, R. R., Burke, M. J., & McLendon, C. L. (2003). Cohesion and performance in groups: A meta-analytic clarification of construct relations. *Journal of Applied Psychology, 88*(6), 989–1004. https://doi.org/10.1037/0021-9010.88.6.989

CaseyCampbell, M., & Martens, M. (2009). Sticking it all together: A critical assessment of the group cohesion–performance literature. *International Journal of Management Reviews*. https://doi.org/10.1111/j.1468-2370.2008.00239.x

Grossmann, R., et al. (2021). The team cohesion–performance relationship: A meta-analysis exploring measurement approaches and the changing team landscape. *Organizational Psychology Review, 12*(2). https://doi.org/10.1177/204138662110411

Heim, E., & Sardar-Drenda, N. (2020). Assessment of employees' attitudes toward ongoing organizational transformations. *Journal of Organizational Change Management, 34*(2), 327–349. https://doi.org/10.1108/JOCM-04-2019-0119

McKinsey. (2024). *A new future of work: The race to deploy AI and raise skills in Europe and beyond.* https://www.mckinsey.de/~/media/mckinsey/locations/europe%20and%20middle%20east/deutschland/news/presse/2024/2024%20-%2005%20-%2023%20mgi%20genai%20future%20of%20work/mgi%20report_a-new-future-of-work-the-race-to-deploy-ai.pdf

"Execution Excellence" Demonstrated by Boehringer Ingelheim BioPharma RCV: Vienna/Austria

20

Tilman C. Rock, Teresa Brenner, Christoph Schwarz, Daniela Ruehrig, Joachim Guger-Hoeller, Tobias Kutscher, and Alexander Krauland

Abstract

The journey towards "execution excellence" at Boehringer Ingelheim BioPharma, Vienna, demonstrates the power of process simplification and the strategic integration of digitalization and AI to achieve operational excellence. First, the dedicated, cross-functional PEOPLE team has established a set of measures to address human errors. Additionally, the team facilitates the implementation of these measures by using change formulas to enhance the understanding of the value of organizational changes. Secondly, a use case was presented that demonstrates how machine learning techniques were utilized for troubleshooting activities, thereby supporting the optimization of a manufacturing process for a commercial biopharmaceutical product. Ultimately, "execution excellence" is about fostering the right mindset and focusing on employees, who are the heart of our Boehringer Ingelheim's mission statement.

Boehringer Ingelheim's so-called LEITBILD (mission statement) serves as the starting point for the unique company culture; moreover, it is instrumental in defining the framework of what, why, and how we execute (excerpt as follows):

T. C. Rock (✉) · T. Brenner · C. Schwarz · D. Ruehrig · J. Guger-Hoeller · T. Kutscher · A. Krauland
Boehringer Ingelheim RCV GmbH & Co KG, Vienna, Austria
e-mail: tilman.rock@boehringer-ingelheim.com; teresa.brenner@boehringer-ingelheim.com; christoph_2.schwarz@boehringer-ingelheim.com; daniela.ruehrig@boehringer-ingelheim.com; joachim.guger-hoeller@boehringer-ingelheim.com; tobias.kutscher@boehringer-ingelheim.com; alexander.krauland@boehringer-ingelheim.com

T. Friedli et al. (eds.), *Striving for Operational Excellence in the Pharmaceutical Industry*, Future of Business and Finance,
https://doi.org/10.1007/978-3-032-26287-5_20

- *"We are independent, family-owned, and intend to remain so"*:

Driven by the desire to serve humankind by improving human and animal health, we feel responsible for our communities and respectful of resources; we plan in generations and focus on long-term performance.

- *"We create value through innovation for our customers"*:

We develop breakthrough therapies; we excel in innovation and deliver highest quality.

- *"We are powered by our people"*:

We foster a collective journey to elevate performance, strengthen collaboration, and shape a future where we consistently go beyond.

In this respect, the customer/patient and our people/employees play a pivotal role in living up to the LEITBILD; hence, a key strategic priority for Boehringer Ingelheim BioPharma Vienna is the element of "Execution Excellence," especially in operations.

Friedli and Belim defined Operational Excellence as a continued pursuit of improvement of a production plant in all dimensions (Friedli & Belim, 2013).

Building on this definition, we use the term "execution excellence" which in practical terms is demonstrated by the BioPharma Team in Vienna: At the Boehringer Ingelheim's Vienna site, around 1800 colleagues are active in the manufacture of biologic drug substance (BDS). The development and manufacturing of biopharmaceuticals based on microorganisms (e.g., *E. coli*) and by cell culture/mammalian cells (e.g., CHO) are well established in cutting-edge facilities and structurally organized in mainly two so-called focused factory departments (cell culture and microbials). Engrained in our mission: our continuous drive for "execution excellence" and cost efficiency is fueled by our passion, industry leading experience, and patient focus.

Specially, manufacturing processes are critical to the success of a production plant and offer therefore significant opportunities for optimization. In the context of manufacturing processes, this translates to enhance productivity and efficiency, improving product quality, minimizing cost, and waste. Consequently, working in operations (i.e., BioPharma manufacturing), the task of "execution excellence" is clear to "demonstrate that you are doing ordinary things extraordinarily well," namely:

(i) Ensure uninterrupted supply of quality products for patients—on-time-in-full (OTIF).
(ii) Unlock growth by continuous improvement with (manufacturing) data and digitalization/AI in machine learning.

Both imperatives are mutually dependent on each other: our manufacturing processes are ultimately executed by humans to define, implement, and steer the corresponding processes, and in turn we are generating ever-growing process data that makes it difficult to sort relevant from irrelevant data in order to run the process robust and stable—causing a dilemma, generally seen by typical symptoms like "human error rate" and backlog.

It is a common theme across the industry that the so-called human error issues are attributed to non-robust processes or "mura" (irregularity in processes) and "muri" (overburden on people in operation processes).

Not surprisingly, high human error rates lead to extensive investigations, delays, replanning, and potentially rejected batches, consequently leading to high costs (rejections) and negative customer perception. In summary, manufacturing is on a downward spiral, fueling a vicious cycle! Therefore, the mission is obvious: identifying the underlying root causes for (human) errors and establishing an environment and culture of error prevention and continuous process improvement. In the following paragraphs, we wanted to share the approach towards "execution excellence" that we have taken by the deliberate decision to concentrate on:

- Pushing "execution excellence" and tackling human error by radically simplifying (and therefore stabilizing) processes
  → There is a huge potential in our workforce to deliver effective CAPAs—listen to them about the underlying root cause and focus on simplifying their work and success will follow.
- Using digitalization/AI in machine learning to optimize production processes
  → Machine learning substantiates execution excellence: machine learning enables data insight and is therefore instrumental in the rapid detection of potential influencing factors within the complex manufacturing environment of BioPharma and enables for data-driven decision-making. It is not about the volume of data, but rather how it is used. When the data is properly aligned, prediction models can be quickly set up and implemented, leading to more efficient and effective operations.

On a site note: both imperatives can only be realized through and by the above-mentioned people focus: "execution excellence" is also about mindset. It lives in the choices we make each day, the ownership we take, and the energy we bring to turn plans into impact. It is about each one of us, across every function.

20.1 Pursuing "Execution Excellence" by Tackling Human Error and Maintenance Coordination Issues in BioPharma Operations by the "PEOPLE" Team

To address "execution excellence" and human error, a temporary internal consulting team—the PEOPLE team—was established. The core task of the team is to systematically reduce complexity in BioPharma operations by simplifying processes, fostering accountability, and driving sustainable improvements. This directly addresses the human error challenges outlined above, developing sustainable solutions for error prevention in production processes. In addition, measurable improvements in other domains, such as maintenance coordination, can be highlighted as a success story of this internal consulting program.

P E O P L E stands for:

→ *P*rocess simplification
→ *E*fficiency enhancement
→ *O*utcome optimization
→ *P*erformance elevation
→ *L*earning embracement
→ *E*rror reduction

Together, these principles form the foundation for a cross-functional, dedicated team (consisting of 6–8 members) operating under the guiding principle: "easing your work, enhancing our success."

Accountability and ownership are strengthened by selecting team members from within the organization, ensuring a wide spectrum of technical expertise. Appointment to the PEOPLE team requires individuals to demonstrate leadership qualities such as self-motivation, assertiveness, and openness to change.

20.2 Setup, Roles, and Responsibilities

The PEOPLE program is initiated and governed by the Site Leadership Team Austria (SLTA), which provides ongoing sponsorship for the PEOPLE team. This support plays a critical role in breaking down departmental silos, enabling cross-functional collaboration, and cultivating an environment that drives continuous improvement.

To maximize impact, team members work exclusively for the program, temporarily stepping away from their previous roles. This full-time commitment ensures that the primary workload during the project phase is managed by the PEOPLE team, allowing smooth implementation without burdening the affected areas. The team remains engaged until the change is fully embedded, the target state is achieved, and processes operate reliably.

A hybrid methodology was introduced, combining agile values and principles for the team's working mode with a classic stage-gate program governance structure. Under this approach, the PEOPLE team reports to the Product Owner Team (POT), composed of key stakeholders from the SLTA. Every four weeks, the PEOPLE team and POT review the previous sprint and agree on a sprint contract for the next cycle, outlining required deliverables. For each focus topic, a POT member acts as a sponsor, working closely with a dedicated topic leader from the PEOPLE team to provide support and address challenges (e.g., integration of focus topic goals into performance expectations for first level and middle management).

20.3 Project Management

The team's mandate to deliver quickly within a strict framework introduced planning challenges under the new methodology. To overcome these, agile project management methods were evaluated. The decision was made in favor of the hybrid STAGILE® framework from ALEAS AG (see Fig. 20.1), which integrates agile project management into the traditional stage-gate process, combining established

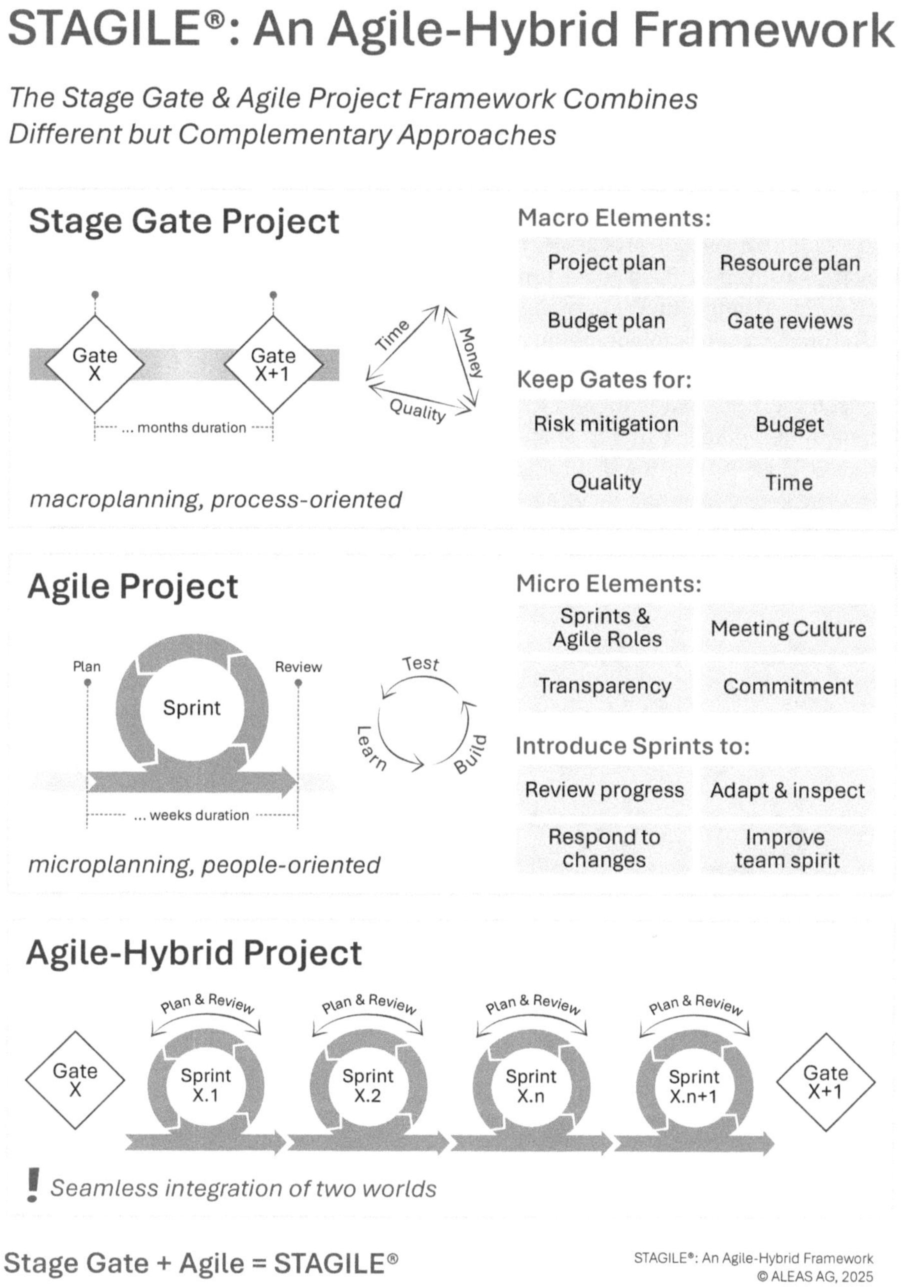

Fig. 20.1 STAGILE® Agile hybrid framework

project management practices with the new needs of the initiative. An online collaboration platform (ALEAS board) was implemented to further support project management and the self-organized way of working.

The role of the scrum master is distinct from that of the topic leaders, allowing the latter to concentrate on stakeholder management, communication, and scoping. Scrum masters are responsible for ensuring that sprint goals are met, and all assigned tasks are completed on time.

To optimize various processes, the project was organized in a sequence of individual focus topics. Each focus topic is processed in a series of three sprints, with each sprint having a duration of 4 weeks (see Fig. 20.2).

Before each sprint planning meeting, in a POT meeting, the top targets and risk focus of the upcoming sprint are agreed on. This is the so-called sprint contract between the PEOPLE team and the POT. Based on this contract, the activities for the sprint are planned by the team in a joint planning session. The success of the sprint is closely monitored through daily scrum meetings. The sprint cycles are linked by sprint planning meetings consisting of a sprint review and a sprint planning part. Additionally, risk and opportunity reviews, masterplan reviews, and lessons-learned sessions are conducted on a rolling basis.

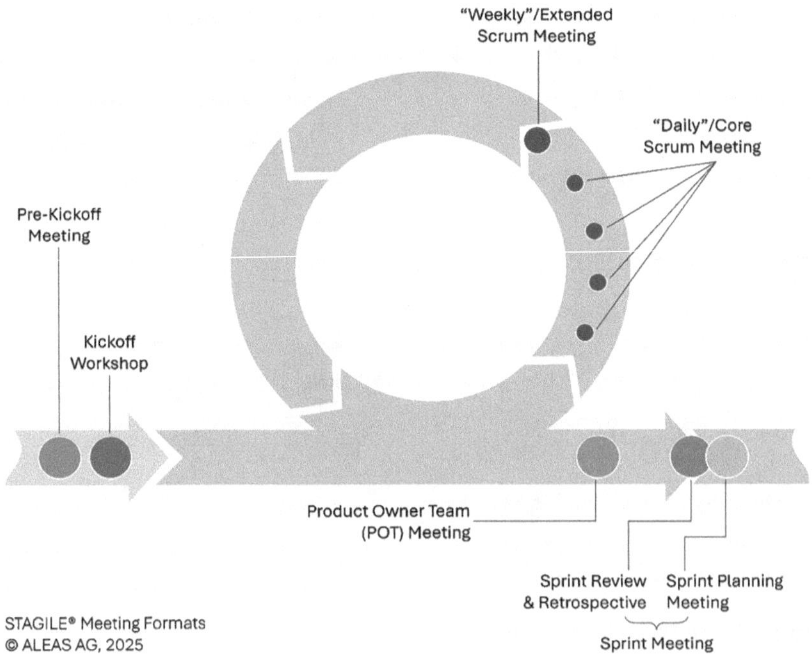

Fig. 20.2 Scrum circle according to ALEAS AG

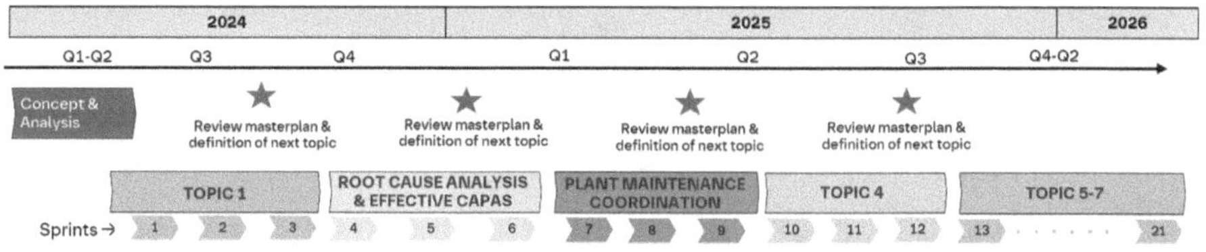

Fig. 20.3 Program timeline

20.4 Program Timeline

Prior to the first sprint of each topic, a pre-analysis of the problem field is performed, usually leveraging a voice-of-the-customer approach.

Typically, the first sprint covers an analysis phase to gain deeper understanding, structure the topic, and identify potential challenges and improvements. The second sprint focuses on developing improvement measures—new or optimized processes, development of IT tools, adapted meeting structures, etc. These measures are rolled out in the third and final sprint of the topic. Once the subsequent topic is started, implementation and control of previous topic measures are still supported during a "hyper care" phase.

The entire program is scheduled for 24 months, starting with a gap analysis followed by seven individual focus topics and finally a fade-out phase to secure the knowledge gained (see Fig. 20.3).

20.5 Working Mode

Initially, a data analysis was performed to identify potential focus topics, based on main drivers for errors, "hot spots," and improvement areas. Over 40 interviews were conducted, with interviewees selected for their opinion leadership, openness, and ability to provide candid feedback. Following transcription, a keyword analysis was performed, and a shortlist of topics was compiled based on team experience and interview insights.

Each topic begins with a comprehensive stakeholder analysis. Daily project work is conducted in close collaboration with subject matter experts. To minimize the workload of the line organization, most development and rollout activities are done by the PEOPLE team. A significant portion of their efforts are dedicated to communicating the program's objectives and ongoing activities.

To foster bottom-up and top-down communication (see Fig. 20.4), two committees were formed:

- Sounding Board: Selected first-level management meets biweekly to discuss processes, solutions, and measures and to provide feedback.
- Middle Management Steering: Selected middle management also meets biweekly, with the goal of facilitating involvement and easing implementation within respective reporting lines.

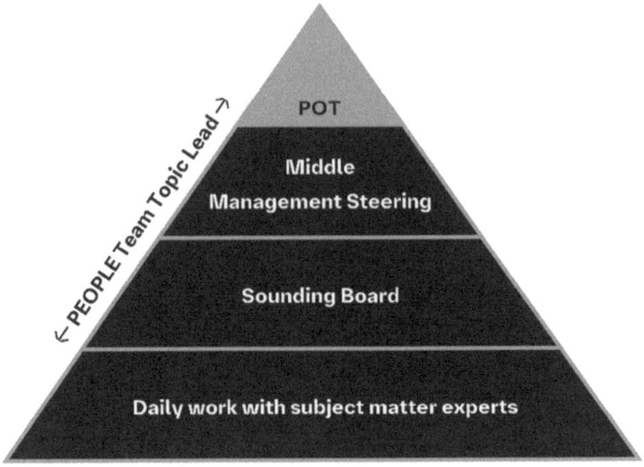

Fig. 20.4 Communication

Members of the PEOPLE team promote the program in various committees to raise awareness across the organization, including the global BioPharma network meeting, senior management meetings, all-staff meetings, and department meetings. Since PEOPLE team members originate from diverse professional backgrounds and consistently communicate outcomes within their former reporting lines, this practice ensures broad and effective dissemination of information throughout the organizational structure.

20.6 Change Management

During one of the first sprint retrospectives, the team identified limited acceptance of implemented measures as a major challenge. Further discussions with key personnel revealed underlying causes: certain leadership levels were not involved in developing the measures, and some experts failed to communicate planned activities within their respective lines.

The team experienced significant variations in acceptance, often aligned with departmental boundaries. To address this, professional change management support was engaged. Key learnings emphasized the need for consistent, comprehensive communication across all management levels—from the POT through middle management to sounding boards—ensuring uniform messaging throughout the organization.

From a set of change tools, the "change formula" (see Fig. 20.5) was selected to clearly illustrate the need for change, its urgency, initial effort, available capacity, and expected benefits. A distinctive feature is its consideration of the personal "perceived cost" associated with change. Presented as a product of these factors, the formula emphasizes that success requires attention to every element, avoiding the risk of "multiplying zero."

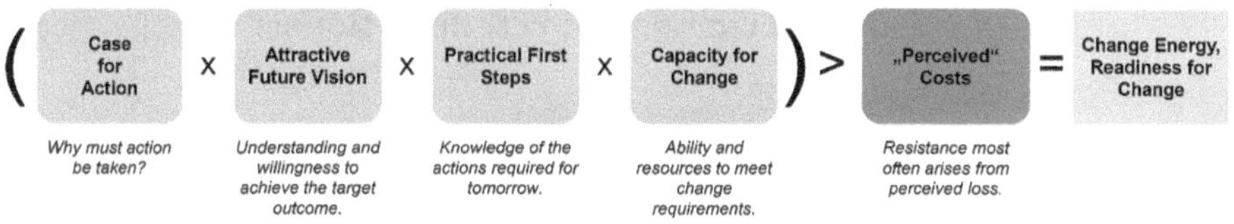

Fig. 20.5 Change formula according to osb international, adapted from Organization transition: Managing complex change Richard and Harris (1987) and The change agenda Kamphuis (1992)

20.7 Success Story A: Reduction of Human Error and Effective CAPAs

Persistent weaknesses in root cause analysis have become evident, particularly the tendency to attribute deviations to "human error" (see Fig. 20.2). While this explanation may seem straightforward, it often conceals deeper systemic issues. Historically, time pressure and ingrained procedural habits have resulted in corrective and preventive actions (CAPAs) that were largely superficial. Common measures, such as revising standard operating procedures or scheduling retraining, created an illusion of resolution without addressing the true underlying causes.

The outcome was a cycle of ineffective measures. Procedures are multiplied, adding complexity instead of clarity. Training requirements expanded, consuming resources without delivering meaningful improvements. Additional control mechanisms were introduced, not to resolve issues, but to compensate for unresolved risks. These patterns highlight the urgent need to move beyond superficial fixes and adopt strategies that target the true drivers of deviation.

To resolve these challenges, a comprehensive set of measures were implemented across the organization:

- An *investigation toolbox* was introduced to support structured and consistent root cause analysis.
- A *risk-based cost-benefit CAPA tool* was deployed, enabling transparent and evidence-based decision-making.
- *Clear criteria for categorizing human error* were defined and embedded into investigative practices.
- A *standardized CAPA cross-check procedure* was established to ensure quality and consistency across departments.
- The *training framework was revised*, incorporating "tool buddies" to reinforce practical application.
- *Leadership training on human error perception* was conducted and has become part of ongoing development programs.
- A *cultural shift was initiated and embraced*, positioning human error not as a conclusion but as a starting point for deeper investigation.

These measures, now fully integrated into daily operations, have significantly strengthened CAPA effectiveness, minimized procedural inefficiencies, and

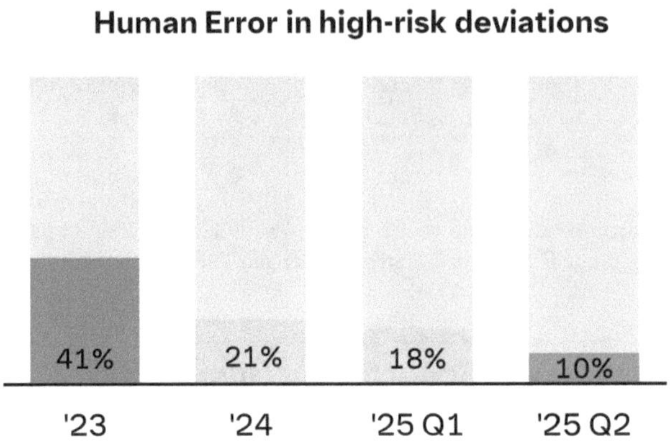

Fig. 20.6 Percentage of high-risk deviations attributed to human error from 2023 through Q2 2025. Bars represent annual and quarterly values: 41% (2023), 21% (2024), 18% (Q1 2025), and 10% (Q2 2025). The KPI target is ≤15%

fostered a culture of accountability and continuous improvement across the organization.

Figure 20.6 illustrates the reduction in human error rates in high-risk deviations over time, the main driver for CAPA effectiveness increase.

20.8 Success Story B: Plant Maintenance Coordination

To shorten the annual maintenance shutdown of facilities, the organization performs a significant portion of maintenance activities throughout the year during ongoing production. This approach requires close collaboration between operations and engineering. Initially, the planning process lacked robustness and transparency, resulting in idle time for technical maintenance staff as maintenance tasks were frequently rescheduled due to equipment unavailability.

Together with specialists from operations and engineering, the PEOPLE team developed specific measures:

- *An equipment occupancy cockpit*
- *Updated meeting framework*
- *KPIs for all departments/areas involved*

The equipment occupancy cockpit was introduced as the central platform for planning and communication. This tool consolidates detailed production equipment utilization based on the production schedule and required maintenance activities with respective timelines. Built in MS Excel with macro automation, it serves as the interface between production scheduling software and SAP PM. All planning and coordination are managed within the cockpit, ensuring transparent visualization of scheduled activities.

To enable smooth communication with clearly defined responsibilities, a structured alignment process was introduced, covering three planning horizons:

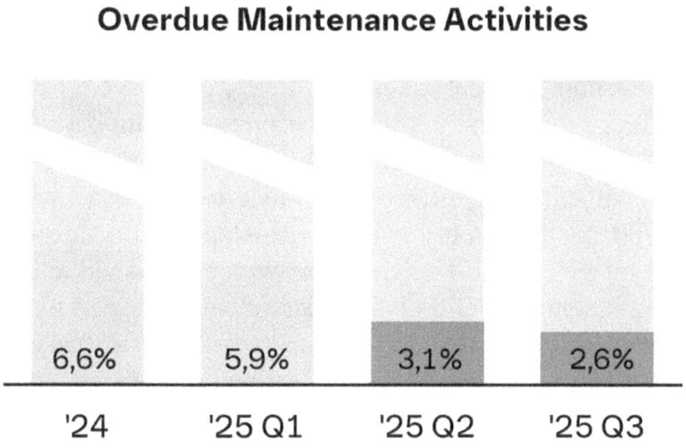

Fig. 20.7 Percentage of overdue maintenance activities from 2024 through Q3 2025. Bars represent annual and quarterly values: 7% (2024), 6% (Q1 2025), 3% (Q2 2025), and 3% (Q3 2025). The KPI target is ≤5%. Overdue work orders were calculated as the number of overdue work orders for GxP equipment, divided by the total number of monthly planned work orders for GxP equipment, multiplied by 100

- *Long-term planning:* Identifies equipment for preventive maintenance during the annual shutdown
- *Midterm planning:* Maintenance activities performed throughout the year
- *Short-term planning:* Activities that need to be rescheduled

The primary KPI for maintenance is the percentage of overdue activities (Fig. 20.3). Initially, this metric was reported in aggregate for the entire production site. To enhance transparency, regular monitoring was extended to each production area and engineering department, enabling faster responses to negative trends. Additional performance indicators were incorporated into performance reviews. All measures, along with background information and practical guides, were published on an internal SharePoint for easy access.

Within 12 weeks, the PEOPLE team developed and implemented these measures, fostering a positive work atmosphere and strengthening trust among departments. Department heads reinforced the change by jointly communicating their commitment to collaboration. After tools and processes were handed over, the PEOPLE team continued to provide troubleshooting support and monitored planning and communication activities. As collaboration and trust grew, KPIs quickly stabilized at an excellent level (Fig. 20.7).

20.9 Success Factors

Improvement programs often depend on external consultants, yet the primary workload remains with the sponsoring organization. A critical success factor is the full dedication of PEOPLE team members, enabling them to manage high workloads, execute strategic tasks, and perform essential operational work, such as revising documents, developing tools, and testing measures before implementation. By assuming responsibilities typically handled by the target organization, the initiative accelerates progress and achieves strong acceptance, as improvements do not solely rely on local effort.

The team was recruited from multiple departments, creating a combined network that spans nearly the entire organization. This broad connectivity enhances communication and enables rapid, effective conflict resolution. Challenging issues, whether with shop floor operators or senior decision-makers, are easier to address when approached by a PEOPLE team member with established relationships. Team members bring both seniority and diverse perspectives shaped by their respective areas. Their collaboration is characterized by openness rather than competition, fostering high trust and strong team spirit. This enthusiasm extends to the wider organization, where the PEOPLE team is seen as a real benefit.

Summary of success factors:

- *Full team commitment:* Ownership of both strategic and operational tasks accelerates progress and ensures high acceptance.
- *Strong internal networks:* Cross-departmental team constitution enables fast communication and effective conflict resolution.
- *High trust and team spirit:* Seniority and diverse expertise foster mutual respect, openness, and enthusiasm.
- *Proven impact:* Early KPI improvements build credibility, turning initial skepticism into active support.
- *Project management framework and tool:* The hybrid project management approach provides long-term orientation while maintaining short-term focus.

20.10 Learnings

Throughout the PEOPLE program, continuous learning is essential. A key insight is, even with a well-defined governance structure, broad and inclusive communication is critical for success. Keeping all management levels informed about program objectives and progress fosters trust and strengthens acceptance of proposed solutions. Initiatives like PEOPLE identify that a significant portion of effort goes into maintaining a steady stream of communication across all organization layers.

20.11 Using Digitalization/AI to Optimize Production Processes

Recognizing the power of smart data usage, we as Boehringer Ingelheim have been taking steps to standardize its data management across all manufacturing sites to increase the use of big data. This move is expected to further streamline manufacturing processes and increase their predictability and robustness, enabling us to continue to deliver high-quality biopharmaceutical products to our customers. It proves the biopharmaceutical paradigm that the process is the product, and its quality is determined by how the process is conducted.

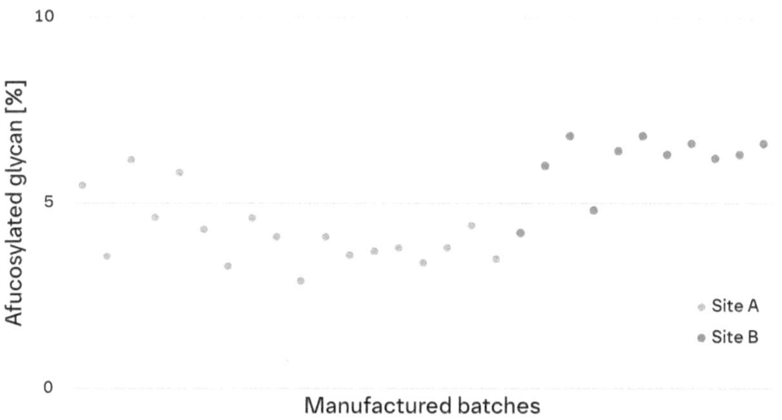

Fig. 20.8 Afucosylated glycan level of manufactured batches at sending site A and receiving site B

AI is an emerging and rapidly growing field in digitalization and is steadily increasingly adopted within the pharmaceutical industry. At Boehringer Ingelheim, artificial intelligence, specifically machine learning techniques, is regularly applied to support troubleshooting activities and to optimize production processes in the manufacturing environment.

The following presents a successful use case demonstrating how machine learning techniques were utilized for troubleshooting activities and consequently supported the optimization of a manufacturing process of a commercial biopharmaceutical product. More in detail, machine learning techniques identified quickly potential influencing factors for an unexpected impurity increase in a complex biopharmaceutical manufacturing process following a product transfer between two sites.

Technical (and analytical) product transfers within a given production network (e.g., from site A to site B) are a key activity in the biopharmaceutical industry to ensure a stable supply of prescriptive medicines. Ultimately, the process performance qualification (PPQ) confirms that the critical quality attributes (CQAs) along with set process and product parameters are comparable and reliable within a given range. It is normal and indeed necessary to make process adjustments. These adjustments ensure that the transfer process aligns with the unique environment and maintains the required quality and efficiency.

Hence, a higher than usual so-called afucosylated glycan level after the transfer to the receiving site B was investigated to ensure Boehringer Ingelheim's highest quality standards (see Fig. 20.8).

20.12 Utilizing Data Science

The in-house data science team of Boehringer Ingelheim, Vienna, performed an analysis of all available production data by accessing the different data sources of the distributed control system (DCS) and in-process control (IPC).

The first step after the initial data collection is always to prepare the data by converting it to the same units (e.g., PSI into bar) to allow for a direct comparison between the two manufacturing process datasets. Diligently transforming and harmonizing data are crucial to enable the analysis of the underlying factors and dependencies using machine learning algorithms.

To transform the data and align all collected production parameters of the previous runs performed at the two manufacturing sites, a method known as "tsfresh" (time series feature extraction based on scalable hypothesis tests) was used. Tsfresh is a powerful tool for automatically extracting and selecting variables from time series data. Tsfresh generates numerous statistical features and uses hypothesis testing to retain only the most relevant ones, creating a structured feature matrix for machine learning tasks (Christ & Braun, 2018).

With the data now successfully aligned, the data science team was able to delve into the root cause of the higher-than-usual glycan level. This was accomplished using correlation algorithms, which allowed for a comprehensive breakdown of the data. To facilitate this process, a web application was implemented to analyze the correlation of all 812 computed variables, providing a powerful tool for data investigation. The use of this web application led to a rapid overview of the most related variables and influencing factors, which revealed some insightful findings. It was discovered that osmolality, ammonia, and glucose are correlated to the afucosylated glycan level. This correlation is exemplarily illustrated for day 9 of the upstream process in Fig. 20.9.

To confirm the findings of the correlation analysis, a machine learning approach was used. The goal was to predict the afucosylated glycan level using regression and to predict the production site via classification. For both regression and classification, a machine learning model was trained on all available variables. Glycan level and production site were not included as input variables for either model, as they were the target variables for regression and classification, respectively. A high-level illustration of the machine learning approach applied here is shown in Fig. 20.10.

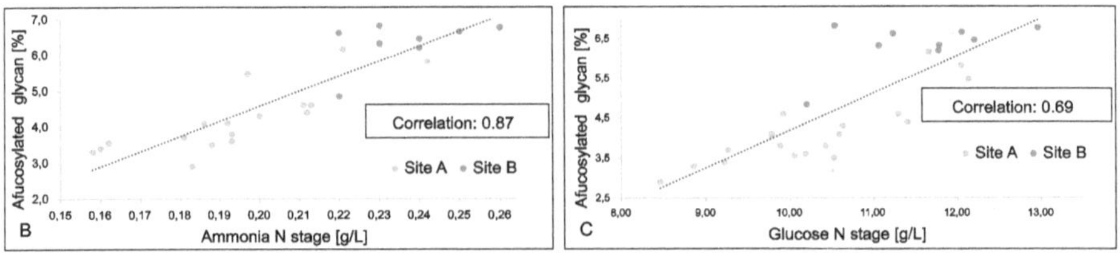

Fig. 20.9 Afucosylated glycan correlations to osmolality (A), ammonia (B), and glucose (C) on day 9 of the upstream process for previous runs at the two manufacturing sites

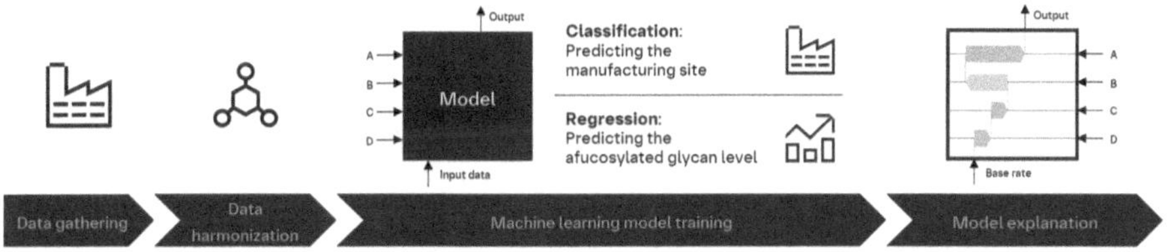

Fig. 20.10 High-level illustration of the machine learning approach applied in the afucosylated glycan case study

After collecting and harmonizing the data using the same data sources as for the correlation analysis, the machine learning framework XGBoost was implemented for its powerful capabilities with classification and regression tasks. XGBoost is an ensemble learning technique that utilizes decision trees that integrate the predictions of numerous base models to come to a final overarching prediction. Two XGBoost models were trained on the data provided by the two sites, one to achieve the classification goals and the other for the regression task (Cloudthat.com, 2024).

To explain the output of the two machine learning models, the game theoretic approach Shapley Additive exPlanations (SHAP) was implemented. Utilizing the Shapley values from game theory and their related extensions, the approach combines credit allocation with local explanations (Shap.ip, 2024).

After analyzing the data of the machine learning algorithms for the 12kL fermentation, we supported the findings from the correlation analysis. To then transition from correlation to causation, diving into literature research was necessary. The literature research supported the correlation that high ammonia concentration and osmolality can elevate the afucosylated glycan level (Rish et al., 2023).

Upon review of the data, it was found that the viable cell density (VCD) at site B was lower than at site A at a comparable glucose consumption rate. Consequently, it was suspected that the root cause of the higher ammonia, osmolality, and glucose levels, and as a result, higher afucosylated glycan levels, could be caused by "overfeeding." To tackle the root cause of "overfeeding" the cells and therefore reducing afucosylated glycan levels in site B, the feed flow rate was adjusted.

As can be deducted from Fig. 20.11, the first batches produced at site B after this adjustment showed significantly reduced glycan levels. The results of the now lowered afucosylated glycan levels highlight a successful investigation using data science and machine learning in the biopharmaceutical manufacturing environment.

20.13 Executive Summary

In summary, the journey towards "Execution Excellence" at Boehringer Ingelheim BioPharma Vienna demonstrates the power of process simplification and the strategic integration of digitalization and AI to achieve operational excellence. First, the

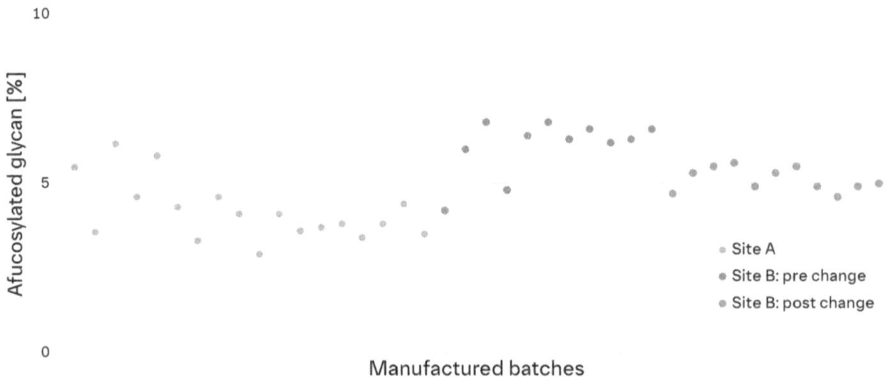

Fig. 20.11 Afucosylated glycan level of manufactured batches at sending site A and receiving site B. Post-change batches at site B represent runs after data science investigation was concluded and the adjustments to the process variables were made

dedicated, cross-functional PEOPLE team has established a set of measures to address human errors. Additionally, the team facilitates the implementation of these measures by using change formulas to enhance understanding of the value of organizational changes. Secondly, a use case was presented that demonstrates how machine learning techniques were utilized for troubleshooting activities, thereby supporting the optimization of a manufacturing process for a commercial biopharmaceutical product. Ultimately, "Execution Excellence" is about fostering the right mindset and focusing on employees, who are the heart of our Boehringer Ingelheim's mission statement.

References

Christ, M., & Braun, N. (2018). Time series FeatuRe extraction on basis of scalable hypothesis tests (tsfresh—A python package). *Neurocomputing, 307*(7).

Cloudthat.com. (2024). *XGBoost: A Powerful Machine Learning Framework* | last accessed 21.11.24, www.cloudthat.com/resources/blog/xgboost-a-powerful-machine-learning-framework

Friedli, T., & Belim, D. (2013). *OPEX: A definition; in leading pharmaceutical operational excellence—Outstanding practices and cases* (1st ed.). Springer-Verlag.

Kamphuis, A. (1992). *The change agenda*. Addison-Wesley.

Richard, B., & Harris, R. T. (1987). *Organization transition: Managing complex change*. Addison-Wesley.

Rish, A., Siddiquee, K., Huang, Z., Xu, J., Anderson, C., Borys, M., & Khetan, A. (2023). Strategies for controlling afucosylation in monoclonal antibodies during upstream manufacturing. *Biotechnology Journal, 18*(7).

Shap.ip. (2024). *SHAP documentation* | last accessed 21.11.24, www.shap.readthedocs.io/en/latest

OEE Improvement in Sterile Operations of Dechra Pharmaceuticals

21

Radovan Biloglav

Abstract

The COVID-19 pandemic put a lot of pressure on manufacturers, making it harder for many sites to keep up productivity and quality. At Dechra Pharmaceuticals, these broader impacts became very real for one of its small manufacturing sites. Despite a history of reliable supply, the site saw drops in efficiency and mounting operational bottlenecks. This triggered a thorough benchmarking assessment, revealing strong customer service but highlighting urgent needs to improve productivity, reduce changeover times, and streamline deviation management—setting the stage for a focused Operational Excellence (OPEX) transformation. This chapter shows that sustained success comes from building capability, recognizing effort, and making continuous improvement part of everyday operations.

21.1 Intro to Dechra Case Study

Dechra Pharmaceuticals is a company that provides a variety of medicines for the veterinary industry and has a network of internal and external manufacturers across the world. One of the key sites in the network, responsible for sterile products, is a small-sized manufacturer with about 100 employees that supplies products to over 20 countries. Historically, the site had a good track record in reliably supplying the market, but the COVID-19 pandemic put a strain on the site operations. At the start of 2021, the decision was made from global functions to analyze the strengths and weaknesses of the site and improve its performance.

R. Biloglav (✉)
Dechra Pharmaceuticals, Rakov Potok, Croatia
e-mail: Radovan.Biloglav@dechra.com

T. Friedli et al. (eds.), *Striving for Operational Excellence in the Pharmaceutical Industry*, Future of Business and Finance,
https://doi.org/10.1007/978-3-032-26287-5_21

The site enrolled for the St.Gallen operational excellence (OPEX) benchmarking in mid-2021. The St.Gallen OPEX research team preformed comprehensive interviews and gathered performance data over several months. The usual benchmark analysis of that type takes a few weeks, but the challenges in the management team prolonged the conclusion of the study, which was also a signal about the ability of the site to meet special requests. Additionally, the site did not have a dedicated OPEX lead that could support the benchmarking process. The preliminary data showed that the site performed well in terms of customer service measures, but they had room to improve in productivity measures and conversion costs.

21.2 Preliminary Findings

Operations on the site consist of sterile and powder manufacturing, with quality control and supply chain departments that include control of materials, microbiological testing, and warehousing activities. The sterile operations themselves include first a formulation phase, followed by two filling lines performing primary packaging in the sterile area, after which materials are stored before visual inspection on a single line and final packaging happening on two packaging lines. The flow of operations was restricted due to the small dimensions of the site, so there was a lot of movement of materials and personnel observed. The St.Gallen report suggested to commence a continuous improvement program focused on understanding both operational losses and cultural challenges and start OPEX improvements in the sterile manufacturing to increase competitiveness of the site.

The main causes of concern were high deviation rates due to frequent interventions in aseptic processing, long durations of deviation closures, and high changeover times leading to lower utilization of the equipment. The site was still doing well during regulatory inspections, but the concerns of the global quality team were growing that something might go wrong soon. Detailed analysis of the OEE data was done separately for formulation and primary and secondary packing. The results are shown in Table 21.1 (compared to St.Gallen Benchmarking Data):

From the data analysis, a couple of key findings could be drawn:

- OEE percentages were at low quartile levels, with significant room for improvement, especially in setup and cleaning duration and frequency of changeovers.
- Loading was very low, which means if any improvements are achieved, one of the filling lines could either be eliminated from operations or alternatively be used for significant expansion of the capacity.
- A deeper analysis of secondary packaging made clear that it is better to leave two packaging lines in operation due to noninterchangeable packing capabilities related to bottle size.
- In general, it was concluded that through resource optimization and process standardization, a lot of benefits could be achieved. Therefore, plans were made for the introduction of OPEX techniques and tools, starting in the manufacturing area and later expanding it to quality, supply chain, and engineering.

Table 21.1 Detailed analysis of OEE data

Metric	Process step	Industry mean	Site data
Setup and cleaning [%]	Formulation, primary and secondary packaging	19.69	30.50
Unplanned maintenance [%]	Formulation, primary and secondary packaging	19.62	16.80
Availability [%]	Formulation, primary and secondary packaging	63.48	52.70
Performance [%]	Formulation, primary and secondary packaging	78.28	72.30
Quality [%]	Formulation, primary and secondary packaging	96.43	95.70
OEE [%]	Formulation, primary and secondary packaging	49.47	36.50
Changeover time [min.]	Formulation	117.49	63.75
Loading = scheduled time/ calendar time [%]	Formulation	62.05	27.50
Loading = scheduled time/ calendar time [%]	Primary packaging	64.32	38.70
Loading = scheduled time/ calendar time [%]	Secondary packaging	61.96	32.20

21.3 Beginning of the OPEX Deployment

The start of OPEX activities was initiated through site-wide communication, linking the OPEX activities to the strategic site goals. Together with the management, communication documents were prepared and continuously displayed on the screens located on the factory floor and canteen. The communication announced the start of gemba walks by management, in which leaders of different departments were formed into two-man teams and visited different departments on the site according to the defined schedule visit. The purpose of the gemba walks was to increase awareness of inefficiencies on the site, as well as give the employees an opportunity to share their problems directly with management.

This was followed by an awareness training, in which 90% of employees got basic information about the method of lean improvement. The training was done in the classroom format, where employees were first made aware of the types of waste they encounter in daily work and then were introduced to simple lean tools and methods that could be used to reduce this waste and to drive the improvement process. Examples from other Dechra sites were used to ensure that employees understand the practical aspects of OPEX methodology. Global functions spent some time introducing the site to One Dechra Production System (ODPS) that became a mainstay knowledge and experience source, a repository of tools and techniques that the site personnel could rely upon. ODPS was organized into six chapters, covering sets of improvement methods, such as workplace organization, standard work, and process confirmation, as well as chapters describing performance management

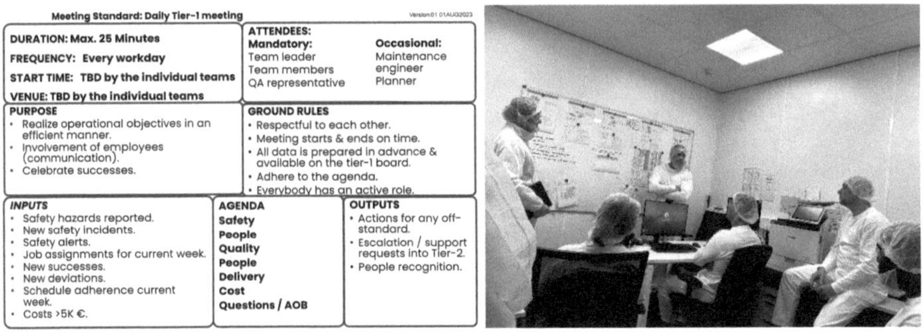

Fig. 21.1 Tier 1 meeting standard (left), tier meeting in action (right)

routines. Special attention was given to mindset, values, and behaviors chapters that placed a focus on creating transparent communication by mandating the presence of the management on the shop floor. In that time, there was also a change of site management, with the new site head being very keen on supporting the introduction of OPEX across the site.

In the next step, strategic site goals were discussed with employees during the monthly employee meetings and inquired about the ideas of how to achieve them. Employees were encouraged to discuss the barriers they encounter during normal work, where continuous improvement crystallized as a best method to overcome them. Q&A sessions were organized to allow employees to raise any questions and concerns that remained unanswered.

A first milestone was achieved by establishing the tier performance management system across the plant, with designated meeting flow, cadence, and outcomes, where every issue that was discovered on shop floor tier 1 meetings was discussed and reported to higher tier meetings on the same day (see Fig. 21.1). The goal was to ensure that all problems are made visible and are brought to the level of management that can solve them. At the start of the process, the tier 3 site meetings were overwhelmed by the number of reported issues, but through careful management of shop floor employees, empowerment, and support, more and more issues were resolved already on tier 1 level, and only real critical issues that needed site management support were brought up to the plant leadership level.

A second milestone was a site-wide problem-solving training, in order to enable and empower all employees to take an active role in resolving issues that surfaced out through performance management. Shop floor employees got full support to take ownership, find solutions, and improve the process by themselves through creating a strong collaboration between production, maintenance, and quality teams, which speeded up documentation modification that would capture all suggested improvements. In this phase, several of the most critical processes were captured in detail and described in standard work documentation, including workplace organization and work instructions. Visual examples were used wherever possible to ensure that the employees quickly recognized if any processes were not performed according to the standards.

Fig. 21.2 Daily quality inspection check

The site director also introduced daily quality/workplace organization inspection checks that would take 5 mins to complete and would have measurable outcome (see Fig. 21.2). The process of the checks was that employees would randomly take one of the ten available cards that describes the content of the quality/workplace organization check, and they would perform it. In case everything was good, they would place the card into "completed" row showing green side. If the check discovered some discrepancies, they would return the card into "completed" row showing red side and open an action item on the performance board with observations. At the beginning, quality assurance led the checks, but as the time progressed, employees from more departments were involved and quality inspection became a norm for all teams, even without the presence of quality employees. Every month, the best scoring department was recognized, which fostered healthy competition between the departments.

21.4 Mid-Stage of OPEX Deployment

About 6 months after the start of the OPEX deployment, the site finally hired a dedicated OPEX lead who was quickly introduced to One Dechra Production System through visits to other Dechra sites and establishing a coaching program. As Dechra does not have internal coaching capabilities, it enrolled key professionals in the external functional core leadership program. Coaching topics were related to management routines such as leading change, developing cultural intelligence, leading diverse teams, and influence and stakeholder management. The program was organized in monthly sessions and was updated as more people started with a new cycle (see Fig. 21.3).

The next step was to start improving the changeover process. As was previously established, OEE data displayed long change over times contributing to the time loss, which is due to small batch sizes and numerous countries that the product is delivered which requires frequent changes of the packaging materials. Hence, it was decided to do a deep dive into the changeover data, with kind support of St.Gallen

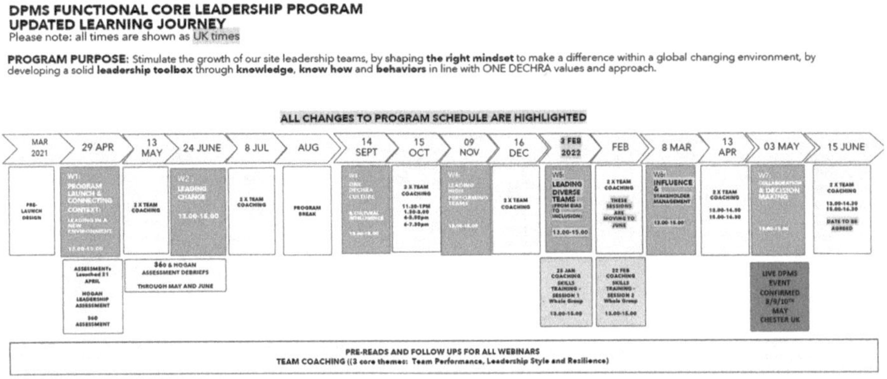

Fig. 21.3 Functional core leadership program

team that supported with industry statistics, which showed that the average duration of different types of changeover in Dechra was 51% longer than industry standard.

In order to improve the duration of the changeovers, several workshops were held where the team measured each step of the process, carefully reviewed work content, and discussed best practices for every step. It was interesting to note that even though all operators involved in the process were adequately trained and had considerable experience in working on the machines, each of them considered his way of working as the "best practice." It was very useful to bring all that knowledge together and ensure that the agreed standard became really the "best practice of all." After the new standard was defined, the following changeover runs were made supported by the site OPEX lead to ensure that the new standards are feasible and fine-tuned. The operators found the introduction of visual cues and signs very helpful, because it reduced the possibility of unintentional mistakes and errors. As a result, changeover duration was reduced by 47% overall within 4 months.

The next priority topic of analysis was to understand the frequency of changeovers. Data was gathered based on production planning routines, accuracy of forecast, production adherence, and usual scheduling methods. Data was gathered across few months of operations and was compared with industry standards provided by St. Gallen.

Through careful analysis of market trends and practices, it was agreed to apply a rhythm wheel planning methodology. Based on the frequency and volume of the product requirements, three main categories of products were defined (runners, stragglers, and strangers). Runners would be made with monthly and bimonthly frequency, stragglers would piggyback on the runners depending on the current market situation, and strangers would have quarterly or twice-a-year production runs. In conclusion, the redesign of planning methods and better information exchange had reduced the number of changeovers by 12%, but the biggest benefit was in the achieving of a more robust production plan and increased schedule adherence from about 60% to over 85%.

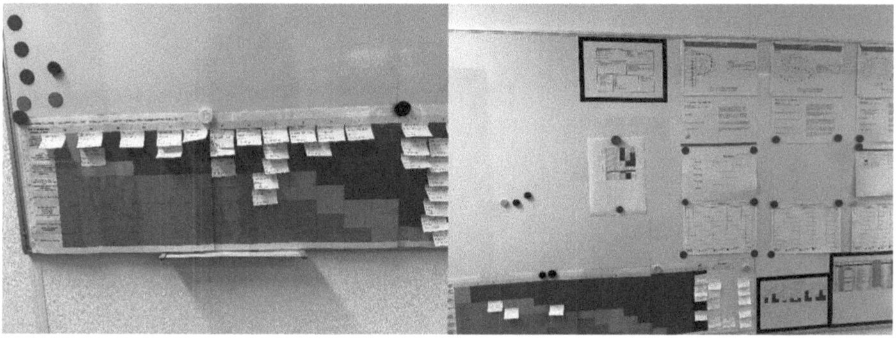

Fig. 21.4 Deviation management on the quality board

There was also an effort put into increasing throughput of the deviation handling process as long batch release times had a considerable impact on production schedule stability (see Fig. 21.4). Deviation life cycle was agreed in advance, and visual management was established, so each team could easily understand the stage of open deviations. The status was reviewed daily on the tier 2 board of the quality department, and the teams became also aware that if a specific stage of deviation became overdue, they could react before the overall deadline of 28 days for deviation closure was reached. In just a couple of months, the number of open deviations was reduced by 80%, which was appropriately celebrated.

21.5 Conclusion

After 2 years of OPEX activities on the site, the site has improved their OEE results by almost 50%. Inspection readiness was also increased, which was recognized by global quality. The decision was also made to automate OEE collection so that further improvements could be made. All the effort of the employees and leadership was officially recognized by achieving maturity level 2 of the One Dechra Production System maturity assessment, a result that would usually take 3 years to achieve. The site was able to create enough free capacity to eliminate one of the filling lines that had been in operation since 1990. Additional space was created for material and personnel movement that reduced complexity of walkways and pathways, bringing additional reduction of waste. Next steps are aimed to further improve the flow of the products within the site so that overall lead time of the site can be reduced, therefore freeing up more finished goods inventory. New Dechra strategy places more emphasis on short, focused improvement events, which can foster interdisciplinary collaboration, so that problems can be measured and analyzed and solutions proposed without the input of the managers. Also, as Dechra is a continuously learning organization, efforts are made to streamline training approaches through the introduction of electronic and visual aids that could make the adoption of new knowledge and onboarding new employees quicker.

Data Analytics as an Enabler of Quality and Operational Excellence in the Pharmaceutical Supply Chain

22

Michael Grischeau

Abstract

This chapter explores how data analytics is a foundational enabler of quality and operational excellence within the pharmaceutical supply chain. It discusses the growing complexity of ensuring compliance, innovation, and patient safety amid rapid technological change. The author outlines four key pillars for achieving analytics and AI readiness: data, technology, people, and process. Each pillar is crucial for building sustainable capabilities, improving data accessibility, and overcoming operational silos. Practical use cases are presented, such as automating supplier performance scorecards, digitizing Certificates of Analysis, and streamlining post-market complaint coding with artificial intelligence (AI). These real-world examples show how analytics eliminate manual workflows, reduce errors, and empower faster, more informed decisions across the supply chain. The chapter also underscores the importance of governing data quality and maintaining human oversight, especially in regulated environments. Advancing data maturity is linked to quality management maturity (QMM), which is promoted by regulators like the U.S. Food & Drug Administration (FDA) to move beyond basic compliance towards continuous improvement. Ultimately, the chapter argues that harmonized analytics practices drive sustainable operational excellence, support regulatory and patient safety goals, and strengthen the industry's ability to respond to future challenges.

M. Grischeau (✉)
AbbVie, Chicago, IL, USA
e-mail: michael.grischeau@abbvie.com

T. Friedli et al. (eds.), *Striving for Operational Excellence in the Pharmaceutical Industry*, Future of Business and Finance,
https://doi.org/10.1007/978-3-032-26287-5_22

22.1 Introduction

In today's pharmaceutical landscape, ensuring operational excellence is more nuanced and complex than ever before. The industry faces continuous pressure to innovate, maintain compliance, and ultimately safeguard patient safety while adapting to rapid technological change. At the heart of these challenges lies a key component: data analytics.

Data analytics is transforming how pharmaceutical companies design, manage, and monitor their supply chains. Modern organizations are increasingly asked not only to deliver safe and effective products but also to protect the integrity of the supply network from supplier to patient. This chapter explores how data analytics acts as an "enabler of quality," equipping pharmaceutical companies to rise to these challenges and drive operational excellence at scale.

Drawing on practical examples presented at the 2025 Parenteral Drug Association (PDA) Regulatory Conference, an industry conference that focused on "Achieving CGMP Excellence: Sustainable Compliance Across the Lifecycle," and leadership experience from AbbVie, this chapter will guide you through the foundational elements of a data analytics and artificial intelligence (AI)-ready organization, strategies for data accessibility, use cases along the supply chain, and advanced concepts like data maturity and quality management. The chapter begins by providing strategic framework and elements needed for an organization to advance its readiness, followed by contextualizing the pharmaceutical supply chain through data inputs and challenges; it then explores three use case examples that have been operationalized across the various elements within the supply chain and finalized with some key considerations for managing, governing, and operationalizing data analytics.

22.2 Foundational Elements of Data Analytics and AI Readiness

Achieving operational excellence through data analytics begins with a solid foundation across four key pillars: data, technology, people, and process (Fig. 22.1). These pillars should be considered as focus areas and critical points to measure and address in business or continuous improvement project structures. These elements do not act in isolation; instead, their integration is essential to build sustainable capabilities and foster enterprise-wide success with business maturity and small- to large-scale business system implementations. These pillars as foundational to data analytics and AI readiness will be explored further in the following section.

22.2.1 Data

Data is the lifeblood of any technology-driven initiative. In pharmaceutical operations, high-quality data underpins everything from product design to post-market

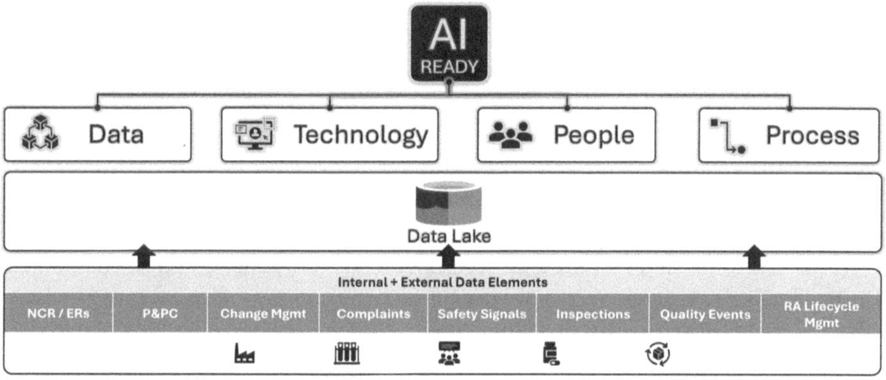

Fig. 22.1 Data analytics and AI readiness framework

surveillance. Yet, the industry struggles with data silos, inconsistent formats, and challenges in access and governance.

AI readiness in data requires organizations to ask critical questions: Is our data accessible? Is it well managed and governed? Addressing these is not just a technical concern, but an operational imperative. Organizations with high data maturity take the time to recognize the importance of clean data and implement controls regarding data lineage, where data can be mapped from originating source systems to any number of target outputs, including system-to-system connections, business intelligence tools, or management reporting mechanisms. Equally important as understanding data sources is assigning data ownership. This can be an independent role accountable for data completeness and quality that works across the organization to close gaps and resolve issues related to master data attributes.

Implementing an operating model around master data that established clear data management and governance frameworks to ensure regulatory compliance, AI model accuracy, and risk reduction is becoming a critical success factor and key to delivering value across the industry.

22.2.2 Technology

Technology is often perceived as the "shiny new object" in any transformation program—but success comes from more than simply deploying the latest platforms or applications. Many organizations find value in enhancing existing solutions with AI capabilities, rather than starting from scratch. Additionally, technology vendors offer AI and other advanced analytics offerings that integrate with these already established solutions. Organizations must assess both internal and external technology road maps to identify the best solutions that align with their business needs.

Technology does not stop after implementation. Monitoring AI tools post-implementation is required to ensure compliance, drive continuous improvement, and enable scalability. Technology must support business goals and be aligned with data and process requirements, not operate in a vacuum.

22.2.3 People

People are the change agents in any digital transformation. From problem definition to solution delivery, cross-functional collaboration is critical. Teams must not only prioritize use cases and resource projects, but also lead education and enablement efforts. Change management is vital; anxiety and resistance often accompany technological change, especially in regulated environments. By engaging stakeholders and understanding user concerns, organizations can drive adoption and maximize return on investment. Complex problems require multidisciplinary collaboration—quality, science, technology, IT, and operations need to co-design models, define metrics, and standardize processes for successful implementation and adoption.

22.2.4 Process

Process readiness is just as critical as data readiness. Organizations must assess their understanding of the AI solutioning process and ensure that standard operating procedures are harmonized for optimal data capture and flow. Subpar processes result in "garbage in, garbage out"—meaning that flawed process inputs inevitably lead to poor analytics and costly interventions. Harmonization and continuous process improvement are nonnegotiable on the pathway to excellence.

22.3 Data Accessibility: Sources, Strategies, and Challenges

The pharmaceutical supply chain spans raw material suppliers, contract manufacturers, laboratories, internal manufacturing sites, distribution networks, and ultimately patients (Fig. 22.2). Each node generates vast volumes of data—deviations, lab results, exception reports, complaints, and more.

22.3.1 Data Sources

Core systems like deviation monitoring, laboratory information management, and regulatory information management (RIMS) platforms generate both internal and external data. Examples include exception reports (internal signals) and production results or customer complaints (external signals).

For robust analytics, organizations must make these data elements accessible—using data lakes, warehouses, shared files, or federated extracts. The strategy depends heavily on the company's digital maturity, existing IT landscape, and business priorities.

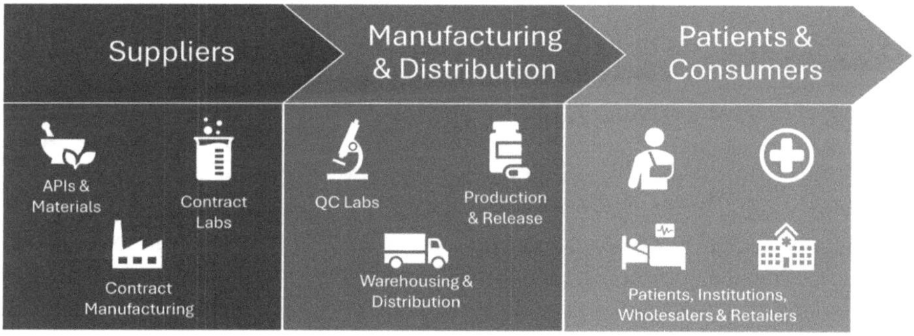

Fig. 22.2 Pharmaceutical supply chain

22.3.2 Accessibility Strategies

A robust data lake architecture provides scalable and secure access to data, enabling disparate information to be harmonized and analyzed in real time. A data lake can be used as a central storage system where data is held and maintained from multiple sources. The data can be structured or unstructured and accessed without having to organize it in advance. This makes it easy for organizations to store different types of data from various sources and analyze them later as needed. Data standardization and establishing data definitions are key to this process, as harmonizing data elements across systems and partners is essential for global adoption and accelerating onboarding. Automated ingestion further enhances efficiency by extracting data from formats such as PDFs or paper documents—like Certificates of Analysis or batch records—and loading them into structured repositories. Throughout the data life cycle, strong governance ensures data quality, security, and regulatory compliance.

22.3.3 Challenges

Organizations often face barriers due to legacy systems, varying formats, and differing standards across partners. Supply chain complexity means careful mapping of data flows, ownership, and stewardship is required. Overcoming these challenges positions the business for advanced analytics and continuous improvement.

22.3.4 Analytics and Technology Planning

Establishing an analytics, data, and technology strategy begins with aligning objectives to the broader business goals of the organization and ensuring that key stakeholders are engaged early to drive buy-in and relevance. The process should start with a review of the current state of data assets, technology platforms, and analytic capabilities to identify both strengths and gaps. With this understanding, a clear future vision should be defined, including the specific business questions analytics should answer and the intended data and technology architecture that supports organizational strategic direction. A phased road map should be constructed, covering initiatives such as data governance, infrastructure improvements, and skill development, while prioritizing efforts that deliver the most value and are feasible within the business context. It is important to establish robust governance frameworks for data quality, security, privacy, and stewardship responsibilities and to invest in change management through training and transparent communication to foster adoption and cultural alignment. Finally, key performance indicators should be set to track progress, feedback should be regularly solicited, and the strategy should be refined as the organization evolves. Confirming and revisiting the strategic road map should line up with periodic planning cycles and be reviewed, adjusted, and set annually as business strategies are adjusted along with financial budgets. Success depends on strong alignment between business and technical priorities, cross-team collaboration, robust governance, and a commitment to continuous improvement.

22.4 Use Case Introductions: Practical Applications Across the Supply Chain

To illustrate the power of data analytics, this section will consider three real-world use cases, each targeting a different stage in the pharmaceutical supply chain.

22.4.1 Use Case 1: Supplier Performance Management—The Automated Scorecard

Many companies today leverage a vast network of partners to perform services that can include the manufacture or supply of finished products. These companies must monitor the quality performance of these contract manufacturing organizations (CMOs) using available indicators such as nonconformities, exception reports, changes, complaints, inspections, and quality events. Traditionally, compiling these inputs has been manual and error prone.

To address the challenges of manual data collection, inconsistent inputs, and difficulty interpreting results, the sponsoring company implemented an automated CMO performance scorecard (Fig. 22.3) that leverages risk-based inputs, standard scoring and weighting, thresholds, and relationship factors based on the CMO type. The final use case deliverable is a globally accessible dashboard for external product QA managers, elimination of all manual data management and processing, and a trusted tool operationalized for all internal and external business reviews.

Implementation of this new approach to performance management has made a positive impact on the organization across multiple business factors. Common definitions and transparent scoring methods have led to objective performance reviews between partners and enabled accelerated decision-making with clear escalation

Performance Scorecard

Metric	CMO 1	CMO 2	CMO 3	CMO 4
Total	92	100	70	85
Product	0	0	-10	-10
Compliance	0	0	-10	-5
Timeliness	-8	0	-10	0

- Scoring starts at 100
- Deductions based on each subgroup
- Subgroup is a compilation of elements (e.g., Product = Initiated NCRs, Effectiveness, etc.)
- Subgroups assigned maximum deduction

Fig. 22.3 Performance management automation

paths. Elimination of manual data processing has allowed the company to reallocate resources on value-added tasks and improve compliance through error proofing. Transparency and common language have improved relationship management with CMOs, allowing manufacturers to focus on delivering a higher-quality product.

22.4.2 Use Case 2: Manufacturing Process and Product Performance Data—Digitizing Certificates of Analysis

Manufacturing oversight requires the ability to analyze process and product data from CMOs to ensure release criteria and adherence to specifications. This data is often received in varied, nonstandard formats, most of which is provided in PDFs or paper formats. Certificates of Analysis (CoAs) are especially challenging to ingest and analyze because document structure and format are not uniform across suppliers or manufacturers. While some CMOs have created standardization using printed text or manufacturing system output, no two CoAs are the same and some are still received as handwritten documents.

To overcome the challenges of non-digitized formatting and eliminate the need for manual data transcription, this sponsoring company formed a cross-functional team to design, develop, and implement an AI technology solution to digitize CoAs that could be used for automated trending. The approach ultimately provided access to data within the company's unified data lake, accelerated ingestion cycle times, and standardized product data across critical CMOs.

It is important to note that while the use of AI and advancing technology solutions as tools to drive efficiencies and accelerate decision-making, human verification is still required for any decisions that may impact product quality and/or patient safety. In this specific use case, a "human in the loop" (HITL) interface is maintained to ensure that data extracts are verified before entering the data lake repository. The term "human in the loop" refers to processes or systems in which human involvement is required to make decisions or oversee automated actions, especially in the context of AI or machine learning. HITL can mean that a human reviews, approves, or supervises outputs generated by an algorithm or automation system to ensure accuracy, address edge cases, or apply ethical judgment. This approach helps balance automation efficiency with human oversight, particularly for tasks where errors could have significant consequences. See Fig. 22.4 for a high-level process

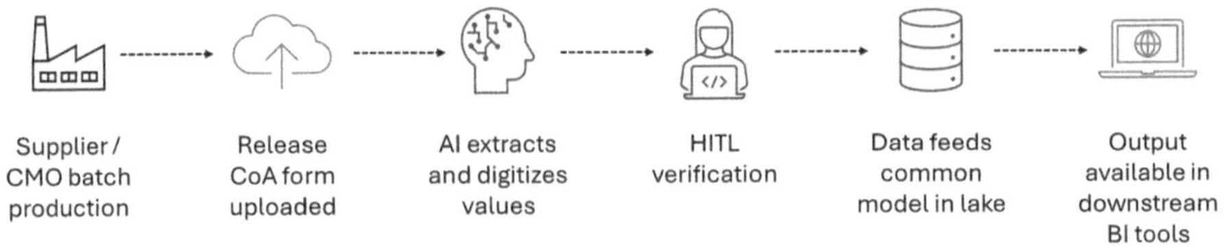

Fig. 22.4 Digitized Certificate of Analysis (CoA) flow

flow from the receipt of the CoA, HITL verification, and final use of data in downstream business intelligence (BI tools).

As a result of the updated solution and process, the company has standardized data points across both internal and external sites by implementing a common set of parameters. Manual transcription tasks have been eliminated, allowing the transition from a two-person workflow—one person transcribing and another verifying—to a streamlined, single-person verification step. This approach accelerates the onboarding of new manufacturing partners and enables the company to scale the solution to additional use cases within operations and quality.

22.4.3 Use Case 3: Post-market Surveillance—Coding and Investigation of Complaints

The third and last use case focuses on the final step in the pharmaceutical supply chain, where the product is used by patients and consumers. A critical input to product quality includes data that is received via post-market surveillance. This data can be received in many formats including via call center operations, online portals, emails, and word of mouth. Ultimately, patient safety and satisfaction drive the need to react quickly to customer complaints and adverse event data. The company set out to address current process challenges that involved manual coding of complaints and prioritization of potential high-risk events that were hampering efficiencies and responsiveness.

The proposed solution was an AI-powered process that automated complaint coding at intake to rapidly identify and assign the correct code and risk-rank priority complaints directly within the company's electronic quality management system (eQMS). The process was enhanced by HITL review to ensure proper code selection and, much like the digitized CoA use case, streamlined the overall process workflow (Fig. 22.5).

As a result, the company realized significant improvements to complaint cycle times with the coding process reduced from minutes to seconds at intake. Internal operational metrics were also improved for right-first-time coding, delivering value not only in efficiency but also in compliance standards. The ability to quickly flag high-priority complaints has also accelerated the investigation process and allowed for rapid engagement with internal and external partners to address emerging concerns and deliver higher-quality products.

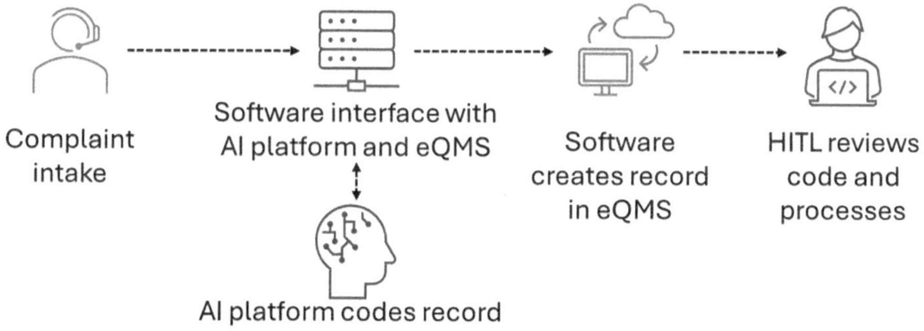

Fig. 22.5 Complaint coding automation flow

22.5 Data Maturity and Linkage to Quality Management Maturity (QMM)

Data maturity refers to the organization's ability to manage, govern, and leverage data strategically. In regulated industries, advancing data maturity closely aligns with the principles of QMM.

QMM is increasingly recognized by regulators as a means of ensuring robust, resilient, and quality systems that look beyond compliance and towards continuous improvement and cultural excellence. The U.S. Food & Drug Administration's Center for Drug Evaluation and Research (CDER) is establishing a program to promote QMM at drug manufacturing establishments. The QMM program aims to encourage drug manufacturers to implement quality management practices that go beyond current good manufacturing practice (CGMP) requirements (1 and 2). An additional white paper was published by the CDER Office of Pharmaceutical Quality in July 2025 titled "Quality Management Initiatives in the Pharmaceutical Industry: An Economic Perspective" (3) that describes how strategic investments in quality management initiatives in the pharmaceutical sector have the potential to yield returns for companies and for public health.

Mature data practices support QMM objectives by enabling:

- Proactive risk identification and mitigation
- Continuous monitoring and improvement
- Standardization and harmonization of quality processes
- Transparency in reporting and decision-making

Transforming from manual, fragmented practices to automated and standardized data analytics is a strong indicator of both operational and quality maturity. Change management underpins this transformation and enables organizations to build capabilities, educate stakeholders, and foster a quality-focused culture.

22.6 Operationalizing Analytics: Moving from Manual to Automated Solutions

Operationalizing analytics involves more than technology deployment; it requires re-engineering processes, aligning stakeholders, and embedding intelligence into daily workflows.

22.6.1 Scorecard Implementation

The automated CMO performance scorecard exemplifies this journey. Previously, data extraction and verification were manual, time-consuming steps fraught with risk. Transitioning to automated, nightly-refreshed dashboards led to faster business reviews and escalation; objective, data-driven insights for internal and external partners; and reduction in process failures and error rates.

22.6.2 Digitizing Data Flows

Digitizing CoAs increased speed and accuracy, eliminated bottlenecks, and brought formerly disparate data into a unified analytical environment. Solutions were designed for scalability and ease of adoption, aligning core operations with strategic objectives.

22.6.3 Accelerated Investigations

Automated complaint coding revolutionized post-market surveillance, decreasing response times, improving compliance, and enabling near-real-time prioritization and investigation.

Key lessons across these implementations—focus on data and process harmonization, collaborate cross-functionally for impact, and embed monitoring and governance from day one.

22.7 Sustainability and Governance: Continuous Improvement and Compliance

Operational excellence is not a destination, but a journey. Post-deployment, organizations must continuously monitor analytics tools, track compliance with internal and external standards, and drive continuous improvement.

22.7.1 Monitoring and Metrics

Instituting monitoring plans with defined business value metrics allows organizations to track tool performance, impact, and compliance—providing feedback loops for refinement.

22.7.2 Governance

Robust governance frameworks safeguard data quality, integrity, and privacy. These frameworks must adapt with evolving regulations and business needs, ensuring that solutions remain fit for purpose in a changing landscape.

22.7.3 Change Management

Sustaining change requires ongoing training, education, and communication. By supporting the human element inherent to transformation, organizations prevent regression to manual practices and foster a culture of excellence.

22.8 Integration with Operational Excellence and QMM Concepts

Operational excellence (OPEX) and quality management are inextricably linked—the former focusing on process efficiency, and the latter on product and compliance. Advanced data analytics integrates both driving strategic value across the enterprise.

22.8.1 Core OPEX Principles

Analytics supports OPEX principles by enabling standardization of processes and outputs, measurement and transparency of performance, rapid identification and remediation of issues, and optimization of resource allocation and workflow.

22.8.2 Quality Management Maturity

Data maturity feeds QMM by creating evidence-based insights for decision-making, underpinning compliance with regulatory and patient safety standards, and facilitating scalable and adaptable quality systems.

22.8.3 Enterprise-Wide Impact

By operationalizing and sustaining analytics solutions, organizations elevate quality and efficiency across their supply and value chains. The examples presented—automated scorecards, digitized CoAs, and complaint coding—demonstrate the power of analytics as an enabler of quality.

22.9 Conclusion

Data analytics is no longer optional; rather, it is an essential capability for pharmaceutical organizations committed to operational excellence and patient safety. The foundational pillars of data, technology, people, and process must be harmonized for success. Practical use cases along the supply chain demonstrate the transformative impact of analytics in supplier management, manufacturing oversight, and post-market surveillance.

True operational excellence demands continuous improvement, robust governance, and enterprise integration. Mature data practices support both OPEX and QMM, propelling organizations forward in an increasingly complex and regulated environment.

Embracing data analytics means not only protecting the integrity of product and process but also creating resilient, innovative organizations capable of meeting tomorrow's challenges. Let us continue to collaborate, learn, and advance our practices—ensuring that our ultimate responsibility, to deliver safe and effective products to patients, remains our guiding principle.

References

CDER Quality Management Maturity Program. https://www.fda.gov/drugs/pharmaceutical-quality-resources/cder-quality-management-maturity

CDER's Office of Pharmaceutical Quality White Paper on Quality Management Maturity. (April 2022). https://www.fda.gov/media/157432/download

Quality Management Initiatives in the Pharmaceutical Industry: An Economic Perspective. (July 2025). https://www.fda.gov/media/187861/download?attachment

Unlocking Business Value Through Quality Management System Transformation

23

Lynn Bottone, Deepti Sharp, Amy Butters, Bina Nair, Steve Stears, and Tarangi Patel

Abstract

In the fast-changing world of pharmaceuticals, having an effective quality management system (QMS) is essential—not just for meeting regulations, but for supporting business success. This chapter illustrates how standardizing, simplifying, and digitizing QMS processes can help organizations respond quickly to new challenges, reduce wasted effort, and deliver better results for both business and patients. Using Pfizer's experience as an example, the chapter highlights practical ways to streamline operations, clarify roles, and ensure that teams use shared tools and best practices across a diverse network.

23.1 Introduction

In today's rapidly evolving and highly regulated pharmaceutical and life sciences landscape, a robust quality management system (QMS) is no longer just a compliance necessity; it is a strategic enabler of business value. This chapter explores how organizations can transform their QMS to drive operational efficiency, financial impact, and measurable success.

L. Bottone · D. Sharp (✉) · A. Butters · B. Nair · S. Stears · T. Patel
Pfizer Inc., New York, NY, USA
e-mail: Lynn.A.Bottone@pfizer.com; Deepti.Sharp@pfizer.com; Amy.Butters@pfizer.com; Bina.Nair@pfizer.com; Steven.Stears@pfizer.com; Tarangi.Patel@pfizer.com

T. Friedli et al. (eds.), *Striving for Operational Excellence in the Pharmaceutical Industry*, Future of Business and Finance,
https://doi.org/10.1007/978-3-032-26287-5_23

23.2 Overview

Like the greater world around us, the biopharma industry is experiencing rapid and continuous change. Regulatory requirements are constantly evolving, at times in seemingly divergent ways. For companies that supply lifesaving medicines to patients around the world, meeting those demands with data-driven decisions is essential, no matter the size of your enterprise or the number of patients you serve.

And the bottom line is that patients desperately need the industry to supply safe medicines and therapies 100% of the time, to meet whatever health challenges arise quickly and efficaciously, and to always challenge the status quo with an innovative mindset.

Pfizer is a well-known pharmaceutical company that has a long-standing legacy of 176 years and iconic products such as PFIZERPEN® (Penicillin), citric acid, VIAGRA®, Prevnar 20™, and COMIRNATY®, Pfizer's COVID vaccine. In 2024, Pfizer impacted >414 million patients with their medicines and vaccines.

Reflecting on how to support delivery of the Pfizer purpose—breakthroughs that change patients' lives—it became evident that the quality management system (QMS) for GMP operations presented significant improvement opportunities to ensure that it is fit for purpose and fit for future and enables advanced predictive performance monitoring through the use of digital tools including artificial intelligence (AI) allowing us to become more proactive, innovative, and digitally driven, with a focus on simplified/optimized business processes, risk-based solutions, and predictive analytics at our fingertips.

This chapter outlines the strategic transformation of the QMS for GMP operations to drive operational efficiency, financial impact, and measurable success. At its core, the story focuses on a simplified, standardized, and robust QMS (see Fig. 23.1).

Fig. 23.1 QMS Revolution

It presents a realistic value proposition, discusses ways to quantify financial benefits to make a compelling business case for investment, and defines measures to realize value. The story is grounded in real-world case studies and is supported by visuals that illustrate the required mindset, tools to execute, and business impact.

23.3 Vision

23.3.1 Is This Where You Are Today?

Like many large pharma companies, Pfizer is an amalgamation of many different companies, formed over the years through acquisitions and mergers. While this has produced one of the world's largest portfolios, it has also resulted in a diverse network of sites—or to borrow a term from Baskin Robbins®, "flavors." This applies to technology, business practices, digital enablement, ways of working, site culture, local governing regulations, you name it! Simply speaking, this means that for every business process we have, there are multitudes of ways that process is run.

As you might imagine, all those "flavors" can create layers of complexity. With a limited global, standard way of operating, our above-site operations were built to deal with every possibility and every different way of working, often unable to take advantage of our network's size and scale. We were largely reactive and had few common business processes or digital solutions that might provide a global look at trends or areas needing proactive attention.

In addition, the above-site quality teams supporting our GMP operations had a reputation often associated with one of "consultative oversight."

Beginning with an analysis of our "current state" (see Fig. 23.2), we found a complex and segmented QMS, resulting in largely reactive approaches, fragmented business processes, and underutilization of advanced technologies and tools. Inconsistent standardization of business processes and operations often bogged down by unnecessary details reducing overall flexibility were also identified.

To address these challenges, transforming the QMS was essential. This would require listening to the specific needs and priorities of each team, ensuring effective engagement, meaningful progress, and driving global business process ownership. We organized the work around three key principles—standardize, optimize, and digitize—under the umbrella concept of the "QMS Revolution," which signified the need for change and comprehensively contemporizing the QMS.

"Revolutionizing" our work would mean moving beyond traditional, often reactive processes and embracing a proactive, innovative, and digital approach that benefits our colleagues, our business, and the patients we serve. By aligning around a shared vision and answering the critical question, "What's in it for me?", inspiring engagement at every level became a key goal. Achieving this transformation would require leadership mandate, clear direction, a strong call to action, and a focus on the measurable benefits of improved effectiveness, efficiency, and quality value indicators. We realized that our success depended on how well these changes were communicated, implement cost-saving measures, and consistently track

Fig. 23.2 Where we started

performance improvements and, along the way, transform the quality organization from a reactive, consultative oversight arm to a strategic business partner.

23.3.2 Where Are We Going?

"QMS Revolution" describes how we are transforming into a contemporary QMS. The intent is for Pfizer to adopt global standardized processes, assessments, and structures and remove local processes with site specificity that sometimes drove site-to-site inconsistencies, saving time, resources, and money. This will empower leaders to drive consistency and support rapid deployment of colleagues and solutions to address our largest sources of opportunities. The newly revolutionized QMS is designed to be fit for purpose, fit for patients, and fit for future.

By implementing the newly revolutionized QMS, GMP business processes will be streamlined worldwide, allowing different facilities and functions to adhere to the same global standards and eliminating inefficiencies that stem from regional variation, except where required by regulation. This harmonization not only improves operational effectiveness but also strengthens compliance and regulatory alignment in a highly competitive environment. With these improvements, global process owners (GPOs) and local process owners (LPOs) *(roles explained later in this chapter)* are better equipped to oversee business processes, ensuring that best practices are consistently applied throughout the organization. This framework makes it possible to quickly mobilize teams, introduce innovative solutions when faced with new challenges or opportunities, and fully realize the value of digital applications through standard deployment. Ultimately, this provides real business value since business process owners are not only subject matter experts but also "owners" responsible for the "health" of their business process from a compliance and performance perspective and owners of the digital solutions that drive compliance, effectiveness, and efficiencies (see Fig. 23.3).

The emphasis on adaptability and forward-thinking design of its QMS for GMP operations positions the organization to meet patient needs with agility while remaining prepared for future advancements in regulations (see Fig. 23.4).

One of the first steps in the QMS Revolution was to create a comprehensive framework for organizing QMS business processes that supports safety, efficacy, regulatory compliance, and one global approach. This framework, or taxonomy, begins with the Quality Manual and "Level 1" (L1) global policies. Business processes are then categorized under each L1 as Level 2 (L2) Global Standard Operating Procedures (Global SOPs). These global SOPs are then supported by Level 3 (L3) Global Work Instructions (GWIs) that explain how the L2 process is executed. The Global SOP and/or GWI may be further supported by Local Work Instructions (LWIs) based on local regulatory requirements or necessary product or business differences (see Fig. 23.5).

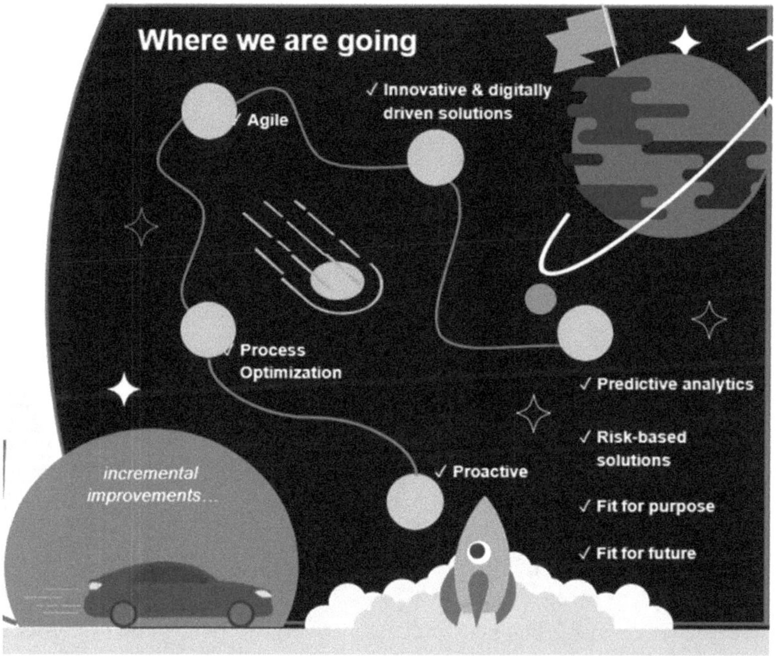

Fig. 23.3 Where we are going

Fig. 23.4 Quality taxonomy

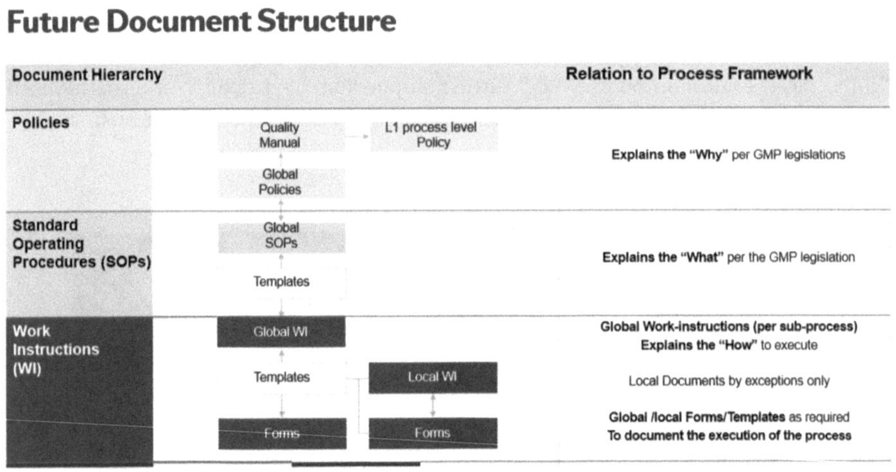

Fig. 23.5 Future document structure

The next step was to create our ways of working that would not only revolutionize the QMS for GMP operations and each business process but also provide the structure for long-term sustainability.

Accountability for each QMS business process lies with a network of colleagues, referred to as the Global Process Network. The roles within the Global Process Network are summarized below.

L1 Processes: Managed by an executive sponsor, responsible for policy documents.

L2 Processes: Each L2 business process is overseen by a global process owner (GPO), accountable for execution, Global SOPs, and Work Instructions. The GPO has end-to-end responsibility for the health of the relevant L2 business process, including:

- Measuring impact and performance through comprehensive metrics and analysis.
- Stakeholder collaboration.
- Change leadership and coaching.
- Training oversight.
- Continuous improvement, including scanning the external environment for best practices and evolving regulations.
- Sustainability and standardization.

In addition, each of these L2 business processes also engages an additional network of colleagues providing structured collaboration between global and local leadership and expertise, designed to sustain and improve the health of the L2 business process:

- *Local process owner (LPO)*: LPOs are required based on site/team needs. They are responsible for the health of the business process locally as well as collaborating with the GPO on continuous improvement initiatives.
- *Subject matter expert (SME)*: Core representatives across sites or functions.

23.4 Why Should This Be a Business Priority?

Centralizing all QMS business processes under a global QMS framework offers significant opportunities for value creation through optimized workflows and digital transformation. By consolidating these critical operations within a unified framework, the organization can drive greater consistency and reduce redundancies not only in execution but also in the types of digital applications and systems that are used to operate a process. It also facilitates seamless data sharing across different sites and geographies. This centralization improves compliance with regulatory requirements and enhances operational efficiency, allowing teams to focus on innovation rather than administrative tasks. Furthermore, a centralized approach supports continuous improvement initiatives by enabling the collection and analysis of quality data at scale, identifying trends, and proactively addressing potential risks before they escalate (see Fig. 23.6).

Business value has been realized through process simplification, standardization, and digitization efforts, which have led to greater efficiency and effectiveness of change management, reduced investigation cycle times, improved quality in nonconformance management/investigation reports, better handling of product complaints, and culturally important, enhanced collaboration across all sites through the global standard processes.

23.5 Structured Approach Leveraging Operational Excellence

A robust quality management system (QMS) is not just a compliance tool (see Fig. 23.7)—it is a strategic enabler of business value. While its primary purpose must remain uncompromised—providing a framework for achieving quality

Fig. 23.6 QMS Revolution is a key enabler

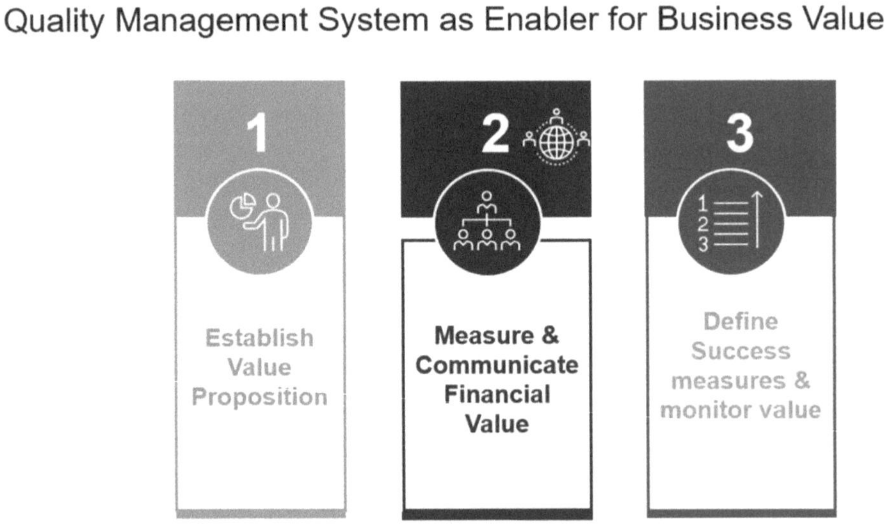

Fig. 23.7 QMS as enabler for business value

excellence and adherence to regulatory and industry standards—it also plays a pivotal role in unlocking operational efficiency and financial impact. By establishing a clear value proposition (see Fig. 23.8), measuring and communicating financial outcomes, and defining success metrics that are continuously monitored, the QMS becomes a driver of tangible business outcomes. Simplification and digitization are key accelerators in this journey, streamlining processes and reducing complexity, thereby enhancing both compliance and quality performance. When implemented with these principles, the QMS evolves from a control mechanism into a catalyst for sustainable value creation.

Fig. 23.8 Establish value proposition

23.5.1 Establish Value Proposition

A persuasive value proposition for transforming the QMS for GMP operations begins with clearly communicating the necessity of change and outlining its anticipated benefits. Organizations should evaluate their existing landscape—typically marked by reactive, compartmentalized processes, minimal standardization, and insufficient utilization of digital tools—while envisioning a proactive, agile, and digitally empowered future.

Essential components of the value proposition include the following:

- Vision alignment: Integrating the transformation with overarching organizational objectives such as patient outcomes, innovation, and scalability.
- Opportunity identification: Identifying critical business processes (e.g., investigations, change controls, training management) that present significant opportunities for enhancement.
- Stakeholder relevance: Customizing communication to different groups to foster engagement and secure stakeholder commitment.

23.5.2 Measure and Communicate Financial Value

Quantifying the financial impact is critical to making a compelling business case for investment by executive leadership to support and fund the program (see Fig. 23.9). The process includes the following:

- Baseline financial metrics: Estimating potential business value/savings resulting from process improvements, digital integration, and standardization.
- ROI communication: Articulating value in terms of reduced rework, minimized deviations, accelerated cycle times, and improved customer satisfaction, tying this value to expected business outcomes.

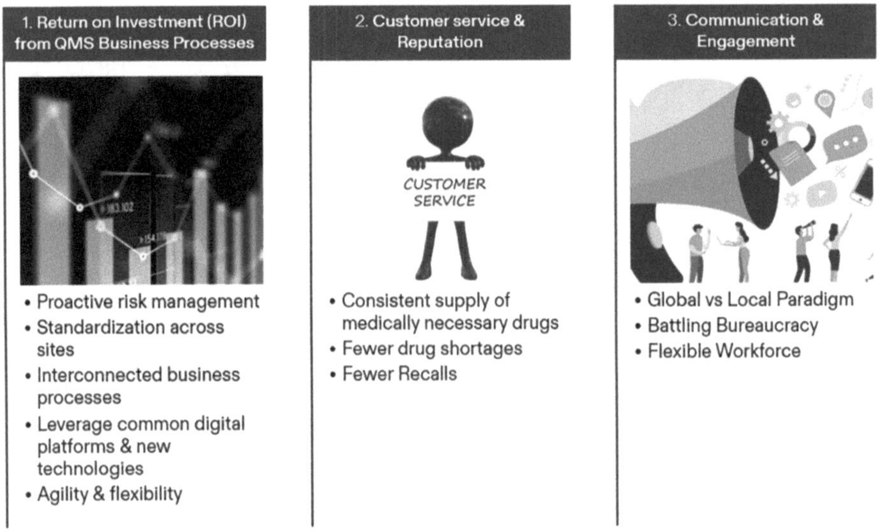

Fig. 23.9 Measure and communicate financial value

23.5.3 Define Success Measures and Monitor Value

Business performance metrics are monitored through value indicators such as retesting; rework; cycle time for but not exclusive to deviations, investigations, batch record reviews, and customer complaints; batch losses; and CAPA effectiveness. Efficiency and productivity costs are calculated to quantify the business value.

To ensure robust business monitoring, these metrics are tracked using digitally enabled visual management dashboards, enabling real-time analysis and identification of trends. For example, tracking retesting rates can highlight areas where initial quality standards/test methods may need enhancement, while analyzing rework statistics helps pinpoint process inefficiencies.

A methodology for incremental value statements assesses efficiencies achieved through phased implementation, following simplification, standardization, and digitization, as well as tracking changes by business process and each end user site or function.

This methodology is implemented systematically, starting with a thorough baseline analysis of current processes at each site. Simplification efforts might entail removing redundant steps or automating manual tasks, whereas standardization focuses on aligning workflows across teams and locations for consistency. Digitization introduces tools such as electronic batch records and automated reporting systems, significantly reducing cycle times. Progress is measured incrementally to demonstrate tangible value added at every phase.

Efficiency improvements are determined by sites based on measured impact at each location.

Each recommendation is tailored to the unique challenges and opportunities present at individual sites. Feedback loops ensure that local teams can adapt proposed improvements to their specific contexts, encouraging ownership and sustained adoption. These recommendations often address both quick wins—such as minor process tweaks—and long-term strategic initiatives, like investing in advanced predictive analytics to forecast demand and optimize resource allocation.

A tiered governance process ensures that the program delivers to the defined ambition or goals. It incorporates visual management and escalation processes with defined roles and responsibilities, structured meeting agendas and charters, comprehensive reviews of project plans and Gantt charts, and diligent follow-up on actions. This offers clear visibility into active projects and outstanding issues. Defined escalation pathways facilitate swift resolution of bottlenecks or risks. Each tier within the governance model, from project teams to executive sponsors, has distinct responsibilities, ensuring that critical decisions are made efficiently and communicated broadly. Structured meetings are held regularly, following established charters and agendas to maintain focus and drive actionable outcomes.

This governance framework (see Fig. 23.10) provides comprehensive management oversight, addresses operational challenges, supports escalation processes, defines success metrics, and monitors value realization. This robust structure ensures that decision-making is guided by transparent criteria and real-time performance data, facilitating both short-term corrective actions and long-term strategic planning. Another key element of our successful governance model is a continuous improvement loop through stakeholder feedback and lessons learned from previously deployed processes.

Ultimately, this integrated approach promotes sustained progress toward organizational goals and seeks to strengthen trust among internal and external stakeholders, ensuring continued support and buy-in from executive leaders and colleagues across all tiers.

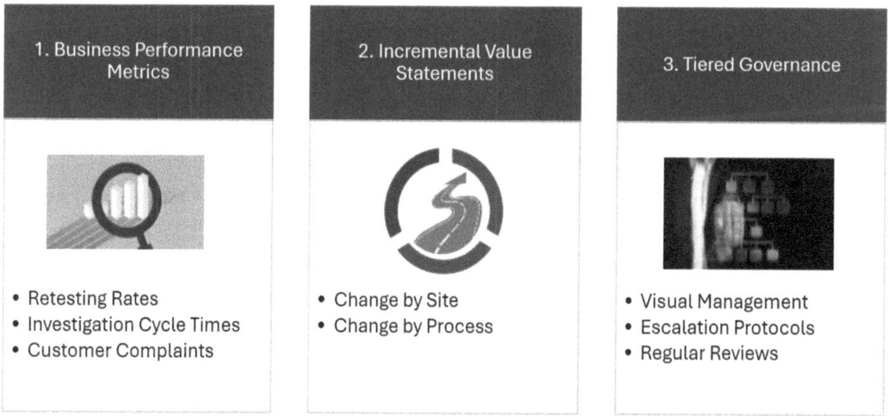

Fig. 23.10 Define success measures and monitor value

23.6 Driving Performance and Continuous Improvement

Periodic performance reviews enable the identification of best practices and lessons learned, which are subsequently disseminated to all operational units via centralized communication platforms. This drives alignment across the network and fosters a culture of shared learning and continuous improvement.

Continuous improvement is driven using feedback from Voice of the Customer (VOC) and lessons-learned surveys conducted after each revolutionized business process is effective. VOC surveys collect input directly from customers, helping organizations align their processes with user needs and expectations. The insights gained from these surveys systematically analyze project execution post-implementation, informing future strategy adjustments and corrective actions. Focusing on both customer experience and operational excellence, this integrated approach sustains momentum toward ongoing performance improvement and innovation.

23.7 Change Management and Business Adoption

Depending on the size and complexity of an organization, the magnitude of change from programs such as "QMS Revolution" and the impact it has on colleagues cannot be underestimated. This warrants thinking creatively and collaboratively to design and implement effective stakeholder management methods at all levels of the organization. These include the following (see Fig. 23.11):

Fig. 23.11 Change management and business adoption QMS Revolution

- *Stakeholder identification*: Identify key stakeholders early for each L2 process to enable targeted, GPO-led business adoption.
- *Clear messaging*: Ensure that stakeholders understand "What's in it for me?" in the context of the change initiative. Include critical information like from-to, long-term benefits, and the "why" behind the change.
- *Quick wins*: Celebrate early successes to maintain stakeholder engagement.
- *Standardization*: Develop a standardized business adoption playbook for LPOs and enablers at sites.
- *Leadership clarity*: Define roles, responsibilities, and accountability for leaders in the initiative, and update as the program evolves.
- *Feedback loop*: Improve feedback mechanisms (e.g., stage gates, surveys conducted during site visits by leadership, pulse checks, Gembas).
- *Adoption agents*: Empower local influencers (key leaders/experts) to drive engagement and outcomes within sites and teams.

23.8 Sustainability

Sustaining the transformation of the QMS for GMP operations requires embedding continuous improvement (CI) into everyday governance. CI plays a central role in ensuring that simplified, standardized, and digitized processes remain effective and responsive over time. GPOs use structured feedback loops, performance dashboards, and team-based problem-solving to identify inefficiencies and drive innovation. Governance frameworks—tiered across global and local levels—support decision-making, escalation, and alignment. By integrating CI into business-as-usual (BAU) operations, companies can create a culture of accountability and agility, enabling their QMS for GMP operations to better evolve with changing business needs and regulatory expectations. This approach not only reinforces sustainability but also empowers teams to continuously improve how quality is delivered and compliance is sustained.

23.9 Conclusion

Transforming a QMS for GMP operations is a journey that requires strategic vision, financial rigor, and operational discipline. By establishing a strong value proposition, quantifying business/financial impact, and defining robust success measures, organizations can unlock the full potential of their quality systems—not just for compliance, but as a competitive business advantage.

The Six Traits of Quality Business Leadership

24

Anders Vinther

Abstract

Quality business leadership (QBL) represents a transformative approach where quality leaders function as strategic business partners rather than reactive compliance functions, driving sustained product quality, reliable supply, and financial performance. Moving beyond traditional cGMP compliance, QBL masters six core traits:

1. Visibly demonstrate responsibility for quality: Leading the quality management system as an integrated system using systems thinking, transforming it from passive control to active leadership tool.
2. Advocate that quality is owned by all: Building senior management partnership and organizational alignment so quality becomes everyone's responsibility, not just the quality department's.
3. Engage employees to continuously improve quality: Fostering psychological safety and genuine engagement through balanced leadership that addresses both rational and emotional dimensions of improvement.
4. Make risk-informed decisions that benefit patients: Using evidence-based, structured decision-making that balances patient safety with reliable product supply.
5. Promote quality as a value-driven financial advantage: Translating quality activities into financial impact using frameworks like the quality value index and financial risk number to demonstrate competitive advantage.
6. Lead the path to one global quality regulatory framework: Driving science-based global alignment to accelerate innovation and strengthen quality systems beyond fragmented national requirements.

A. Vinther (✉)
Quality Business Administration LLC, Burlingame, CA, USA
e-mail: anders@qbaleaders.com

T. Friedli et al. (eds.), *Striving for Operational Excellence in the Pharmaceutical Industry*, Future of Business and Finance,
https://doi.org/10.1007/978-3-032-26287-5_24

24.1 Introduction

Quality leadership is at a turning point. Traditional models are no longer sufficient to keep pace with rapid advances in science and technology in an increasingly complex and resource-constrained world.

Quality business leadership (QBL) represents a new vision in which quality leaders operate as strategic business partners rather than reactive support functions. QBLs move beyond a narrow focus on cGMP compliance to drive sustained product quality, reliable product supply, and strong financial performance, making quality a true competitive advantage.

At the core of QBL is the progressive mastery of the six traits of quality business leadership (see Fig. 24.1):

Quality leaders who have transitioned to QBL consistently demonstrate value through improved financial performance, faster and more patient-focused decision-making, timely adoption of new technologies, and creation of engaged cultures that sustain continuous improvement.

Becoming a QBL is not trivial. As Mahatma Gandhi observed, "You must be the change you want to see in the world." This change begins with the individual and requires acknowledging that existing ways of working and ingrained mindsets, particularly those limited to basic cGMP compliance, are no longer sufficient.

The starting point for QBL is a clearly articulated personal ambition to integrate quality into the broader business. Becoming a true business partner then requires deliberate unlearning of entrenched habits and development of new capabilities over time.

Fig. 24.1 Traits of quality business leadership

With this foundation established, we now examine each of the six traits of quality business leadership.

24.2 Trait: Visibly Demonstrate Responsibility for Quality

An objective of "meeting cGMP expectations" sets the quality bar at the minimum, or "the floor," leaving the organization vulnerable to noncompliance and failing to use quality as a lever for improved business performance. Organizations that adopt this minimal compliance mindset often develop their quality management system (QMS) narrowly, focused on meeting basic regulatory requirements rather than integrating it into broader business processes. If the QMS slows rather than accelerates the introduction of new technologies, or if its stakeholders are limited to quality and manufacturing operations alone, it is time to rethink the QMS.

Visibly demonstrating responsibility for quality means taking ownership of how the QMS performs in practice, not just how it is designed or documented. When quality leaders lead the QMS as a true system and apply systems thinking to complex challenges, responsibility for quality becomes observable, measurable, and inseparable from overall company performance. Despite the name QMS, or pharmaceutical quality system (PQS), and Deming's decades-old teachings on the system of profound knowledge, many organizations still manage the QMS as a collection of procedures rather than as an integrated system. As a result, they fail to realize its potential not only to improve cGMP compliance, but also to enhance patient satisfaction and drive continuous improvement in quality and financial performance.

Taking a true systems approach to the QMS enables organizations to demonstrate "PQS (QMS) effectiveness," a topic receiving increasing attention from FDA, EMA, and ICH as a means of reducing regulatory burden. Viewing the QMS as a system also changes the nature of the problems quality leaders must address. Many challenges are no longer purely technical or procedural, but systemic. QBLs therefore distinguish between complicated problems, which have solutions and can be addressed through linear cause-and-effect thinking, and complex problems, which do not. Applied to complex problems, linear approaches and addition of more data often make matters worse. By mastering systems thinking, including broad stakeholder involvement, pattern recognition, and strategic intervention, QBLs make complex problems more manageable and transform the QMS from a passive control mechanism into an active leadership tool.

24.3 Trait: Advocate That Quality Is Owned by All Through Developing Senior Management Partnership

Deming's 14th Point of Management states, *Put everybody in the company to work to accomplish the transformation. The transformation is everybody's job* (Deming: Out of the Crisis 1982, MIT Press). In short, quality is everybody's responsibility.

Yet in practice, quality and quality-related issues are often owned primarily by the quality organization.

The idea that quality is solely the responsibility of the quality department is often reinforced by the shared name of the attribute "quality" and the organizational function of "quality." Actions by quality leaders can also unintentionally reinforce this perception of sole ownership, for example when quality operates in a policing rather than collaborative manner, when regulators primarily engage with quality during inspections, or when quality reports quality metrics on behalf of other functions instead of those functions owning and presenting their own metrics. While cGMPs do not explicitly state that every employee owns the quality of their own work, they do establish quality's role as *oversight* to ensure the quality of drug products provided to patients. Where this distinction is unclear, QBLs must actively clarify it.

For quality to be truly owned by all, QBLs understand that the most senior leaders must regularly communicate the importance of quality to employees at every level. Everyone across functions regulated by the cGMPs should have a quality goal, and the operations and quality organizations should share key objectives for the year and operate with a common vision. Active engagement from the CEO as sponsor, together with strong partnership between senior leaders and the head of quality, is critical. As Juran observed, "*It is most important that top management be quality-minded. In the absence of sincere manifestation of interest at the top, little will happen below*" (Juran: The Quality Control Handbook 1951, McGraw-Hill).

QBLs deliberately build ownership of quality across the organization by articulating its value in financial, supply, and reputational terms; ensuring alignment and ownership across cGMP-regulated functions; and integrating deeply into the business through a strong understanding of its technology and operations.

24.4 Trait: Engage Employees to Continuously Improve Quality

Operational Excellence (OPEX) was initiated by **Juran** and other quality thinkers in the 1970s to drive continuous improvement through problem-solving, teamwork, and leadership, with the goal of meeting customer needs by reducing waste, improving efficiency, and adding value. At its core, OPEX was intended to actively engage employees in improving quality.

Over time, OPEX has moved out of the quality organization in many companies and, in some cases, has become a short-term cost-cutting exercise led by employees with limited quality and cGMP expertise. This shift increases risk to long-term cGMP compliance and runs counter to Deming's fifth Point of Management, which calls for continually improving systems to enhance quality and productivity and thereby reduce costs. QBL research data supports that most organizations today prioritize cost reduction over quality improvement (Data from Certificate in QBL Program).

For employees to move beyond compliance-driven participation toward meaningful contribution, two conditions must be present: psychological safety and genuine engagement.

While motivation may exist on its own, sustained improvement depends on employees feeling safe to speak up and actively involved in shaping outcomes. Understanding the distinction between motivation and engagement is therefore central to how QBLs create the conditions for continuous improvement.

Psychological Safety Full engagement requires psychological safety, which cannot be assumed and is demonstrated through behavior: how failures are handled, whether employees speak up about risks, and whether they suggest improvements. Leading people differs from leading technical or scientific work. While manufacturing improvement reduces variation, effective people leadership encourages diversity of thought and a strong speak-up culture.

Motivation vs. Engagement Motivation, as described by Daniel Pink (Pink: Drive 2009, Riverhead Books), is intrinsic and driven by purpose, autonomy, and mastery. Engagement goes further, emerging from motivation combined with involvement and reflected in enthusiasm and connection to work. Because individuals differ, engagement cannot be created through a one-size-fits-all approach, requiring QBLs to adapt their leadership.

The distinction between motivation and engagement is central to effective change leadership. **John Kotter**, a leading authority on change and creator of the eight-step change model, identifies four principles critical to organizational success (Kotter: Accelerate, 2014, Harvard Business review Press):

24.4.1 Management and Leadership

Management and leadership are often conflated but represent different activities. Management emphasizes reliability, efficiency, and execution, while leadership focuses on direction, opportunity, and change. Under time pressure, leaders tend to default to urgent management tasks (firefighting) at the expense of leadership. Quality business leaders (QBLs) deliberately protect time for leadership, recognizing that without it they cannot fully lead the business or develop employees as business leaders.

24.4.2 Select Few and Diverse Many

This distinction influences how leaders engage employees. When forming task forces or project teams, leaders frequently rely on familiar experts with proven results. While effective short term, this limits broader involvement. Inviting participation from employees who volunteer out of interest expands perspectives and can surface valuable alternative approaches.

24.4.3 Head and Heart

Sustained engagement requires addressing both rational (head) and emotional (heart) dimensions of change. The head is engaged when employees understand priorities, logic, and outcomes. The heart is engaged when they feel connected to purpose, believe their contributions matter, and see themselves in the change. Opportunities that appeal to both understanding and meaning broaden and sustain participation.

24.4.4 Have to and Want to

Change driven by obligation "have to" may achieve compliance, but lasting engagement occurs when employees "want to" contribute. Consistent recognition of individual and team achievements, including small but meaningful wins, helps shift improvement from a requirement to a choice, reinforcing commitment over time.

QBLs reverse current patterns by reclaiming OPEX as a quality-led, people-centered system rather than a short-term cost initiative. They foster psychological safety that enables speaking up, learning from failure, and contributing ideas, recognizing that engagement extends beyond intrinsic motivation alone. By balancing management and leadership, engaging both the select few and the diverse many, appealing to head and heart, and shifting improvement from have-to to want-to, QBLs actively involve employees in continuous quality improvement. In this model, cost reduction follows from stronger systems, better decisions, and sustained engagement, rather than serving as the primary objective.

24.5 Trait: Make Risk-Informed Decisions That Benefit Patients

A structured approach to decision-making, combined with disciplined follow-through, is essential for quality leaders as many decisions can be critical to quality and cGMP compliance, including batch release to patients.

24.5.1 Defining Risk

Explicitly defining risk tolerance is critical. When leaders implicitly set tolerance at zero risk, decisions may become overly conservative and, in some cases, unintentionally contribute to drug shortages of safe and effective products. A practical test of decision robustness is to consider whether the decision could be confidently explained and defended to both a patient and a regulatory agency FDA investigator during an inspection. This framing reinforces accountability to both patient outcomes and regulatory expectations.

24.5.2 Assessing Risk

Critical decisions require balancing risk and benefit to ensure patient safety while maintaining reliable product supply. ICH Q9(R1), *Quality Risk Management*, provides important guidance by clarifying risk-based decision-making and emphasizing reduced subjectivity. As the guideline notes, higher levels of uncertainty, importance, or complexity may require more formal risk management approaches.

Within this framework, risk assessments should be evidence based, justified by data, supported by science and technology advancement, bounded by compliance guardrails, guided by experience-based judgment, and benefitting patients. Together, these elements enable consistent, transparent, and defensible decisions.

24.5.3 Making the Final Decision and Sharing the Rationale

The decision itself is not the endpoint. Communicating both the decision and its rationale across relevant functions reinforces transparency and strengthens organizational capability in quality decision-making.

24.5.4 Applying Learnings

QBL research shows that while many organizations conduct AARs after major deviations or project completion, learnings are often not systematically embedded into processes and systems. This creates a missed opportunity to close the loop and institutionalize new knowledge. Effective quality business leaders deliberately acquire, analyze, store, and disseminate knowledge related to products, processes, and components, ensuring that lessons learned are ingrained in the system to help inform future decisions and continuously strengthen the overall quality system.

24.6 Trait: Promote Quality as a Value-Driven Financial Advantage

A primary barrier to quality leaders fully integrating into the business is language. Quality professionals speak in cGMP terms, while executive management operates in financials. Quality success is typically measured through timeliness, reduced CAPAs and deviations, and avoidance of inspection observations. While these measures matter, they must be translated into financial impact for quality to influence decisions and lead the business. Reframing quality from a necessary cost to a source of value and competitive advantage is therefore essential.

This shift begins with terminology. The word "cost," as used in cost of quality, implies reduction and can unintentionally diminish the perceived value of quality activities. In contrast, "value" conveys optimization and strategic importance, something to be increased rather than minimized.

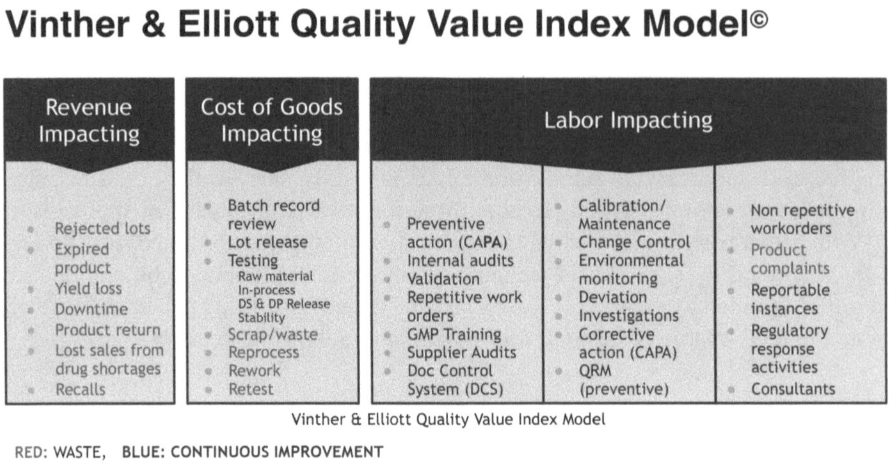

Fig. 24.2 The QVI model

Demonstrating quality value requires understanding the financial impact of quality-related activities, including deviations, batch release, batch failures, and prevention efforts to mention just a few. Cost of quality (CoQ) models, pioneered in the 1950s and refined in the 1970s, categorized activities into prevention, appraisal, internal failure, and external failure to quantify waste. Despite their historical importance, CoQ models have fallen out of favor in the pharmaceutical industry due to their focus on cost reduction and poor alignment with corporate financial reporting.

Compounding this issue, many organizations have not implemented an alternative framework, leaving quality activities weakly quantified in financial terms. To address this gap, we redefined both categories and language through the Vinther & Elliott Quality Value Index (QVI) model, which directly links quality activities to revenue, cost of goods sold, and labor costs, as illustrated in Fig. 24.2.

For each of the activities, the coding reflects if it is a cost that ideally should approach zero (red color) or offers an opportunity to continuously improve (blue) value wise.

Risk registers are now common and usually are developed around risk to patients. They are scored based on severity (S), probability (O, occurrence), and detectability (D), which when multiplied generate a risk priority number (RPN = SxOxD). The higher the score, the higher the risk to patients.

We suggest adding a risk to business dimension to the individual high RPN scores. We call it the financial risk number (FRN), which is calculated as FRN = SxOxC/P. C is the cost of correction in case the risk happened, and P is the cost of preventing the risk from happening.

Incorporating FRNs into your risk register enables quality professionals to engage finance partners and executive leadership into the work of quality by framing quality risks into financial terms they are familiar with, rather than relying solely on compliance-based language.

The adoption of FRNs fosters meaningful dialogue with senior leadership, facilitating proactive budget allocation for risk mitigation to prevent costly corrective actions that could impact profitability (the P&L).

Patient safety should always remain the primary focus for prioritizing risk mitigation, although financial considerations should also guide sound decision-making. Companies should implement a structured, systematic approach to prioritize risk mitigation strategies, balancing patient safety with financial prudence.

24.7 Trait: Lead the Path to One Global Quality Regulatory Framework

Diseases know no borders, but quality regulatory frameworks do. Post-approval changes (PACs) illustrate this challenge. Regulatory expectations for reporting levels, documentation, and timelines vary by country and region, often spanning months or years rather than days or weeks. This fragmentation strains global supply chains and slows the adoption of new technologies, even when they benefit patients. Yet science is global, and the pharmaceutical industry operates at global scale, producing fundamentally equivalent products.

Quality leaders have an opportunity to move beyond reacting to national regulatory requirements and instead lead the path toward more aligned, global quality regulatory frameworks. This leadership is grounded in science and industry quality expertise and driven by the need to advance innovation while strengthening the robustness of quality systems for both patients and companies.

If the pharmaceutical quality profession establishes a clear, shared vision supported by science- and data-based benchmarks for best practices, global standards can be defined and implemented across key quality topics. Achieving this requires quality business leaders to intentionally invest time in thought leadership beyond their own organizations, remain focused and consistent, build trust with regulators and peers, and resist the pull toward narrow, company-specific solutions.

24.8 Conclusion

Quality leadership is at a turning point, reflecting a fundamental redefinition of quality's role in organizational leadership. Quality business leaders operate as strategic business partners, shifting quality from reactive oversight to proactive leadership and from reporting outcomes to shaping decisions. This transition requires deliberate mastery of the **six traits of quality business leadership**, which provide a prescriptive framework for integrating quality into core business decision-making. Collectively, the six traits position quality beyond compliance and as a driver of sustained performance for patients, organizations, and the broader healthcare system.

OPEX and System Engineering: Advancing Healthcare Through the End-to-End Supply Chain Management

25

Jim Cafone

Abstract

This chapter explains how using operational excellence (OPEX) and systems engineering can improve the way healthcare companies manage their supply chains, from research all the way to getting medicine to patients. By standardizing how work is done, tracking performance, and always looking for ways to improve, companies can work more efficiently and handle problems better. The chapter highlights how new technology—like digital data systems, automation, and artificial intelligence—helps Pfizer connect different steps in their supply chain. This makes it easier to see what is happening in real time, spot risks before they become serious, and keep everyone focused on common goals. The main message is that combining strong processes with right systems and the right data foundation digital tools creates a supply chain that is more reliable, flexible, and better at meeting patient needs.

25.1 Context Setting

The chapters of this book collectively illustrate how adopting operational excellence within manufacturing and quality operations significantly transforms traditional work methodologies. We also believe that operational excellence (OPEX) is relevant to the broader end-to-end supply chain process—from discovery through to patient delivery. In this chapter, we will discuss the advantages of applying *operational excellence* and *systems engineering principles* throughout the entire supply

J. Cafone (✉)
Pfizer, New York, NY, USA
e-mail: jim.cafone@pfizer.com

T. Friedli et al. (eds.), *Striving for Operational Excellence in the Pharmaceutical Industry*, Future of Business and Finance,
https://doi.org/10.1007/978-3-032-26287-5_25

chain ecosystem. Core OPEX principles including standardized work, performance measurement, and continuous improvement are fundamental. But the journey toward excellence is radically changing with the addition of new digital toolsets that must be harnessed. These include the use of unified master data libraries, digital automation, and advanced artificial intelligence solutions (such as large language models and agent-based frameworks). Through the application of these tools, one can extend OPEX benefits from conventional manufacturing to comprehensive end-to-end supply chains. Now let us explore this thesis.

25.2 System Engineering: Background Context

Systems engineering is an interdisciplinary field focused on the design, integration, and administration of complex systems. Unlike traditional engineering disciplines that target individual components, systems engineering utilizes a comprehensive approach to ensure that *all elements work seamlessly toward a common objective.* A real-life example of systems engineering is automotive antilock brake system. An antilock braking system comprises several integrated components designed to prevent wheel lockup and maintain steering control during braking. These are:

- ***Caliper***: Utilizes hydraulic pressure to press brake pads against the rotor.
- ***Wheel speed sensors***: Monitor the rotational velocity of each wheel.
- ***ABS control module***: Processes sensor input, identifies potential wheel lock scenarios, and modulates hydraulic pressure as required.
- ***Hydraulic control unit (valves):*** Adjusts brake pressure on individual wheels to optimize performance.
- ***ABS pump***: Restores hydraulic pressure following ABS activation.

Effective coordination among these components is essential to achieve safe and controlled braking, particularly in demanding driving conditions.

In the business environment, effective integration of enterprise processes is essential for establishing an optimized design and deployment of strategic objectives and quantitative outcomes. If not, the enterprise will be suboptimized at the expense of serving the interests of individual functions. Realizing this level of optimization is inherently challenging, as it requires careful management across four key dimensions: process, information, technology, and people. Without a unified approach to integrating process, information, technology, and people, organizations will, at a minimum, utilize a lot more kinetic energy than needed for a customer output. In extreme situations, an enterprise sees a complete degradation of their outcomes and customer service. Why is this? Several examples of these failures are provided below:

- ***Process decay***: Previously effective standard operating procedures (SOPs) become outdated, prompting employees to develop informal workarounds and uncoordinated processes that lack management oversight.

- ***Information silos***: Communication becomes fragmented, leading to situations where customer demand data fails to inform appropriate supply chain actions. This results in the erosion of a single, reliable source of information.
- ***Bureaucratic friction***: Organizational growth is accompanied by additional approval layers and administrative barriers, diverting effort from value-added activities for customers.
- ***Goal alignment decay***: Over time, individual departments prioritize their own objectives at the expense of the enterprise's overarching mission, reducing overall organizational cohesion.
- ***Business entropy***: In thermodynamics, entropy refers to the degree of molecular disorder, randomness, or inefficiency in energy utilization within a system. When applied to business, maintaining an integrated and structured approach to executing work is essential to effectively channel organizational energy toward fulfilling the company's purpose, mission, measurable objectives, and stakeholder expectations. Absent deliberate efforts to organize and optimize processes, disorder naturally increases over time, often resulting in significant costs as resources and energy are expended without generating commensurate value.

25.3 System Engineering the BioPharma Industry

So using a systems engineering approach to the biopharmaceutical sector necessitates the integration and coordination of multiple processes to optimize patient outcomes and efficiently align the end-to-end (E2E) value chain. The E2E value chain in the biopharmaceutical industry encompasses the following eight macro processes:

1. ***Research***: Rigorous investigation designed to enhance scientific understanding, develop innovative technologies, and identify emerging patient needs.
2. ***Development***: Transformation of research findings and customer requirements into validated, reproducible product designs.
3. ***Marketing***: Strategic identification of target audiences, establishment of brand awareness, and implementation of demand generation initiatives.
4. ***Selling***: Conversion of recognized needs into substantive customer interactions and sales transactions.
5. ***Planning***: Incorporates demand forecasting; resource allocation for materials, labor, and equipment; inventory management; as well as scheduling for manufacturing, release, and distribution.
6. ***Procurement***: Acquisition of necessary raw materials, components, and services essential for production.
7. ***Manufacture***: The process of converting raw materials into finished products through specialized labor, advanced machinery, and technical expertise.
8. ***Deliver***: Management of logistics and distribution operations to ensure prompt and reliable delivery of finished goods to customers.

25.4 System Engineering the BioPharma Supply Chain

The pharmaceutical supply chain encompasses planning, procurement, manufacturing, and delivery. \Pfizer \ has established a unified process governance framework within its supply chain, and adopted integrated work practices to optimize operations and ensure the entire organization is constantly monitoring and improving the "integrated work". As stated previously, this has been realized through the deployment of the ***Integrated Manufacturing Excellence production system (see Chaps. 4 and 16)***. Process centric teams at all levels, from frontline personnel to executive leadership, adhere to standardized processes and procedures (across plan, procure, manufacture, and deliver), apply recognized performance indicators, and participate in ongoing training focused on continuous improvement to uphold strong process control. As a result, process-centric teams take full ownership of results and continuously enhance performance, integrating work both vertically and horizontally through enterprise management systems. This system forms the foundation of our operating model for planning, procurement, manufacturing, and delivery. The comprehensive catalogue of end-to-end supply chain processes is organized into three hierarchical tiers: ***(1) frontline execution***, ***(2) strategic long-range planning***, and ***(3) enabling capabilities***. A brief description of each of these processes is as follows:

1. Frontline Supply Chain Processes
    1. ***Demand planning***: Utilizes epidemiological data, market access information, pricing strategies, and marketing initiatives to accurately forecast patient and market demand.
    2. ***Production planning***: Translates demand projections into capacity-aligned manufacturing production plans throughout the network.
    3. ***Material requirements planning (MRP):*** Identifies and schedules material needs to facilitate compliant and uninterrupted production.
    4. ***Site material procurement***: Secures critical materials at manufacturing sites to maintain operational continuity and regulatory compliance.
    5. ***Detailed scheduling***: Refines production sequencing to enhance throughput and service reliability.
    6. ***Manufacturing and release***: Oversees production and quality release of products in full compliance with regulatory, safety, and supply obligations.
    7. ***Warehouse management***: Manages secure and compliant storage, handling, and movement of both materials and finished products.
    8. ***Global logistics***: Coordinates the compliant, temperature-controlled transportation of products throughout the global supply network.
    9. ***Customer experience/in market logistics***: Facilitates prompt, dependable delivery of medications to patients via efficient localized distribution and service channels.

2. Strategic Supply Chain Processes

   - ***Integrated business planning***: Establishes a cohesive, cross-functional framework for business planning and operations in the short, mid, and long terms, ensuring alignment in decision-making and outcomes.
   - ***Product life cycle management***: Enforces robust governance from inception through retirement of products, optimizing value, regulatory compliance, and operational performance.
   - ***Strategic network design and sourcing***: Strategically configures both internal and external networks to achieve an optimal balance among cost, service quality, resilience, and risk mitigation.

3. Enabling Capabilities

   - ***Data Management, analytics, and reporting:*** Design and deployment of a unified data foundation encompassing material master records, BOMs, vendors, customers, and more. This enables the operational execution of work and transformation of data into actionable insights through advanced analytics and standardized reporting mechanisms.
   - ***Supply chain parameter automation***: Automates the creation, maintenance, and optimization of supply chain parameters, ensuring that plans are continuously aligned with actual business conditions. Parameters such as cycle time, shelf life, lot size, and safety stock are effectively managed through this approach.
   - ***Operational excellence*:** Leveraging harmonized processes, information, and data, organizations can drive continuous improvement by refining processes, controlling costs, and enhancing performance. The focus remains on maximizing efficiency, productivity, and value delivery across end-to-end operations.
   - ***Risk management***: Through integrated processes, information, and data, potential operational and strategic risks are proactively identified, assessed, and mitigated. This strategy helps safeguard business continuity, bolster resilience, and support informed decision-making.
   - ***E2E command center*:** Delivers comprehensive visibility and control throughout the value chain by monitoring performance, identifying issues promptly, and coordinating swift, data-driven responses.

25.5 IMEX Enables End-to-End Supply Chain Performance

A simplified frontline process framework is given below (see Table 25.1).

During operational execution, each process-centric team in the process framework operates under a standardized set of work practices, a standard set of metrics, as described below (see Fig. 25.1):

When deviations occur, the PCT initiates a continuous improvement cycle to restore process integrity. This approach fosters a cohesive end-to-end framework by

Table 25.1 Frontline process overview

Frontline process	Standard work	Traditional metric example
Demand planning	• Frontline *demand planners* validate and endorse forecasts	• Forecast accuracy/bias
Production planning	• Frontline *supply planners* drive E2E horizontal product planning across multiple frontline supply nodes	• Plan attainment • Break in causality
Manufacturing and release	• Frontline *operators* and *lab analysts* execute manufacturing conversion and quality release	• Right first time • Cycle time
Global logistics	• Local *logistics planners* and *logistics operators* execute origin to destination transport execution	• Air-to-Ocean Transformation • Freight cost • Right first time • Cycle time
Customer experience	• *Customer order management teams* drive the management, delivery, and ultimate satisfaction with the customer	• On-time in-full • Customer segmentation • Cost to serve

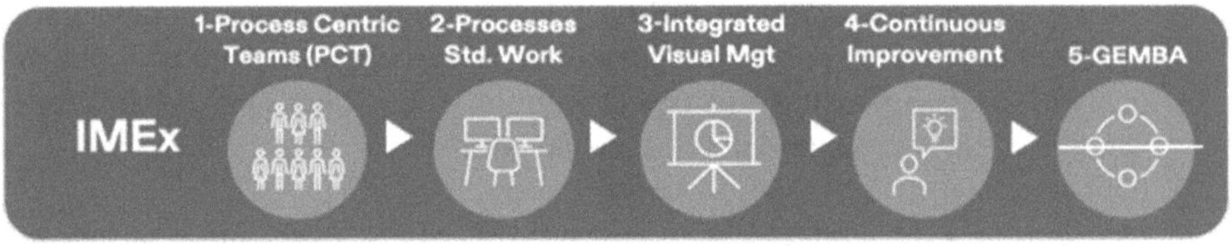

Fig. 25.1 Process-centric team standardized set of work practices

integrating individuals, practices, and leadership within a unified execution system, thereby promoting transparency, accountability, and continuous learning for sustainable high-level performance. The implementation of IMEX throughout the supply chain has resulted in substantial advances in operational execution. Nevertheless, as time progresses, opportunities for further enhancement may diminish due to *inherent constraints of traditional human-centric methods*. At Pfizer, we have acknowledged this challenge and introduced advanced digital methods—termed intelligent IMEX—to redefine productivity and performance standards, as outlined below (see Fig. 25.2):

The integration of *human and digital methods* has facilitated notable progress in operational effectiveness. Human approaches remain essential in scenarios requiring discernment, judgement, and interdepartmental collaboration, particularly when confronting bias and data deficiencies. Digital methods are crucial for enabling rapid and consistent analytical processes, especially where decision complexity exceeds human capability for speed and efficiency. However, optimal outcomes are realized when human expertise is applied alongside digital systems. This collaborative framework enhances velocity, precision, and comprehensive insights, which are fundamental for identifying inefficiencies and leveraging emerging opportunities.

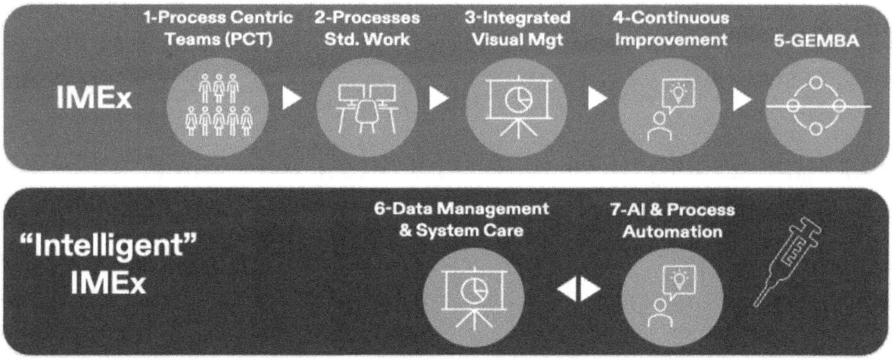

Fig. 25.2 Process-centric team standardized set of work practices with advanced digital methods

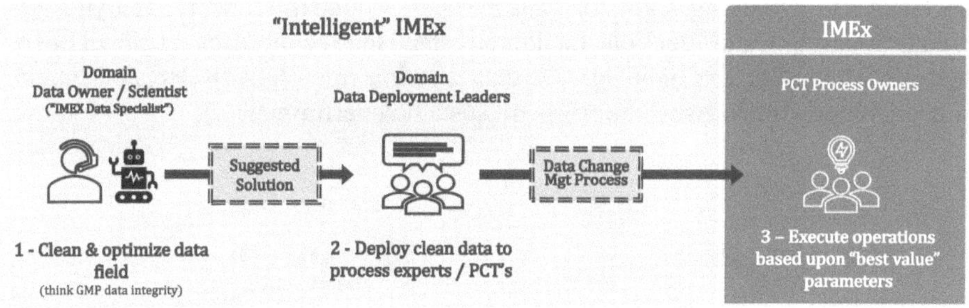

Fig. 25.3 Intelligent IMEX

25.6 New Approach: "Digital Intelligent" IMEX

Intelligent IMEX augments the traditional IMEX framework by deploying unified data management capabilities, process automation, and rigorous enterprise system oversight (see Fig. 25.3). This methodology enhances operational efficiency, ensures regulatory compliance, and enables scalable solutions throughout the organization.

Within the current, multifaceted landscape of global supply flows, the transition toward intelligent IMEX is fundamentally transforming value generation within the supply chain. Through the incorporation of real-time intelligence, predictive analytics, and embedded compliance measures, intelligent IMEX effectively streamlines complex operations, facilitates expedited decision-making, bolsters proactive risk mitigation, and optimizes the efficiency of trade flows. This paradigm provides greater transparency and reliability for our teams, clients, and partners, equipping them to function optimally in a dynamic environment.

This capability aligns with Pfizer supply chain's broader strategy of evolving toward a fully digital operating model. Agentic AI is poised to be an influential advancement, progressing beyond traditional automation to intelligent systems capable of learning, adapting, and taking action. By embedding advanced artificial intelligence and machine learning into our IMEX processes anchored by robust, standardized enterprise practices, we are constructing a supply chain that is increasingly resilient, adaptable, and scalable.

25.7 Identifying Opportunity and Loss in the Digital Supply Chain

Within Pfizer's supply chain, significant opportunities emerge at the intersection of scale, speed, and regulatory requirements. By integrating demand, planning, manufacturing, quality, and logistics data into unified digital platforms, Pfizer achieved real-time visibility that enabled early identification of bottlenecks and dynamic rebalancing of capacity (see Fig. 25.4). This transformation replaced reactive problem-solving with proactive coordination. Leadership played a critical role in setting priorities and managing regulatory and partner complexities, while analytics and digital "control tower" functions facilitated effective execution across global operations. The combination of integrated data and informed human decision-making converted operational risks into resilient, sustained performance.

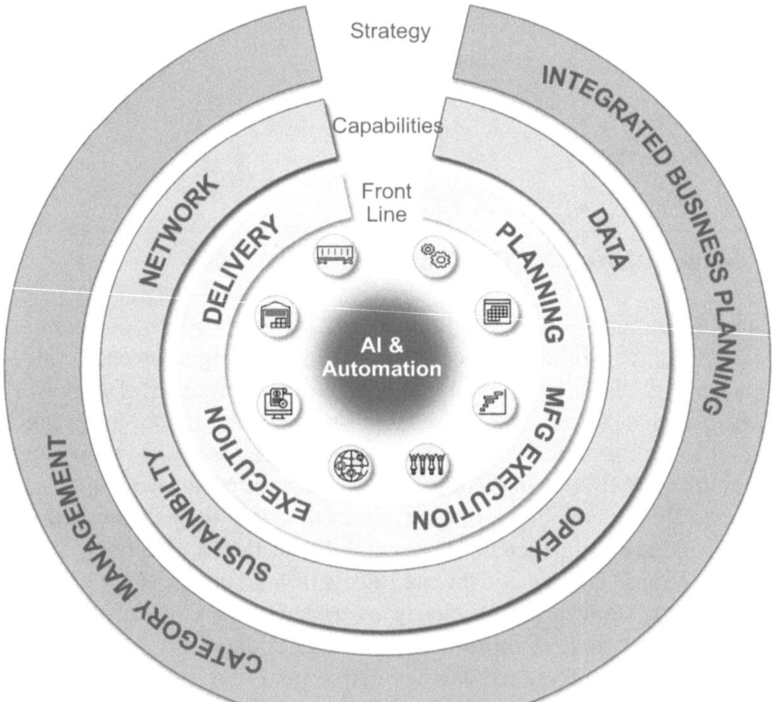

Fig. 25.4 Unified digital platforms

25.8 Key Enabler: Unified Data

Using an airplane example, unified and integrated data streamlines the linkage between flight sensors, central computing systems, and corresponding wing and flap actuator configurations. This optimization ensures that wing positions, flap settings, and fuel consumption are consistently managed throughout aircraft operation to prioritize passenger safety and deliver a seamless flight experience. Moreover, this methodology enhances schedule adherence and contributes to overall optimal performance in every phase of flight.

Within BioPharma supply chains, comprehensive data integration across suppliers, manufacturing, logistics, and demand planning enables real-time visibility, supports timely decision-making, and promotes efficient material flow. In both sectors, integrated data functions as a cohesive control framework: without it, organizations encounter heightened risks of delays, inefficiencies, and misalignment; with it, they achieve improved precision, reliability, and informed decision-making capability.

25.9 Key Enabler: The New Agentic Process-Centric Team

Unified data has allowed Pfizer's supply chain to advance its digital transformation by incorporating agentic AI within process-centric, human-led teams. These agents equip colleagues to work more effectively and efficiently and deliver greater value. The agents optimize operational parameters, reduce manual intervention, and support faster, more informed decision-making. With the implementation of a new process-centric team model (see Fig. 25.5), human expertise works in close partnership with autonomous AI agents to improve agility, accuracy, and innovation while maintaining robust governance and oversight. This evolution not only enhances efficiency but also fundamentally transforms workflows, ultimately achieving improved outcomes for patients and the organization.

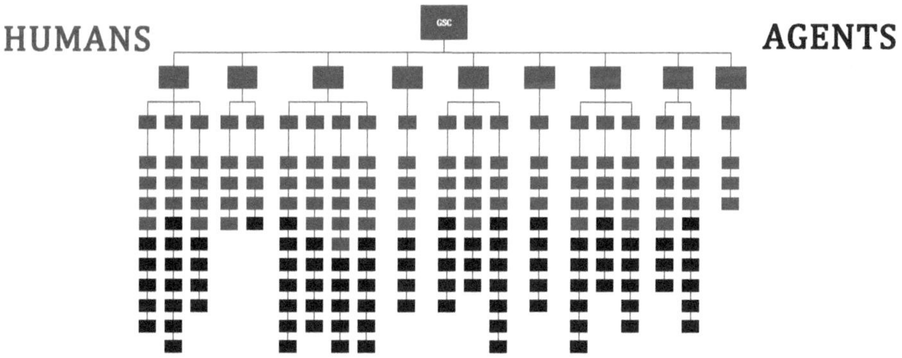

Fig. 25.5 Agentic process-centric team model

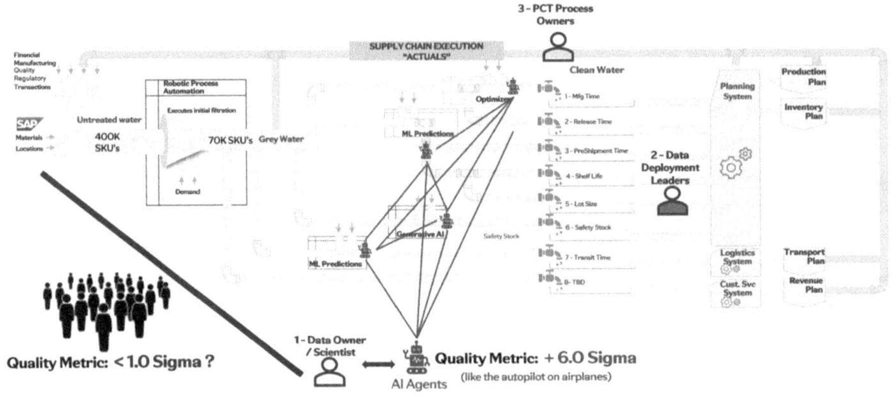

Fig. 25.6 Multi-agent network-wide AI system

25.10 Key Enabler: The Supply Chain Parameter Data Factory

At Pfizer, we implemented a "data factory" that utilizes advanced analytics, machine learning, and optimization to transform historical data into robust supply chain planning parameters. This approach establishes an integrated and dependable data repository accessible by all systems, thereby supporting precise and consistent decision-making, and ensures that the supply chain operates under optimal planning parameters.

Pfizer is also piloting a multi-agent AI system to distribute these analytical insights throughout its global network (see Fig. 25.6). This initiative promotes transparency with explainable models, scenario analyses, and structured governance. Collectively, these innovations strengthen consistency, trust, and confidence in data-driven decisions enterprise-wide.

25.11 Key Enabler: The Supply Chain Command Center

The Pfizer Global Command Center, powered by agentic AI, serves as the central hub for global supply chain operations. Autonomous agents systematically monitor real-time signals to detect potential disruptions at an early stage, evaluate their impacts, and coordinate prompt, integrated responses. By proactively identifying and mitigating incidents before escalation, the Command Center bolsters operational resilience, expedites recovery, and enables Pfizer supply chain to respond adaptively to changes throughout its worldwide network.

25.12 The Outcome: A Supply Chain in a Constant State of Visibility, Control, and Optimization

A modern business system should operate as a dynamic ecosystem that remains visible, strategically managed, and continuously refined. At Pfizer, this is achieved through an integrated approach: the planning data factory transforms enterprise data into reliable, actionable intelligence; the global command center leverages real-time monitoring and advanced agentic AI to identify, address, and mitigate operational disruptions; and fusion teams combine human expertise with artificial intelligence for rapid and accurate execution. Collectively, these capabilities drive a comprehensive transformation, establishing a closed-loop system designed to detect changes, coordinate effective responses, and foster ongoing improvement.

25.13 The Future: From Planning to Orchestration—Enabling a Faster, Data-Driven End-to-End Supply Chain

Volatility has materially compressed planning horizons and exposed the limits of fragmented, sequential planning models. In response, Pfizer has embarked on a deliberate orchestration journey initially anchored in unified data, data factory, agentic process-centric team, and E2E visibility across nodes, to move from disconnected planning activities to an integrated, enterprise-wide approach. These early foundations established the connective tissue across data, processes, and decision rights, creating the conditions for scalable orchestration.

Building on this foundation, Pfizer addresses the planning gap through true orchestration: connecting signals, decisions, and actions across the enterprise through a unified data foundation, AI-enabled capabilities, and redesigned strategic processes.

AI-driven orchestration represents a fundamental shift in how planning is executed. Multi-agent architectures now automate data preparation, model selection, explainability, and forecast alignment at scale accelerating decisions while increasing trust in outcomes. Realizing this value, however, requires disciplined data ownership and governance. Orchestration is not a technology upgrade; it is an operating model transformation anchored in data accountability, decision clarity, system visibility, and speed at scale.

Part IV

Conclusions from the Book

What Are Deep Structures? 26

Thomas Friedli and Fabian Klinkner

Abstract

In this chapter, we define deep structures, apply the concept to OPEX programs, and elaborate our perspective on this phenomenon. We further deepen our conceptual understanding alongside selected examples, providing the foundation for the concluding remarks of the book.

26.1 Introduction

In literature, *deep structures* are usually seen as a barrier to revolutionary change in organizations (cf. Gersick, 1991; Heracleous & Bartunek, 2020). In parts, this perception is compatible with our understanding of deep structures. However, we do not view deep structures as a hurdle that must be overcome to enable change. We have indeed experienced cases where structures in Operational Excellence (OPEX) have been punctuated and finally broke, but these very experiences made us see the functionality of deep structures to sustain OPEX. In this chapter, we will define deep structures, apply the concept to OPEX programs, and elaborate our perspective on this phenomenon. We will further deepen our conceptual understanding alongside selected examples, providing the foundation for the concluding remarks of this book.

T. Friedli (✉) · F. Klinkner
Institute for Production and Supply Chain Management, University of St.Gallen, St.Gallen, Switzerland
e-mail: thomas.friedli@unisg.ch; fabian.klinkner@unisg.ch

T. Friedli et al. (eds.), *Striving for Operational Excellence in the Pharmaceutical Industry*, Future of Business and Finance,
https://doi.org/10.1007/978-3-032-26287-5_26

26.2 Deep Structures Defined

When analyzing models from six different domains dealing with organizational system development and change, Gersick (1991, p. 12) identified a common paradigm: "… relatively long periods of stability (equilibrium), punctuated by compact periods of qualitative, metamorphic change (revolution)." A deep structure from this perspective is "… a highly durable underlying order …" (ibid.). She continues to state that "this deep structure is what persists and limits change during equilibrium periods, and it is what disassembles, reconfigures, and enforces wholesale transformation during revolutionary punctuations" (ibid.).

26.3 What Does This Mean for OPEX Programs

T. H. Netland (2012, p. 31) defines an OPEX program "as a corporate process improvement program with the goal of driving competitiveness" (cf. Chap. 5).

Therefore, the OPEX program is not—and should not be—subject to radical change, as its very essence is to *drive* change, not to be changed itself. Recent research has examined why and how OPEX programs change (cf. Grothkopp, 2023; Netland et al., 2026). However, the cases discussed typically involve situations where the change was detrimental to the company or where the program was poorly designed from the outset. In these examples, organizations failed to build deep, embedded structures; their programs remained superficial, which made later adaptations possible—and in some cases unavoidable.

The real challenge during periods of disruption does not necessarily lie at the management or leadership level. It emerges where continuous improvement actually occurs: on the shop floor and within core processes. When OPEX programs falter, credibility erodes quickly, and engagement levels deteriorate just as fast. Rebuilding trust requires significant time and effort.

26.4 Observed Punctuations from Our Experience

Through our long-term collaborations with selected companies in our research on pharmaceutical OPEX, we have witnessed firsthand situations where OPEX programs could not be sustained:

1. In a global pharmaceutical company, a power struggle occurred when the very successful OPEX program, launched initially based on a Six Sigma philosophy, was intended to be complemented by traditional lean methods, like value stream mapping. As a senior manager wanted to reap the benefits of lean, he applied pressure to set up a parallel structure for the lean organization to report to him. Even though this could be avoided, the clash left no winners and had a negative impact on the existing organization, weakening previously built problem-solving capabilities.

2. In another case, the CEO invited all his top managers to a meeting to transport his vision for OPEX. In his first address, he stated: "We will do a lot of OPEX going into the future." In his subsequent address, he talked about the savings in the range of billions that would come from the OPEX efforts. All the managers present therefore perceived these efforts as a cost-cutting exercise going forward. The term OPEX had thus received a negative connotation and was subject to significant prejudice thereafter.
3. In still another instance, a case company bought a license for an off-the-shelf OPEX program from a consultancy and tried to implement it as it was without adaptation. For the employees, this program never became *their* program. It was seen as external; there was no buy-in at all.
4. A major pharma company that was an early adaptor of lean, already defining measure to become best in class, saw its head of production and sponsor of the OPEX program leave the company after 10 years. The new head of production was not supportive of a seamless continuation and made employees question everything and rethink from scratch. The company has not recovered from this setback until today.
5. In another company, the OPEX program was for some time seen as the very future of the company. This attracted a lot of talent to the OPEX organization and became part of its success. Initially, there was a rather aggressive approach taken, copying what they understood as lean from companies like Toyota and other automotive manufacturers without considering pharma-specifics. The program failed when hierarchies at plant level were reduced from seven to three, leaving process unit heads with over 100 direct reports. The confusion at plant level got so high that the program could not be sustained.
6. A global pharma company had been running its OPEX program for multiple years with considerable success. The employees believed in OPEX and continuous improvement. However, an FDA inspection in one of the company's leading plants caused destabilization. From an OPEX perspective, the company was pushed by the FDA to ramp up a new organization for manufacturing science and technology. This new organization was seen as the future, whereas the old OPEX organization became a relic of the past. The program lost its credibility and support; the success factors were gone.
7. A company tried to establish its OPEX program by training selected experts. The reasoning was that we build the capabilities and will reap the benefits. However, people were selected based on the basic knowledge of statistical tools; leadership skills were neither evaluated nor seen as a prerequisite. The training program that led to an expert certification but only when a project was conducted at the plants of the participant had additional not-intended consequences. There were too many projects in parallel that were not based on the real needs of the specific plants but were seen as the fastest way to the certification. This led to a situation where everybody not involved with training developed a poor attitude towards the program.
8. When we met with a global OPEX manager in his office, he asked us if we would like to see the production. We certainly wanted to and followed him out

of the office just to find out that he could not find the entrance to the part of the plant he wanted to show us. Obviously, the time he spent on the shop floor was rather limited. Accordingly, in this company, there was nobody taking the OPEX program seriously. The main spirit was to see this as a PowerPoint exercise with nice slides but no impact.

We could list countless more of such anecdotal examples as evidence but will stop here. To draw some conclusions from these examples, in the next chapter, we will summarize the lessons learned from this book, always deriving consequences for striving towards deep structures.

References

Gersick, C. J. G. (1991). Revolutionary change theories: A multilevel explorations of the punctuated equilibrium paradigm. *Academy of Management Review, 16*(1), 10–36. https://doi.org/10.5465/amr.1991.4278988

Grothkopp, M. (2023). *Managing production systems: A corporate-wide improvement program perspective* [Dissertation]. University of St.Gallen, St.Gallen.

Heracleous, L., & Bartunek, J. (2020). Organization change failure, deep structures and temporality: Appreciating wonderland. *SSRN Electronic Journal*. Advance online publication. https://doi.org/10.2139/ssrn.3602800

Netland, T. H. (2012). Managing strategic improvement programs: The XPS program management framework. *Journal of Project, Program & Portfolio Management, 3*(1), 31–44. https://doi.org/10.5130/pppm.v3i1.2556

Netland, T., Grothkopp, M., Rossini, M., Friedli, T., & Portioli Staudacher, A. (2026). Why and how do corporate lean programs change? A punctuated equilibrium perspective. *International Journal of Production Economics*. Advance online publication. https://doi.org/10.1016/j.ijpe.2026.109980

Towards Deep Structures

27

Thomas Friedli and Fabian Klinkner

Abstract

In this chapter, we review selected statements and insights from the previous chapters to derive consequences for managing OPEX in the pharmaceutical industry. This provides the foundation for our conclusions from the book and an outlook towards deep structures.

"Come back with your shield, or on it."
Plutarch, Sayings of Spartans, 241.

Operational Excellence (OPEX) has the potential to be the deep structural backbone of the organization—not just a set of tools or initiatives, but the underlying architecture that shapes how we work, decide, and improve. If it is embedded in our deepest structures, it remains stable, even if leadership, strategy, or external conditions change. Innovations like digitalization, artificial intelligence (AI), and agile methods can only thrive if they build on this deep, resilient backbone.

For this section, we apply a methodology that is taught in the leadership courses of the Swiss Army and used for situational assessments. This approach helps to transform information (statements taken from the previous chapters) into understanding (insights). Thereby, we derive consequences for the creation of deep

T. Friedli (✉) · F. Klinkner
Institute for Production and Supply Chain Management, University of St.Gallen, St.Gallen, Switzerland
e-mail: thomas.friedli@unisg.ch; fabian.klinkner@unisg.ch

T. Friedli et al. (eds.), *Striving for Operational Excellence in the Pharmaceutical Industry*, Future of Business and Finance,
https://doi.org/10.1007/978-3-032-26287-5_27

structures—not just superficial symptoms—when striving for OPEX in the pharmaceutical industry.

The information is taken directly from the preceding chapters. To ensure readability, we present the three categories, statements, insights, and consequences, in Table 27.1 hereafter.

Based on this condensed review of key content of our book, we provide a core summary of these conclusions:

1. OPEX must be a fully integrated future-proof system, not a project, or a toolbox:

- It must be designed, deployed, and sustained as a holistic system that aligns with all other initiatives and integrates them.
- OPEX has the potential to be the backbone of operations, the way work is conducted.

Avoid the impression of a short-lived program; build it to evolve, progress, and remain future oriented.

2. OPEX must become an inseparable part of the culture:

- Structures and activities impact culture and must be supported by the culture.
- OPEX leaders must work on culture from the start, using hard elements (standards, KPIs, governance) to reinforce soft elements (mindset, engagement, trust).
- Build a beyond compliance mindset.

3. Leadership commitment and communication are key:

- Select the right key people and invest time in enabling them to succeed.
- Top and middle management must communicate constantly and consistently.
- Leaders need to be present, need to interact with people doing the work, and need to be the role models for the behaviors expected.

4. Embed CI into structures and daily routines:

- Standards must be defined, enforced, and continuously improved.
- CI must be natural, systematic, and embedded across all levels and functions—not limited to manufacturing.
- Governance must be cross-functional, with clear responsibilities and performance tracking.

Table 27.1 Statements, insights, and consequences

Statement	Insight	Consequences
Part 1: Introduction		
OPEX is the balanced management of cost, quality, and time while at the same time focusing on customer needs. To achieve this, OPEX comprises structural and behavioral changes, thought to ideally support necessary activities	OPEX is a comprehensive management approach, not a set of tools, not a technical exercise, not a project with an end date	The design, deployment, and sustainment of an OPEX program have to be a reflection of this. Culture and structure have to visibly support the respective activities
OPEX is not only concerned with performance. It also encompasses the journey leading to this superior performance and practices that allow an organization to continuously improve itself and fully engrain OPEX into the organization	OPEX is not a separate organizational unit; it is a fully engrained core element of the entire company	OPEX has to be positioned as what it really is, the way a specific company works and continuously improves itself
Quality culture is a shared commitment, driven by leadership and embraced across the organization, to consistently uphold high standards through accountability, proactivity, and ethical behavior in everyday work	Culture comes from the top but has to be built with, by, and through people	Make culture a focus of your OPEX efforts
Compliance is about adhering to regulations and requirements set by authorities to meet standards, ensure safety, and support the license to operate	It is not either OPEX or compliance; it is OPEX and compliance	Find ways to bring both elements—OPEX and compliance—Together
Beyond compliance means exceeding minimum requirements by embedding continuous improvement (CI), innovation, and quality into every process, striving for excellence and setting new standards	To ensure the future competitiveness, companies cannot "just" rely on compliance alone	Instill a mindset beyond compliance
CI is the ongoing effort to optimize processes, systems, and performance by fostering a culture of excellence, regularly evaluating and refining operations, and eliminating inefficiencies	CI is a must	Ensure that CI is embedded in your structures and your culture
… there must be reasons to engage in serious efforts to become an "operationally excellent company." if this is not given or it is not transparent, the whole program will remain skin-deep and fade away over time	Communicating the strategic relevance and impact on future competitiveness is key	Ensure constant and consistent top management communication
Triggers/forces of OPEX in the first decade: Regulators and, especially the US FDA, requiring more science when it comes to manufacturing; data demonstrating waste and issues across manufacturing, a new process focus going away from "only" looking at the product, the quest for cost improvements	OPEX is and will remain a must	Design, deploy, and sustain it accordingly

(continued)

Table 27.1 (continued)

Statement	Insight	Consequences
We foresaw in the future more integrated approaches, on the one hand bringing together preventive with reactive OPEX (e.g., QbD and OPEX combined) and on the other hand the alignment of all improvement initiatives at the top management level	OPEX is not an isolated, standalone initiative	Design, deploy, and sustain it as a system. Ensure alignment with other programs/initiatives. Make it the very way the company works
Working against deeper and more integrated implementation: Cost pressure leading to the reduction of OPEX resources and travel budgets, experiences during pandemics, digitalization poorly aligned with OPEX, and cutting resources result in lost credibility	OPEX has to be safeguarded against pressure	Make OPEX a matter of course
Working towards deeper and more integrated implementation: The realization that programs must be aligned to have a sustainable impact; a recognition that OPEX first is the way towards impactful digitalization; new US FDA programs like quality management maturity (QMM) recognizing the importance of advanced quality management practices and quality culture	Several trends foster and require full integration	Ensure alignment and understanding
Given the growth in both maturity and performance, we can confirm this hypothesis. However, compared to sectors where OPEX has become ubiquitous (e.g., automotive), the pharmaceutical industry still faces an uphill battle. While maturity levels range between 60% and 70% overall as well as for the building blocks of the St.Gallen OPEX model, respective performance scores rarely exceed the 55% mark	There is a visible impact of OPEX on pharma companies' maturity and performance	Communicate it to further spread excellence in the company
After initial progress during the first decade, the industry has reverted to a rather technical implementation and brute force	There is a limited understanding of the "people" aspects of true OPEX implementation	Take culture and people into consideration
There is already a visible impact of digitalization on enablers. Some traditional approaches for statistical process control lose in importance	Digitalization is and will further become a crucial factor to generate further improvements	Align your OPEX programs with digital
… uncertainties and disruptions have confronted pharmaceutical companies with unprecedented challenges and will continue to do so in the future	There will not be a time to relax in the foreseeable future	Keep up the pace, do not slow

A nonintegrated OPEX strategy enabled the preservation of distinctive cultures, specialized processes, and bespoke workflows. In this case, it was believed that integrating an acquired company into a new production system could jeopardize its core strengths. Additionally, because speed to market is critical and transitioning to a new production system can be time-consuming, we were reluctant to enforce this move	Only standards and relentless drive can support a deep integration across the company	Do not allow for too much freedom when it comes to implementation
A single integrated production system across the end-to-end supply chain allows to enable one common operating model via one language of performance—From the shop floor to the CEO, this one "operating model" is driven via standardized work, shared metrics, and a common set of continuous improvement tools. This also allows for easier oversight, faster best practice transfer, and consistent leadership development	Horizontal and vertical integration is a must	Work over all managerial levels and across functions to make your production system the way the work is done
Companies that excel at OPEX have figured out that it is not just a set of individual tools (lean, six sigma, kaizen, business process management, balanced scorecard, ISO, etc.) but an integrated management system across processes and systems that makes the difference	OPEX is not a toolbox	Manage it as an integrated system
The model is built on the foundational principles: Standardized work, visual management, problem-solving/lean, and six sigma methodologies. Empowerment of teams built around process outcome-focused tracking coaching-centric leadership behaviors	A state-of-the art program encompasses tools, leadership, and broader engagement elements	Build your program as a system
The IMEX harmonization took place both vertically and horizontally across Pfizer global supply. Vertical harmonization connected across tiers from shop floor colleagues to the executive leadership team through shared goals, clear metrics, and a common set of operating standards. Horizontal harmonization connected departments within sites, across sites, and functions globally. IMEX became our new way of working	The production system is the way of working	Position and implement it like that
Part 2: Learnings from two decades of OPEX research		
… OPEX must be managed as a lasting program rather than a temporary project. The TPS, which Toyota began developing in the mid-twentieth century, embodies and exemplifies this perfectly, as it has remained in place ever since	It is possible for programs to sustain over a long period of time	Avoid the impression that the program is only short lived. Make it a fully integrated part of the general management and the way of working

(continued)

[illegible].1 (continued)

Statement	Insight	Consequences
Often emerging in production, OPEX programs are, however, not limited to manufacturing in their applicability. In fact, their purview extends to the entire organization with the intention of transforming processes across all departments and functions	The true potential is the enterprise-wide establishment	Do not limit it to manufacturing even if you pilot it there
Conceptually, OPEX programs may be described in three dimensions, namely content, process, and structure	OPEX is not a set of tools; it is a system	Design, deploy, and sustain it this way
Commonly, the program is owned by a corporate entity, which oversees it globally. Tasked with the local management of the OPEX program, site entities complement this organizational structure. Typically, site OPEX teams report to the general manager of their site, rather than the corporate team. Through this circumstance, these general managers become gatekeepers for the success of an OPEX program and as such key figures to involve. Finally, divisional or regional OPEX teams may also be installed for further support, depending on the company's requirements	There is some complexity involved in a global program Integration horizontally and vertically as well as the inclusion of key people is key	The design of the program must be a reflection of the inherent complexity
Our recent cross-industrial study on the management of OPEX programs shows that companies who update their program frequently and proactively tend to achieve far greater improvements than those who do not	The program is not static but has an always evolving component	Build it in a way that it can evolve and progress It needs a sustaining and a future-oriented part to ensure that the program is an asset
An OPEX program may rest in an equilibrium state, where it has, for example, just been put in place. Certain triggers may now punctuate this equilibrium and cause significant shifts	Even if the program has run for some time, it is not always future proof	Work on full integration and embeddedness to reduce vulnerability
OPEX programs, that would have had the potential to yield great improvements given enough time, fall victim to impatience and short-term thinking	There is a trade-off between short-term gains and long-term success	Position it as long-term; do not compromise and do not sacrifice effectiveness for short-term efficiency gains
OPEX programs represent interrelated, socio-technical system of synergetic practices. In that, their content builds on tools and principles that, when combined and employed correctly, unfold positive effects that lead to operational improvements and competitive advantages	There are soft and hard parts	Be aware of the necessity of a sound culture and work on it from the start Use the hard parts to foster the soft parts

… as a maturity assessment aims to uncover areas of insufficient OPEX implementation, exhaustiveness, and consistency in the approach that are crucial. Companies must therefore be thorough and assess the OPEX implementation across the entire organization and all elements of the program, as well as with sufficient granularity	Maturity must be regularly checked and acted upon identified improvement potentials	Make sure to have the capacity for sound maturity checks Ensure that this is done in a way that the outcomes are accepted
As processes are optimized and inefficiencies eliminated, results translate to key performance indicators (KPIs)	There must be KPI systems in place to measure the outcomes	Define KPI systems that reflect the overall system performance
In the long run, companies do not face a trade-off between quality, cost, and time, but can only excel in all these elements simultaneously if tackled in the right sequence	Sustainable cost improvements are based on sound quality	See quality as the basis for sustainable OPEX and make interdependences visible
HQ resources are limited	The HQ supports the program but must encompass as many employees as possible	Deploy and sustain across the whole company encompassing as many employees as possible Define the role of the HQ resources
HQ OPEX managers should assess their sites' priority, maturity, capacity, and accessibility, as well as the respective context, when determining how to support the sites	Maturity and performance assessments are required	Define and use them to ensure acceptance
Across the sample, OPEX programs are not seen as simple toolboxes what demonstrates that all participating companies were already on a rather high maturity level. Nevertheless, the system thinking is considerably higher in the successful practice group	A mindset beyond tool thinking is an absolute requirement for success	Do not position the program as a toolbox and a training program
Obviously, the successful practice companies were capable of embedding OPEX in a way that it became robust against disturbances like management changes. It is also much more than just a structure or some defined processes; it is the very way to work in the successful practice companies, and it brings a level of engagement with it that goes way beyond the OPEX experts; it embraces all other employees too	Successful programs are all embracing he OPEX program can even become invisible as it is so engrained that it is just the way of working	Make sure that the program becomes the very backbone of operations
When it comes to the question of how production systems are designed, the successful practice companies do this to a higher degree based on the production/business strategy and therefore achieve a better alignment	The why is as crucial as the how	Make it explainable and communicate

(continued)

Table 27.1 (continued)

Statement	Insight	Consequences
Communication seems to be key for a successful implementation	Communication is obviously a success factor	Ensure steady top- and middle-management communication
Tunner was certainly a demanding commander, but at the same time, he was the very role model for everything he expected others to do and to be	There is a need for visible and credible leadership	Select the right key people for your program
Tunner also introduced proper shifts and rest for aircrews	To get the required buy-in, people must feel that they matter	Spend enough time on people and help them to get the chance to be successful
Never, from the very beginning of my command until the end, had I subordinated flying safety to any other phase of operation	Safety and quality are paramount	Do not compromise
I did not want to get this information from squadron commanders or group commanders, men who might conceivably be influenced by what they thought I would wish to hear, but directly from the lips of the men who were flying the route themselves, day in, day out	Unfiltered information is essential for true improvements	Make managers talk to and interact with the people who really do the work
I made constant rounds. I was available to everyone, dozens of times. Making myself available to the pilots and crew members, seeking them out, and listening to their complaints gave me an insight into the true operations of the lift that I could never have gotten sitting at my desk reading filtered and secondhand reports	It is not the single GEMBA walk that makes the difference but the reputation you build around you if you do these walks	It takes presence and time to be effective as an OPEX leader
My meanderings were by no means restricted to the daytime. I would purposely pick late at night or the very early morning hours to check up on what I considered important or vital activities. It was not unusual for a sleepy tower operator to suddenly realize that I was standing there with him. Sometimes, there would be no conversation other than a greeting or a good night; there was not much need for words when things were going properly. The word got around that the general had been there during the small hours. It was a 24-h operation every day, and I wanted everyone in the airlift to remember that	GEMBA is much more effective if it is done without a clear pattern but still being present at all potential key operations	Plan GEMBA, but not in a way that it becomes a pure routine, where everybody knows when it takes place and when not
It is mandatory if operations should be successful to take decisions and make them happen	For OPEX to be successful decisions must be taken	You need people who will lead

The hustle and bustle and excitement of operation Vittles in the early days all comprised a case in point. The last place you should find this type of activity is in a successful airlift. The actual operation of a successful airlift is about as glamorous as drops of water on stone	If the system works, it is not as visible as if it is based on sweat and reactive actions	Structure and design the system
I thus decided that all planes under my command would fly a never-changing flight pattern by instrument rules at all times, good weather or bad, night or day	Standardization was instrumental for success	Standards must be defined and enforced
Tunner made sure that there were always new ambitions and objectives set to avoid negligence and complacency	The system is never static and needs inputs	There is a need for forward planning and continuous management on always higher level of aspiration
Pharmaceutical companies operate in highly regulated environments where efficiency and compliance must coexist	Both perspectives must be taken into account	Ensure alignment
Three basic forms of interaction between global and local levels can be distinguished. Top-down models enable strong strategic control, clear priorities, and rapid, coordinated implementation. Bottom-up models promote initiative, learning-oriented work, cultural anchoring, and employee engagement. Hybrid models combine both approaches and are considered particularly effective because they balance strategic consistency with operational flexibility	The most effective organization is hybrid	Build the program on leadership and engagement
OPEX teams typically promote empowerment, experimentation, and continuous learning, while quality units emphasize discipline, documentation, and compliance	The underlying institutional logics are different	Find ways to align it
When OPEX and quality collaborate, process stability and compliance reinforce each other	Alignment and collaboration are key	Ensure alignment
Integration creates strategic resilience. Companies that align OPEX and quality are better positioned to respond to regulatory changes, market pressures, and technological disruptions. By embedding CI into compliance routines, they ensure that excellence is achieved not in parallel tracks but through a unified path	OPEX and quality can work together	Build joint governance structures, cross-functional routines, hybrid competence building

(continued)

Table 27.1 (continued)

Statement	Insight	Consequences
…, research largely proliferates a complementary relationship presenting both digitalization as an enabler of OPEX and OPEX as an important catalyst for digitalization	OPEX and digital are two sides of the same medal	Integrate it
…, rather than fostering synergies, the OPEX and digitalization relationship is regularly marked by friction and incompatibility. Moreover, it is well known that individual implementation of OPEX or digitalization is already challenging	There is a need for alignment	Ensure shared direction, common language and narratives, integrated governance and structures, trust
Development in OPEX program maturity must be followed to achieve higher digitalization program maturity. A relationship where OPEX program maturity is low and digitalization program maturity is high is not or hardly possible to achieve	Digital maturity built on OPEX maturity—Not the other way round	Establish OPEX as the backbone of the organization
Barriers to CI vary widely across the life sciences industry—From initiative overload to skill gaps to strategic misalignment	The root causes for CI not working must be understood	Strengthen systematic integration, shift from technical to natural CI, balance top-down guidance with bottom-up engagement, broaden the value perspective of CI
Part 3: How it looks in the field		
Culture failures precede operational failures	Culture not tools or structures are most powerful for success	Stress culture wherever possible
The lesson for BioPharma is when purpose and discipline align and where employees feel that their work matters and understand how to succeed; employee engagement accelerates execution without compromising compliance or reputation	People are key	Make sure to make them part of the solution
Patient-focused excellence is not achieved solely through strategy, process, or compliance systems, and purpose statements alone do not build culture	Culture and alignment are key	Make sure that culture aligns behavior to purpose, consistently, visibly, and predictably Leaders need to emphasize the expectation, through their own daily actions and behaviors

Recognition plays a pivotal role in shaping how work gets done and how people feel at work	Focus on people is a requirement	Design recognition thoughtfully, highlighting the behaviors, results, and values that matter most to the organization, clearly aligning day-to-day actions with strategic goals
Our business systems either enable or disable the desired behaviors we seek	Your system can be supportive but also dysfunctional	Analyze the incentives and hurdles for the desired behavior
But true transformative participation is fundamentally rooted in the quality and consistency of leadership across an organization. Developing leadership skills and measuring the quality and impact of those skills are the basis of a culture of excellence	Leaders are key	Select the right key people Focus and measure the ability of our leadership community to identify and support the behaviors that enable positive outcomes
Middle managers, supervisors, and line leaders, through their direct and consistent contact with frontline teams, have the greatest impact on the felt culture in which daily work takes place. They are the engine of any cultural transformation	Middle management is an engine for OPEX	Work with them
Delivering for the business involves the constant interplay between the identification, standardization, and continual improvement of critical work methods and the systems that support them	It is all about the overall system. Holistic thinking is required	These systems must be designed and managed purposefully to enable the behaviors and culture necessary to ensure ongoing and sustainable improvement of the businesses' performance
While power to manage culture is ceded to frontline teams, senior leaders and middle managers must constantly verify and actively support the desired culture through the purposeful design and management of key systems within the business	It is the combination of leadership and bottom-engagement that makes the difference	Consider all levels

(continued)

Table 27.1 (continued)

Statement	Insight	Consequences
IMEX is not merely a framework; it is a philosophy that aligns people, processes, and performance. It empowers teams to work smarter, faster, and more collaboratively, ensuring that Pfizer stays agile and resilient in a rapidly evolving healthcare landscape. From the deployment of digital tools to the standardization of visual management and leader standard work, IMEX represents PGS' commitment to OPEX at every level of the organization	OPEX is a comprehensive, all-encompassing approach	Frame and design it that way
Every ecosystem begins with a PCT, a cross-functional team responsible for executing IMEX elements. The PCT is the operational nucleus, ensuring that standard work, visual management, CI loops, and other elements are implemented consistently. PCTs receive structured GEMBA coaching. They are trained to execute TPM, CI, and standard work routines. Their performance is tracked through leader standard work and visual boards	Ownership is crucial Tracking of outcomes is required	Assign clear responsibilities Set standards for performance tracking and leader standard work
To ensure that IMEX is not only deployed but also sustained effectively, PGS has developed a rigorous IMEX health check and certification process. This process evaluates the progress of IMEX implementation at sites and functions, identifies gaps, and guides teams towards higher levels of operational capability	Maturity checks and continuous improvement are key	Develop sound maturity checks and ensure that there are defined actions based on the results
IMEX is not limited to manufacturing sites; it extends to functional groups such as global supply chain, quality, and external supply, to name a few. Leadership teams at all levels of the organization, from the shop floor to the PGS leadership team, benefit from having IMEX in place	OPEX has the potential to become the way work is done across a company	Make it broad not narrow
One of the most vital enablers of lean and IMEX success is the presence and engagement of leaders. At Pfizer, we have seen that when leaders are actively involved from the start—Helping define the clear "why," supporting teams in building their systems, and guiding understanding of each IMEX element—Momentum builds quickly. Sustaining that momentum, however, requires ongoing leadership presence and support	Leadership is key	Ensure visible presence of leaders
But the significance of mindset in the pursuit of OPEX maturity cannot be overstated. While tools, systems, and methodologies create the infrastructure for improvement, it is the mindset that determines whether these structures thrive or wither	It is only when OPEX becomes the natural way of working that it will sustain over time	Make sure that OPEX is seen as the way work is done

Moreover, GEMBA is essential for embedding CI as an enterprise-wide norm rather than a fleeting initiative. When leaders model curiosity and humility by engaging directly at the source, they signal a commitment to transparency and collaborative problem-solving	GEMBA is essential to have a true leadership impact	Ensure that true GEMBA takes place
Continuous improvement 4.0 integrates classical OPEX disciplines with digital and AI capabilities, enabling organizations not only to sustain improvements but also to achieve exponential gains in performance, agility, resilience, as well as customer/patient and employee experience	New technologies (e.g., AI and digitalization) have the potential to reinforce OPEX	Integrate digital/AI systematically. Keep OPEX as the backbone
OPEX/lean programs, although grounded in process rigor, tend to yield only incremental improvements and struggle to address systemic challenges and disruptive improvements due to limited access to data and digital tools Organizations that fail to integrate OPEX and digital risk underperforming in a world where real-time decision-making, predictive analytics, and autonomous systems are fast becoming the norm	The adaptability of OPEX is key to ensure the readiness for the future	Use OPEX as backbone but keep it adaptable to integrate crucial elements for future competitiveness
… we made a bold decision: To create a fully integrated organization that brings together digital, AI, IT, and OPEX under a single leadership structure, directly embedded within the business	Integration is key	One strategy, one team, one culture, one operating model
When continuous improvement becomes part of the identity of the organization, not a project or an initiative, everything changes. People stop thinking of “doing the work” and “improving the work” as separate activities. Instead, improvement becomes an instinct	CI must become the norm not the exception	Ensure that CI is seen as the norm. Encourage CI
But one leadership trap persists: Viewing the operating system as a toolkit, rather than a system	System understanding is a must	Design and manage it as a system
If you want new behaviors, sometimes you need a new structure	Structures impact behaviors	Use structures to have an impact on culture
We often talk about “safety culture,” “quality culture,” or “CI culture,” as if these are discrete things, but in reality, there is only one: A culture of excellence. The same principles underpin all of them—Discipline, consistency, rigor, and pride in doing things right the first time. A process executed correctly is by definition safer, more compliant, more efficient, and of higher quality	A culture of excellence is required throughout the organization	Build one culture

(continued)

Table 27.1 (continued)

Statement	Insight	Consequences
OPEX is not a destination but an ongoing discipline—One built daily through the choices leaders make, the systems they reinforce, and the behaviors they model	Constant leadership drives OPEX	Be relentless
Our experience demonstrates that strong team cohesion is a critical driver of sustainable results and high performance. By fostering a culture where every team member takes responsibility and embraces continuous learning, we have been able to navigate organizational transformation effectively	Ownership and team spirit are key	Build team cohesion and team spirit
The PEOPLE program is initiated and governed by the site leadership team Austria (SLTA), which provides ongoing sponsorship for the PEOPLE team. This support plays a critical role in breaking down departmental silos, enabling cross-functional collaboration, and cultivating an environment that drives continuous improvement	Real optimization takes place cross-functionally	Build and use cross-functional teams to overcome silos
Four key pillars for achieving analytics and AI readiness: Data, technology, people, and process. These pillars should be considered as focus areas and critical points to measure and address in business or continuous improvement project structures. These elements do not act in isolation; instead, their integration is essential to build sustainable capabilities and foster enterprise-wide success with business maturity and small- to large-scale business system implementations	A comprehensive perspective is required	Take a holistic perspective integrating process understanding with people, technology, and data
QMM is increasingly recognized by regulators as a means of ensuring robust, resilient, and quality systems that look beyond compliance and towards continuous improvement and cultural excellence	Compliance is the minimal requirement	Work on beyond compliance organizations
OPEX and quality management are inextricably linked—The former focusing on process efficiency, the latter on product and compliance. Advanced data analytics integrate both, driving strategic value across the enterprise	OPEX and quality have the potential to reinforce one another	Design OPEX in a way that quality becomes an integrated part
In today's rapidly evolving and highly regulated pharmaceutical and life sciences landscape, a robust quality management system (QMS) is no longer just a compliance necessity; it is a strategic enabler of business value	QMS has the potential to provide business value	Use your QMS to support business and OPEX

"Revolutionizing" our work would mean moving beyond traditional, often reactive processes and embracing a proactive, innovative, and digital approach that benefits our colleagues, our business, and the patients we serve. By aligning around a shared vision and answering the critical question, "what's in it for me?", inspiring engagement at every level became a key goal	A quality professional thinks and talks in business to ensure alignment	Build an OPEX and business mindset in quality leadership
By establishing a clear value proposition, measuring and communicating financial outcomes, and defining success metrics that are continuously monitored, the QMS becomes a driver of tangible business outcomes. Simplification and digitization are key accelerators in this journey, streamlining processes and reducing complexity, thereby enhancing both compliance and quality performance	QMS has the potential to provide business value	Use your QMS to support business and OPEX
Quality leadership is at a turning point, reflecting a fundamental redefinition of quality's role in organizational leadership. Quality business leaders operate as strategic business partners, shifting quality from reactive oversight to proactive leadership and from reporting outcomes to shaping decisions	The state-of-the-art understanding of quality business leadership united quality and OPEX thinking	Integrate quality and OPEX
Core OPEX principles including standardized work, performance measurement, and continuous improvement are fundamental. But the journey towards excellence is radically changing with the addition of new digital toolsets that must be harnessed. These include the use of unified data libraries, digital automation, and advanced artificial intelligence solutions (such as large language models and agent-based frameworks). Through the application of these tools, one can extend OPEX benefits from conventional manufacturing to comprehensive end-to-end supply chains	OPEX must be designed in a way that it can embrace latest technological developments	Use OPEX as the backbone when striving for latest technology adaptions
Intelligent IMEX augments the traditional IMEX framework by deploying unified data management capabilities, process automation, and rigorous enterprise system oversight. This methodology enhances operational efficiency, ensures regulatory compliance, and enables scalable solutions throughout the organization	There are requirements when data should become an integral part of OPEX	Establish capabilities like unified data management to make use of intelligence in OPEX programs
A modern business system should operate as a dynamic ecosystem that remains visible, strategically managed, and continuously refined	Robustness and adaptability are key	Use OPEX as backbone and keep it flexible

5. Align OPEX with compliance, quality, and digital transformation:

- Compliance and OPEX must reinforce each other; quality is the basis for sustainable excellence.
- Digital and OPEX require a shared direction, common language, and integrated governance.

Build capabilities such as unified data management to leverage intelligence for improvement.

6. Make the program people centric:

- Spend time on people, team cohesion, and capability building.
- Recognition systems should highlight behaviors and results aligned with strategic goals.
- Deploy OPEX broadly, involving as many employees as possible.

7. Ensure long-term resilience and maturity:

- Conduct sound maturity checks with accepted outcomes and defined follow-up actions.
- Maintain pace and avoid compromising long-term effectiveness for short-term efficiency.
- Reduce vulnerability by embedding OPEX deeply into the organization's DNA.

OPEX in pharma succeeds when it becomes a long-term, culturally anchored, leadership-driven operating system that integrates compliance, quality, digitalization, and CI into the way the company works every day. It must be built in a way that it results in deep structures.

Based on the conclusions we have drawn from the chapters of this book, we provide suggestions on how to position OPEX in the company to set the foundations for the deep structures required to sustain the gains from initial implementation successes. This will not happen overnight, but it is essential that the vision is set. Either you strive for OPEX seriously or you do not do it at all—or like stated in the opening quote of this chapter: either you come back with your shield or on it. The implementation of OPEX must be serious from the start, and you can never relent.

OPEX must be the backbone of the organization—the structure that keeps everything aligned, stable, but also capable of moving forward. Newer developments like digitalization, AI, sustainability initiatives, and agile ways of working are the limbs and muscles: powerful, flexible, and full of potential. But without a strong backbone, they cannot coordinate, scale, or perform at their best.

OPEX for an industrial company is like the spine for a human body: you cannot run or jump if your spine is not strong enough. OPEX was, is, and remains essential.

This understanding helps to overcome a common flow in many companies. They talk about digitalization, AI, agile, sustainability, etc., as if they were something fully independent from everything else, but OPEX has the very potential to anchor them: these are all crucial topics for future competitiveness, but they only work when the core is healthy, aligned, and disciplined.

When designed, deployed, and sustained, right OPEX will form the deep structure of the organization—the backbone that shapes behavior, decision-making, and performance at a fundamental level. Deep structures are resilient: they persist through leadership transitions and strategic shifts. This is why digital transformation, AI, and agile practices can only scale sustainably when anchored in a strong OPEX backbone.

Kolibri 360 GmbH
Ferdinand-Jühlke-Straße 7
99095 Erfurt, Deutschland
produktsicherheit@kolibri360.de